Optical Sensors Technology and Applications

Optical Sensors Technology and Applications

Editors

Maria Lepore
Ines Delfino

MDPI

Basel • Beijing • Wuhan • Barcelona • Belgrade • Novi Sad • Cluj • Manchester

Editors
Maria Lepore
Università della Campania
"Luigi Vanvitelli"
Napoli, Italy

Ines Delfino
Università della Tuscia
Viterbo, Italy

Editorial Office
MDPI
St. Alban-Anlage 66
4052 Basel, Switzerland

This is a reprint of articles from the Special Issue published online in the open access journal *Sensors* (ISSN 1424-8220) (available at: https://www.mdpi.com/journal/sensors/special_issues/Optical_Sensors_Technology_Applications).

For citation purposes, cite each article independently as indicated on the article page online and as indicated below:

Lastname, A.A.; Lastname, B.B. Article Title. *Journal Name* **Year**, *Volume Number*, Page Range.

ISBN 978-3-0365-9917-5 (Hbk)
ISBN 978-3-0365-9918-2 (PDF)
doi.org/10.3390/books978-3-0365-9918-2

Contents

sensors

MDPI

Editorial

Optical Sensors Technology and Applications

Maria Lepore [1,*] and Ines Delfino [2,*]

[1] Dipartimento di Medicina Sperimentale, Università della Campania "Luigi Vanvitelli", I-80138 Napoli, Italy
[2] Dipartimento di Scienze Ecologiche e Biologiche, Università della Tuscia, I-01100 Viterbo, Italy
* Correspondence: maria.lepore@unicampania.it (M.L.); delfino@unitus.it (I.D.)

Optical methods are non-invasive tools, and their use in various fields, including sensing applications, is continuously increasing, which is thanks to the continuous development of innovative low-cost sources and detectors. Together with the use of nanodevices, these new optical technologies enable the development of new sensing schemes and devices characterized by superior working parameters: optical sensors frequently offer very low detection limits, high specificity, and high sensitivity. In addition, they are very versatile and allow the realization of innovative applicative approaches for quantitatively and qualitatively determining the components of analytes of interest in many fields of application, such as in pharmaceutical research, medical diagnostics, environmental monitoring, agriculture, industry, food safety, and security, just to name a few. The present Special Issue aimed to offer an overview of recent advances in optical sensor technology and applications, including source and detection technologies, sensor architectures, sensor performance, and processing approaches and applications.

The polarization of light represents a formidable tool for the development of innovative methods for the control of products and manufacturing processes. Three of the papers in this Special Issue exploit polarization properties for proposing innovative experimental approaches in very different fields [1–3]. In Ref. [1], the authors prove that individual measurements of the optical polarization parameters of light scattered by suspended particles are a powerful tool to classify the particulate compositions in seawater. This information plays a pivotal role in ecological research and environmental monitoring. A method based on a dense sampling of polarized light pulses is proposed, and the experimental setup is built. Ref. [2] addresses a completely different field of application of polarization. In fact, this paper presents the implementation of an automatic Stokes dynamic polarimeter to characterize non-biological and biological material samples. Experiments demonstrated the efficiency of the proposed optical array to calculate the Mueller matrix in reflection and transmission mode for different samples. A comparison with similar papers reported in the literature validates the approach proposed by the authors. Ref. [3] is devoted to stress detection of the conical frustum window in deep manned submersibles by using polarization imaging. The authors built a Mueller matrix polarimetry, and the examined samples were similar to the typical conical frustum windows in submersibles. The results evidence that the polarization parameters can characterize the stress transfer process and the elastic–plastic transformation process of the window under different pressurization pressures. This proposed method can offer new possibilities for monitoring the window stress and to further ensure the safety of deep manned submersibles.

Interferometry is an optical technique that allows ingenious approaches in very different experimental situations, as occurs for Refs. [4,5]. In Ref. [4], a three sensors-based Mach–Zehnder interferometer is adopted for developing a non-invasive control method for the burning rate of solid fuel in a model solid propellant rocket motor. The results show that the proposed method allows the non-invasive control of solid fuel burnout both by recording the time of arrival of the combustion front to the sensor and by analyzing the peaks of the signal recorded by the available optical fibers. In Ref. [5], the authors propose

Citation: Lepore, M.; Delfino, I. Optical Sensors Technology and Applications. *Sensors* **2022**, *22*, 7905. https://doi.org/10.3390/s22207905

Received: 22 September 2022
Accepted: 26 September 2022
Published: 17 October 2022

the use of a Fizeau interferometer for a self-calibration stitching method for the pitch deviation evaluation of a long-range linear scale. The developed method can represent a significant improvement in the field of optical linear encoders that are widely employed for precision positioning applications, such as semiconductor manufacturing, precision machine tools, and coordinate measuring machines due to their low cost, high resolution, and robustness.

In recent years, the joint use of nanotechnologies and optics has opened new scenarios in the development of sensing schemes. In this framework, the authors of Ref. [6] developed a simple but effective method for the colorimetric detection of antibiotic residues in raw milk by using aptamer-conjugated gold nanoparticles. In fact, the improper use of antibiotics in cattle breeding can cause milk contamination, which has negative effects on the dairy industry and human health. The proposed method could allow the semiquantitative analysis of antibiotic residues in raw milk to be obtained directly from dairy farms. Nanoparticles are also exploited in Ref. [7], in which the authors describe novel flexible plasmonic strain sensors prepared by functionalizing a polymeric transparent substrate with a TiO_2 thin film containing Au nanoparticles. The use of the localized surface plasmon resonance (LSPR) effect due to the interaction of light with the free electrons of noble metal nanoparticles indicated that the LSPR band of these sensors is sensitive to different applied strains. LSPR bands are mainly modified by the changes in the refractive index of the matrix surrounding the Au nanoparticles when uniaxial strains are applied.

Refs. [8,9] are significant examples of the very numerous and various applications of the vibrational spectroscopies, such as Fourier Transform Infrared (FT-IR) and Raman spectroscopy, that span from basic research, medicine, and biology to agrifood, industry, and many other fields.

In Ref. [8], the authors use FT-IR spectroscopy to discriminate different breast cell lines on glass substrates. They aimed to investigate the possibility of using the FT-IR spectroscopy in the clinical diagnostic field to grow cells by adopting the same substrates that are currently used by cytologists and histopathologists. The reported results contribute to promoting the translation of the FT-IR technique in medical practice, representing a complementary diagnostic tool. Ref. [9] addresses a completely different field of application for Raman spectroscopy. In fact, this paper discusses the possibility of using this spectroscopy for analyzing natural gas by taking advantage of Raman spectroscopy's high-speed measurement and the potential to monitor all molecular components simultaneously. The adopted contour fit method to derive concentrations from the spectra of mixtures enables reliable results to be obtained, even when a significant change in the composition of the samples is present. The reported results confirm that Raman gas analyzers can operate without frequent calibrations, differently from gas chromatographs.

In Ref. [10], a new criterion for analyzing the resolution and accuracy of the laser speckle correlation method is presented. This would be of great interest and would be relevant for a variety of applications since the laser speckle correlation method is largely applied to obtain information from vibrating objects. The proposed new criterion combines the mean intensity gradient and frequency spectrum and has been verified by using simulated and real speckle patterns with different speckle sizes, densities, and gray contrasts. The authors use rotation and vibration experiments to confirm the effectiveness of the proposed criterion to improve the performance of the laser speckle correlation method.

The use of neural networks allows for the development of a fringe phase extraction method for optical metrology in Ref. [11]. In this framework, a fringe pattern is usually adopted as output from which the desired parameters can be extracted. The authors propose an end-to-end method of fringe phase extraction based on the neural network. The results of simulation and experimental fringe patterns validate the accuracy and the robustness of the proposed method, which is also characterized by easier operation compare to the other methods that are available.

Ref. [12] is a careful review of the conventional and innovative coating thickness methodologies for application to chromium coatings on a ferromagnetic steel substrate and determines their advantages and limitations regarding in-line measurements. The X-ray Fluorescence (XRF) method is the most common in-line coating thickness measurement method employed in the steel packaging industry, but it is expensive and is characterized by health and safety concerns due to its ionizing radiation. Optical methods can represent a valid alternative. The authors discuss three of these methods: optical reflectometry, ellipsometry, and interferometry. The review presents a detailed references section, and the features of the different coating thickness measurement methods are accurately discussed.

Ref. [13] aims to review the most generally used methods for designing and fabricating optical biosensors and sensors for phenolic compounds that are particularly dangerous due to their toxic effects and their ability to persist in the environment for a long period of time. Their danger is also linked to the fact that they are widely used in many industrial processes and agricultural treatments. The basic principles of the most largely used optical detection techniques are presented, and some selected examples of the most interesting applications of these techniques are also discussed. The large number of bibliographic references demonstrate the ever-keen interest in this research area.

The brief discussion of the papers published in the present Special Issue (SI) attests the diversity of the new developments in optical sensors and their wide dissemination in many applied research fields.

Finally, we would like to thank all the authors who have submitted their work to *Sensors*, specifically to this Special Issue, "Optical Sensors Technology and Applications", all the reviewers for their hard work, and the editors of *Sensors* for their kind and continuous support. All of these efforts made this SI possible. We hope that this SI will help researchers to better understand the state of the art of optical sensors and techniques as well as their wide range of applications. We hope that the 13 published papers will also help the researchers working in the field to disclose future perspectives.

Conflicts of Interest: The authors declare no conflict of interest.

References

1. Guo, Z.; Deng, H.; Li, J.; Liao, R.; Ma, H. Optimized Classification of Suspended Particles in Seawater by Dense Sampling of Polarized Light Pulses. *Sensors* **2021**, *21*, 7344. [CrossRef] [PubMed]
2. Almanza-Ojeda, D.-L.; Rodriguez-Sotelo, D.; Castro-Sanchez, R.; Martinez-Celorio, R.; Ibarra-Manzano, M.-A. Stokes Dynamic Polarimeter for Non-Organic and Organic Samples Characterization. *Sensors* **2022**, *22*, 2155. [CrossRef] [PubMed]
3. Li, H.; Liao, R.; Zhang, H.; Ma, G.; Guo, Z.; Tu, H.; Chen, Y.; Ma, H. Stress Detection of Conical Frustum Windows in Submersibles Based on Polarization Imaging. *Sensors* **2022**, *22*, 2282. [CrossRef] [PubMed]
4. Zhirnov, A.A.; Stepanov, K.V.; Sazonkin, S.G.; Choban, T.V.; Koshelev, K.I.; Chernutsky, A.O.; Pnev, A.B.; Novikov, A.O.; Yagodnikov, D.A. Study of Intra-Chamber Processes in Solid Rocket Motors by Fiber Optic Sensors. *Sensors* **2021**, *21*, 7836. [CrossRef] [PubMed]
5. Xiong, X.; Shimizu, Y.; Matsukuma, H.; Gao, W. A Self-Calibration Stitching Method for Pitch Deviation Evaluation of a Long-Range Linear Scale by Using a Fizeau Interferometer. *Sensors* **2021**, *21*, 7412. [CrossRef] [PubMed]
6. Díaz-García, V.; Contreras-Trigo, B.; Rodríguez, C.; Coelho, P.; Oyarzún, P. A Simple Yet Effective Preanalytical Strategy Enabling the Application of Aptamer-Conjugated Gold Nanoparticles for the Colorimetric Detection of Antibiotic Residues in Raw Milk. *Sensors* **2022**, *22*, 1281. [CrossRef] [PubMed]
7. Rodrigues, M.S.; Borges, J.; Vaz, F. Plasmonic Strain Sensors Based on Au-TiO2 Thin Films on Flexible Substrates. *Sensors* **2022**, *22*, 1375. [CrossRef] [PubMed]
8. Lasalvia, M.; Capozzi, V.; Perna, G. Discrimination of Different Breast Cell Lines on Glass Substrate by Means of Fourier Transform Infrared Spectroscopy. *Sensors* **2021**, *21*, 6992. [CrossRef] [PubMed]
9. Petrov, D.V.; Matrosov, I.I.; Zaripov, A.R.; Tanichev, A.S. Raman Natural Gas Analyzer: Effects of Composition on Measurement Precision. *Sensors* **2022**, *22*, 3492. [CrossRef] [PubMed]
10. Hu, W.; Sheng, Z.; Yan, K.; Miao, H.; Fu, Y. A New Pattern Quality Assessment Criterion and Defocusing Degree Determination of Laser Speckle Correlation Method. *Sensors* **2021**, *21*, 4728. [CrossRef] [PubMed]
11. Hu, W.; Miao, H.; Yan, K.; Fu, Y. A Fringe Phase Extraction Method Based on Neural Network. *Sensors* **2021**, *21*, 1664. [CrossRef] [PubMed]
12. Jones, A.; Uggalla, L.; Li, K.; Fan, Y.; Willow, A.; Mills, C.A.; Copner, N. Continuous In-Line Chromium Coating Thickness Measurement Methodologies: An Investigation of Current and Potential Technology. *Sensors* **2021**, *21*, 3340. [CrossRef] [PubMed]
13. Delfino, I.; Diano, N.; Lepore, M. Advanced Optical Sensing of Phenolic Compounds for Environmental Applications. *Sensors* **2021**, *21*, 7563. [CrossRef] [PubMed]

sensors

MDPI

Article

Optimized Classification of Suspended Particles in Seawater by Dense Sampling of Polarized Light Pulses

Zhiming Guo [1], Hanbo Deng [1,2], Jiajin Li [1,2], Ran Liao [1,3,4,*] and Hui Ma [3,5]

1 Institute for Ocean Engineering, Shenzhen International Graduate School, Tsinghua University, Shenzhen 518055, China; gzm20@mails.tsinghua.edu.cn (Z.G.); dhb19@mails.tsinghua.edu.cn (H.D.); lijiajin21@mails.tsinghua.edu.cn (J.L.)
2 Department of Biomedical Engineering, Tsinghua University, Beijing 100084, China
3 Guangdong Research Center of Polarization Imaging and Measurement Engineering Technology, Shenzhen International Graduate School, Tsinghua University, Shenzhen 518055, China; mahui@tsinghua.edu.cn
4 Shenzhen Key Laboratory of Marine IntelliSensing and Computation, Shenzhen International Graduate School, Tsinghua University, Shenzhen 518055, China
5 Department of Physics, Tsinghua University, Beijing 100084, China
* Correspondence: liao.ran@sz.tsinghua.edu.cn; Tel.: +86-755-869-75-301

Abstract: Suspended particles affect the state and vitality of the marine ecosystem. In situ probing and accurately classifying the suspended particles in seawater have an important impact on ecological research and environmental monitoring. Individual measurement of the optical polarization parameters scattered by the suspended particles has been proven to be a powerful tool to classify the particulate compositions in seawater. In previous works, the temporal polarized light pulses are sampled and averaged to evaluate the polarization parameters. In this paper, a method based on dense sampling of polarized light pulses is proposed and the experimental setup is built. The experimental results show that the dense sampling method optimizes the classification and increases the average accuracy by at least 16% than the average method. We demonstrate the feasibility of dense sampling method by classifying the multiple types of particles in mixed suspensions and show its excellent generalization ability by multi-classification of the particles. Additional analysis indicates that the dense sampling method basically takes advantage of the high-quality polarization parameters to optimize the classification performance. The above results suggest that the proposed dense sampling method has the potential to probe the suspended particles in seawater in red-tide early warning, as well as sediment and microplastics monitoring.

Keywords: classification; suspended particles; polarized light pulses; dense sampling

Citation: Guo, Z.; Deng, H.; Li, J.; Liao, R.; Ma, H. Optimized Classification of Suspended Particles in Seawater by Dense Sampling of Polarized Light Pulses. *Sensors* **2021**, *21*, 7344. https://doi.org/10.3390/s21217344

Academic Editors: Maria Lepore and Ines Delfino

Received: 28 September 2021
Accepted: 3 November 2021
Published: 4 November 2021

1. Introduction

The ocean is the most important resource endowed by nature, which contains the abundant resources necessary for the survival and development of human society [1]. While the ocean continues to create huge benefits for modern society, its own ecology has also encountered great challenges to human life and production [2–4]. As an important and essential component in seawater, the suspended particles significantly influence the optical properties of seawater, as well as the marine ecological environment [5,6]. For example, certain harmful types of microalgae rapidly cause algae blooms in a short period, which is a threat to marine organisms [7–9]. Microplastic has become a prevalent, widespread element of marine litter, threatening marine organisms and human health [10–12]. The accumulation and transportation of sediments such as silts, have a great impact on the stability of the estuarine and seacoast [13,14]. Therefore, the development of effective detection and accurate classification and identification of these different suspended particles is of far-reaching significance [15,16]. At the same time, the detected particle information is also helpful to interpret the data of marine science macroscopic researches, and further, promote the development of remote sensing in marine monitoring [17,18].

Optical methods are currently one of the most popular methods to detect the suspended particles, due to the advantages of high resolution, non-contact, and rich information [19,20]. In recent years, many in situ optical instruments have been applied to the acquisition of suspended particles in seawater such as YSI EXO [21,22], AC-S [23], BB9 [24], and LISST-200X [25]. However, these methods use bulk measurement to obtain comprehensive information of suspended particles in seawater, and cannot obtain information on the morphology and internal structure of individual seaweed particles. The Flow cytometers, such as FlowCytobot, individually measure the scattered intensity and fluorescence or sometimes the images of the particles to classify the particles in seawater. However, its dependency on the pretreatment is based on the hydrodynamic focusing system, limiting its application in the seawater [26,27].

Polarization is the fundamental property of light. Compared with traditional optical methods, polarized light can carry richer information [28]. The polarization information differences can be utilized for the analysis and identification of biological tissue lesions [29,30], and for the identification and classification of atmospheric and marine particles [31,32]. However, the polarization parameters are affected by the comprehensive effects of particles, including size, refractive index, shape, morphological structure, and microstructure. In 2018, Wang et al. obtained the polarized pulses of temporal signals through an experimental setup, and the signal-to-noise ratio of the system was bigger than 5. Then, they used a low-pass filter to suppress the high-frequency noise and used a threshold limit to acquire the polarized pulse signal. Finally, all the samplings in each polarized light pulse are calculated as an average value (PLP-Ave). This method is applied to differentiate the suspended particles of different physical and microstructural properties, which is important for monitoring microalgae, microplastics, and silt concentrations [33]. In 2019, Liao et al. applied the PLP-Ave method and developed a new in situ prototype, whose ability to the classification of the suspended particles in seawater has been demonstrated by field deployments [34]. In 2020, Li et al. used the PLP-Ave method to probe the collapse and regeneration of the cyanobacterial gas vesicles exposed to different static pressures [20]. In addition, Wang et al. chose samples with distinctive microstructural features, and then conducted simulations and calculations to examine how these features affect the polarization of the scattered photons using the PLP-Ave method [35].

This paper will introduce a method for optimizing the classification of suspended particles in seawater by dense sampling of polarized light pulses. The laboratory experiment which is free of the pretreatment of the samples, preliminarily shows that this method has local optimal characteristics and significant classification performance. Through the dense sampling of polarized light pulses, the characteristic information of the suspended particles can be obtained comprehensively, and a large number of local effective characteristics of the suspended particles can be improved to achieve high accuracy. In practical applications, the PLP-All model is more accurate and flexible in the mixed experiment prediction. These results show that using the dense sampling of polarized light pulses is beneficial to the more accurate classification of suspended particles. In addition, it brings a good application method for identifying and classifying suspended particles in seawater.

2. Methods and Materials

2.1. Principle of the Experimental Setup

A polarized light scattering method for differentiating suspended particles of different physical and microstructural properties can detect and classify the suspended particles [33,35]. On this basis, we built an experimental setup using a similar principle, which is free from the pretreatment of the samples. It achieves the classification of suspended particles in seawater by individual particle measurement and machine learning algorithms.

The experimental setup includes four parts: Optical path, photoelectric convertor, analog-to-digital conversion, and the sampling system. The sampling system consists of a sample pool and a flow circulation system. The sample pool has one inlet and one outlet. In addition, the sample is pumped by a pumper to enter the sample pool through the inlet and leave the sample pool through the outlet by another pumper. The particles in the sample are suspended and pass through the scattering volume of the setup, whose polarization parameters can be efficiently and accurately measured.

Figure 1 shows the optical path of the setup. A 520 nm laser is the light source with 2-mm beam size and 0.7 W maximal power. Since most of the microalgae have a low absorption coefficient and high scattering coefficient at 520 nm [36], using the 520 nm laser the light is more conducive to reflecting the structure of microalgae and highlighting the differences between them. A polarization state generator (PSG) converts the light into the desired polarization states. The beam is completely reflected by the equilateral prism to obtain the obliquely illuminating light. Then, the illuminating light passes through Lens 1 and the transparent ceramic window 1, which is focused into a tiny light spot. Once the suspended particle passes through the light spot, it will be illuminated, and the 120° scattered light will pass through the transparent ceramic window 2 and be received by Lens 2. Note that windows 1 and 2 do not change the polarization states of the light passing through them. In addition, the 120° scattering angle has been proven to be sensitive to the microstructure in our previous work, which helps in identifying and distinguishing different suspended particles in seawater [35]. Then, the 120° scattering angle is used to maintain consistency with the previous work [33].

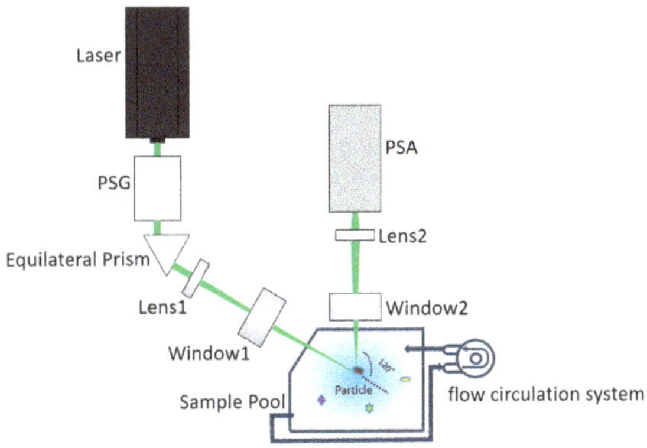

Figure 1. Schematic diagram of the optical system of the experimental setup.

Lens 2 consists of a series of lenses and a pinhole. The size of the pinhole is 100-micron and its position is the imaging point of the light spot by the lenses before the pinhole. The pinhole is followed by a short focal length lens to convert the scattered light to the parallel light beam before entering the polarization state analyzer (PSA). Therefore, the scattering volume as the intersection volume of the illuminating optical path and the receiving optical path, is determined by the pinhole and the light spot. Moreover, in this work, the scattering volume is less than 0.01 microliter. If the volume concentration of the suspended particles is less than 10^5 per milliliter, there is only one particle in the scattering volume at most, based on which the measurement of the individual particle can be realized. Therefore, the requirement for the volume concentration does not depend on the particle size, but depends on the scattering volume, which is determined by the pinhole and the light spot of the experimental device. In the measurement, if a single particle passes through the scattering volume, its scattered light contributes to the signal. When there are no particles in the scattering volume, the electronic noise, environmental light, and the scattering of water contribute to the background, which is smaller than the particle. Therefore, the signals are a series of temporal pulses.

PSG and PSA are the important components of the experimental setup, which realize the key functions of polarized light illumination and detection in the system. PSG is composed of the fixed linear polarizer, achromatic half-wave plate, and achromatic quarter-wave plate. In addition, the independent rotating motors carried by one-half wave plate and one-quarter wave plate change the direction of the fast axis of the wave plate. In our self-written application interface, the operator can set the rotation angle of the motor to obtain the desired polarization of the incident light, thereby obtaining more abundant polarization information of the particulate matter.

PSA is composed of three non-polarizing cube beam splitters, which divide the incident parallel beam into four parts. The two parts are analyzed with 0 and 45° linear polarizers and other analyzers. The other two parts are analyzed by left-hand and right-hand circular analyzers. The left-hand circular analyzer consists of a 135° fast-axis oriented quarter-wave plate and a 90° linear polarizer. In addition, the right-hand circular analyzer consists of a 45° fast-axis oriented quarter-wave plate and a 90° linear polarizer.

Then, the 120° backward scattered light by the suspended particles in the scattering volume is divided into four channels by the PSA and converted into the four voltages by the four independent photoelectric converters. In addition, it is simultaneously digitized into a four-channel signal by a data acquisition card (DAQ). Thereafter, the four-channel signals are transferred into the polarization state of the scattered light by the instrument matrix gained from a polarization calibration procedure [34].

Using the experimental setup, we can measure the polarization state of the light scattered by the suspended particles and obtain a series of temporal pulses. Due to the individual measurement of the setup, each pulse originates from an individual particle. By processing these pulses, we can get the polarization parameters of suspended particles.

2.2. *Samples*

The suspended particles used in this experiment consist of five types of microalgae, two types of microplastics, and one type of sediment. The five types of microalgae (*Dunaliella salina* (DS), *Cryptomonassp.* (CP), *Chaetoceros debilis* (CD), *Phaeocystis globosa* (PG), and *Thalassiosira weissflogii* (TW)) were bought from Shanghai Guangyu Biological Technology Co., Ltd., which carried out a large-scale cultivation of liquid microalgae species based on the production environment including the temperature, nutrient salt formula, and light intensity. In particular, DS has a unique economic value in medicine and health care [37], CD is often used in research studies for oceanography and aquaculture [38], PG is toxic and will cause red tide [39], CP and TW are important components for the phytoplankton ecosystem and productivity [40,41].

In addition, other suspended particles (monodispersed polystyrene microspheres, 2 μm (PS-02) and 10 μm (PS-10), silicon dioxide pellets, 10 μm (SD-10)) were bought from Big Goose (Tianjin) Technology Co., Ltd. They are all white suspensions obtained by dispersing white solid powder in water. The SD-10 mother suspension was prepared by mixing 250 mg of silicon dioxide pellets in 10 mL of 50% ethanol water solution. Moreover, the PS-02 and PS-10 mother suspensions were prepared by mixing 250 mg of monodispersed polystyrene microspheres in 10 mL of deionized water solution.

Filtered seawater is used in this work. We obtained the surface seawater from Yantian Port in Shenzhen, and filtered it in the lab with a 0.2-micron filter membrane. The filtered seawater is prepared in advance to dilute the mother suspensions of the particles.

2.3. *Polarized Light Pulse Processing Algorithm*

Figure 2a shows the temporal signals of a suspension measured by the setup, which is a series of polarized light pulses. The envelope of pulses is obtained by the treatment of low-pass filtering. The polarized light pulses include the information of the particle and the noises. In addition, the filtering can reduce the noisy fluctuation that originated from the electronic noise or environmental noise. However, there is still residual noise in the pulses.

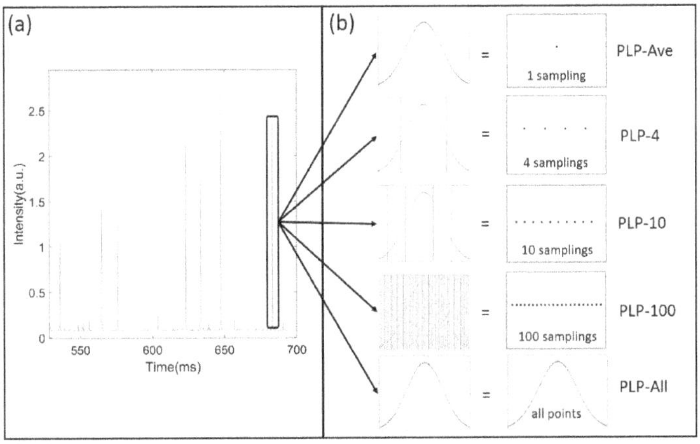

Figure 2. The dense sampling of polarized light pulses: (**a**) The temporal signals; (**b**) polarized light pulse processing algorithm.

When a single suspended particle passes through the scattering volume, it stays for a while during which it is continuously illuminated and its scattered polarization states are measured. The width of the polarized light pulse is the settling time that the particle stays in the scattering volume. In this case, the particle would be sampled many times by the setup. For example, in Figure 2b, a pulse's settling time is about 2 ms and if the sampling rate of the DAQ is 200 K sampling per second, then this pulse consists of 400 samplings and the particle is measured 400 times at one measurement. We have 400 scattered polarization states of the particles. Considering that the particle is moving in the scattering volume, we obtain abundant data of this particle and the method of extracting the information from these data is a serious issue. Due to the noises from the electronic system and the environmental light, in previous researches, we first averaged all of the samplings in each polarized light pulse (PLP-Ave) to reduce the influence of the noises on the signals before classifying the suspended particles.

In this work, we introduce the polarized light pulse processing algorithm and investigate the benefit or loss of the averaging of the polarized light pulses to achieve more efficient and accurate classification than before. We divide the polarized light pulses with specific methods into four samplings (PLP-4), 10 samplings (PLP-10), 100 samplings (PLP-100), and all points (PLP-All), as shown in Figure 2b. For example, in the PLP-4 method, we divide the polarized light pulses into four parts and average each part to obtain four values for one pulse, which is considered as four samplings. Similarly, we get 10 samplings for one pulse in the PLP-10 method, and 100 samplings in the PLP-100 method. For the PLP-All method, all the samplings are considered. Generally, the averaging will suppress the noise and enhance the signal-noise-ratio, and finally help in extracting the information from the polarized light pulse. However, the averaging will omit the detailed information. Definitely, the PLP-All method suffers the most from the noise and the values are most inaccurate. Therefore, this investigation will try to estimate the benefit and loss of the averaging of the pulses using the polarized light pulse processing method.

2.4. Analytical Methods

Stokes vector, S, as shown in Equation (1) is always used to describe the polarization state of light [42].

$$S = \begin{bmatrix} I \\ Q \\ U \\ V \end{bmatrix} \tag{1}$$

where I is the total light intensity, and Q, U, V are the residual 0°, 45°, and right-circularly polarization, respectively.

As shown in Equation (2), q, u, and v are polarization parameters normalized by light intensity I, which can be dimensionless and range from −1 to 1.

$$q = \frac{Q}{I}, \; u = \frac{U}{I}, \; v = \frac{V}{I}. \tag{2}$$

The degree of polarization (DOP) as shown in Equation (3) commonly represents the proportion of polarized light in the total light intensity, ranging from 0 to 1, which is also used to characterize the depolarization ability of particles when they are illuminated by a polarized light [43].

$$\text{DOP} = \frac{\sqrt{Q^2 + U^2 + V^2}}{I} \tag{3}$$

For classification problems, the results of machine predictions and actual values will deviate. The confusion matrix is a standard measure that represents accuracy evaluation, including true positive (TP), false positive (FP), true negative (TN), and false negative (FP) [44]. Each row of this matrix represents an instance in the actual class, and each column represents an instance in the predicted class. It can clearly express the correct classification and misclassification of each category on the visual system. Therefore, the evaluation model's advantages and disadvantages standards introduce accuracy, which is computed from the confusion matrix using Equation (4).

$$Accuracy = \frac{TP + TN}{TP + FP + FN + TN} \tag{4}$$

where TP denotes that the positive class is judged as a positive class, FP denotes that the negative class is judged as a positive class, TN denotes that the positive class is judged as a negative class, and FN denotes that the negative class is judged as a negative class.

As shown in Equation (5), the mean-square error (MSE) is a measure that reflects the degree of difference between the estimator and the true value. $\hat{x}$ is the estimator of the sample, and x is the true value of the sample. When the sample is constant, their distance function is an index used to evaluate the quality of an estimator.

$$MSE(\hat{x}) = \mathrm{E}(\hat{x} - x)^2 \tag{5}$$

2.5. Algorithm Theory

As one of the most traditional neural networks, the main characteristic of backpropagation neural network (BPNN) is that the signal propagates forward and the error propagates backward. Backpropagation is the standard method for training artificial neural networks. This method helps in calculating the gradient of the loss function with respect to all the weights in the network. By fine-tuning the weight of the neural network based on the error rate obtained in the previous period, it can reduce the error rate and improve the generalization and reliability of the model. Therefore, the training and test sets of the BPNN algorithm model that we built use the polarization parameter data, which is measured by the suspended particles in the experimental device. As shown in Figure 3, the input layer of the network training uses $X = [I, q, u, v, \; DOP]$, and there are five input nodes. Through multiple experiments, the hidden layer is set to three layers, which are five nodes, six nodes, and four nodes. Finally, there are four nodes in the output layer, corresponding to four types of suspended particles. In addition, we built the algorithm model using the sigmoid nonlinear transfer function, which minimizes the actual output and expected output error function of the system by revising weights and thresholds repeatedly.

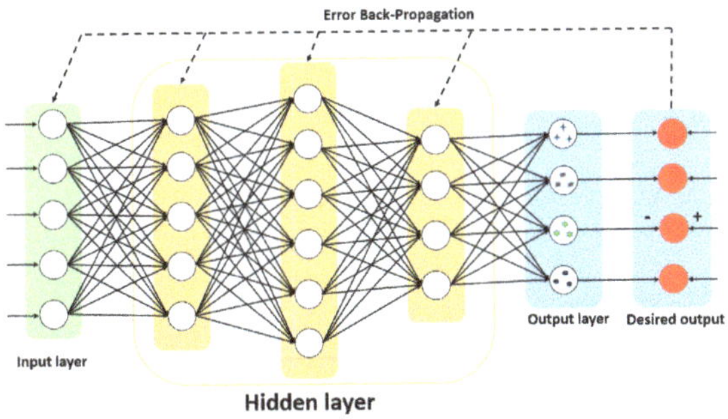

Figure 3. The schematic diagram of backpropagation algorithm structure.

3. Results

3.1. Classification of the Four Types of Microalgae

In this paper, we built the BPNN model which is used to classify the four types of microalgae, DS, CP, CD, and PG for each method of the polarized light pulse process algorithm. The dataset prepared by the PLP-Ave method is used as input data for training and testing and finally, for building the PLP-Ave model. Similarly, we built the PLP-4 model, PLP-10 model, PLP-100 model, and PLP-All model. For each model, the total number of dataset is 11,000. We used the random function of MATLAB to sort the dataset of each model. Then, 70% of the disordered dataset is used for training and 30% is used for testing.

After training for 100 epochs, the model tends to converge and the results are shown in Figure 4. Figure 4a shows the confusion matrix of PLP-Ave model. As can be seen, the classification accuracy for the DS, CP, and PG are less than 80%, and some errors are around 10%. The average accuracy of PLP-Ave model is only 80.77%. In Figure 4b, the classification accuracy of PLP-All model for all the types of microalgae is larger than 90%, and for PG it is larger than 99%. The average accuracy of PLP-All model is 97.32%. Next, we collect the average accuracy of the different methods in Figure 4c. As can be seen, the PLP-All model has achieved the best classification accuracy, and the PLP-Ave model's accuracy is minimal. In addition, the classification accuracy for the suspended particles increases with the sampling numbers in the different methods. Note that the PLP-All model suffers the most from the noises in the data. However, since it contains most of the information about the particles, it achieves the best classification accuracy.

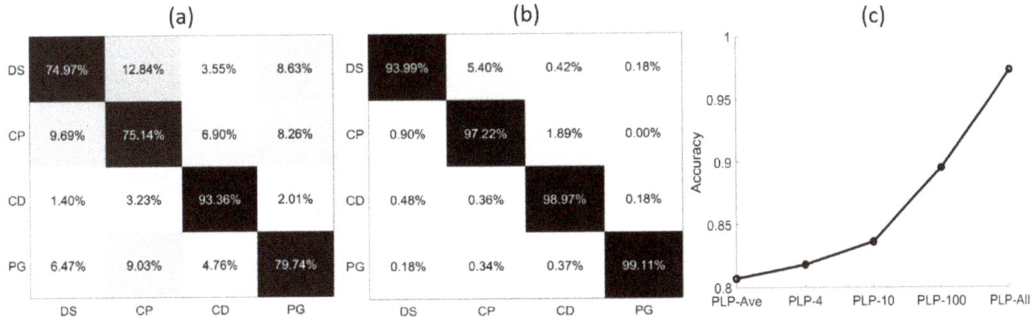

Figure 4. The confusion matrix of different polarized light pulses methods: (**a**) PLP-Ave model; (**b**) PLP-All model; (**c**) average accuracy of five models.

The classification accuracy of the four types of microalgae in general is shown in Table 1. In this case, the classification accuracy of the model trained by the data of PLP-All model is larger by 16.55% than the PLP-Ave model, which indicates that this model has stronger feature extraction capabilities. These high-precision results are consistent in both the training and prediction sets, which also prove that the model we trained has a strong generalization ability.

Table 1. Accuracy of PLP-Ave model and PLP-All model.

Dataset	Training Set	Test Set
PLP-Ave model	80.87%	80.77%
PLP-All model	97.80%	97.32%

3.2. *The Mixed Experiment Prediction*

To check the flexibility and feasibility of the above models, we used them to classify the four types of microalgae cells in the mixed suspensions. First, we measured the four types of microalgae suspensions separately. In order to collect enough particles, we added 2 milliliter of DS to the filtered seawater and measured for 3 min, then recorded the resulting pulse number which is 720 pulses. Similarly, the same steps were performed for CP, CD, and PG, respectively. In addition, we obtained the polarized light pulse numbers, 2160 pulses for CP, 576 pulses for CD, and 648 pulses for PG. Then, we mixed the four microalgae suspensions together, Groups 1, 2, and 3. In each group, the respective volumes were carefully chosen to ensure that the obtained pulse number by the setup for each type of microalgae was consistent with the preset pulse number. Then, the mixed suspensions were separately measured by the experimental setup and each measurement lasted 3 min. The obtained data were fed into the PLP-Ave and PLP-All models respectively to predict the number of microalgae cells in each group. For the pulses of temporal signals information, we need to accumulate enough pulses to obtain most of the particle information. In addition, the machine learning of BPNN algorithm needs enough characteristics of particles in order to correctly identify the category of particles.

Table 2 collects the preset pulse number by the calculated volume and the predicted pulse number by the models for the three mixed suspensions, which can be regarded respectively as the true value and the predicted value by the models. There are obvious differences between the preset pulse numbers and the predicted pulse numbers gained by the PLP-Ave model. The largest error of PLP-Ave model occurs at Group 1, the predicted pulse number for DS is 100, while the preset pulse number is 40, and the relative error is about 60%. However, the largest error of PLP-All model occurs in the same group, the predicted pulse number for DS is 100, while the preset pulse number is 75, and the relative error is about 25%. Generally, the errors between the preset and predicted pulse numbers of PLP-All model are much smaller than the PLP-Ave model. Note that the shape and microstructure of some specific microalgae are not obvious, and an excessive increase in number cannot be well identified by the PLP-All model. Therefore, some specific type of the microalgae worsens the errors of the PLP-All model. However, most other types of the microalgae are good and acceptable, which are quite different from the PLP-Ave model. The results indicate that the PLP-All model is flexible and feasible for use in the classification of the microalgae in mixed suspensions.

Table 2. The predicted results of mixed suspensions.

	Preset Pulse Number	PLP-Ave Model Prediction	PLP-All Model Prediction
Group 1	100, 100, 100, 100	115, 104, 133, 40	115, 103, 99, 75
Group 2	200, 300, 100, 200	292, 258, 80, 194	214, 308, 108, 194
Group 3	200, 400, 100, 300	280, 320, 84, 286	234, 364, 100, 268

3.3. Comparative Analysis

Based on the classification performance of the PLP-All model, we tested a diverse particle group that consists of five types of microalgae, two types of microplastics, and one type of sediment, and the results are shown in Figure 5. Since TW and PG can both cause a red tide, we added TW into the group to confirm that the PLP-All method has a strong generalization ability and strong feature extraction ability. Through the confusion matrix of PLP-Ave model in Figure 5a, it can be seen that the highest classification accuracy for the eight suspended particles is less than 80%, and the lowest classification accuracy is 65.11%. The average accuracy of PLP-Ave model is only 80.20%. In Figure 5b, the classification accuracy of PLP-All model for all types of the suspended particles is larger than 81%, and for PS-10 it is larger than 99%. The average accuracy of PLP-All model is 90.90%. In addition, we noticed that all the suspended particles classification accuracy increase and the maximal increase in the accuracy is about 19.47%. Moreover, for PS-02, which has a rather high accuracy, the PLP-All model still improves the classification accuracy. Meanwhile, for the total effects, half of the particle's classification accuracy increases by more than 10%.

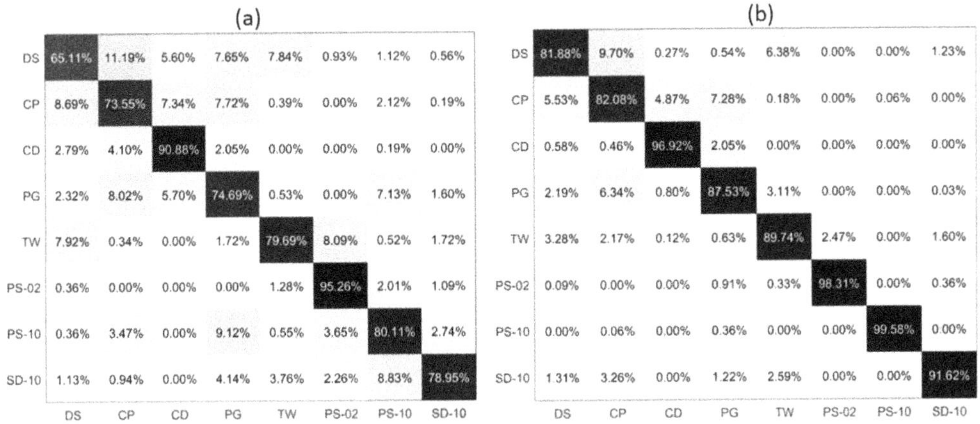

Figure 5. The confusion matrix of eight suspended particles classification: (**a**) PLP-Ave model; (**b**) PLP-All model.

The classification accuracy of the eight suspended particles in the training and test sets using the PLP-Ave and PLP-All models are shown in Table 3. The model's classification accuracy which is trained by the data of the PLP-All model is larger by 10.70% than the PLP-Ave model, which indicates that this model has stronger feature extraction capabilities.

Table 3. Accuracy of PLP-Ave model and PLP-All model.

Dataset	Training Set	Test Set
PLP-Ave model	79.89%	80.20%
PLP-All model	90.80%	90.90%

In general, for the diverse particles classification results in Figure 5, as compared with the PLP-Ave model, the PLP-All model effectively improves the classification accuracy. These results show that the PLP-All model has a strong generalization ability and impressive classification performance.

4. Discussion

4.1. Training Details of Different Models

To further show the classification effects of the polarized light pulse processing algorithm, we provide the training details of all the models for different parts. As the number of epoch increases, the accuracy and MSE of these different models are evidently different, as shown in Figure 6a,b. Of note, when the number of epoch is less than 30, all the models increase sharply, but fluctuate strongly and show little differences in accuracy when compared with each other. However, as the number of epoch increases by more than 40, the PLP-Ave model converges to the stable value at first, but then its accuracy becomes worse. On the contrary, the accuracy performance of the PLP-All model is better than the other four models, but converges most slowly. Meanwhile, we can easily see that the models' accuracy increases with the sampling number of the polarized light pulse. Moreover, these models still have a large difference in MSE. The MSE of PLP-All model reduces the fastest and reaches a lower value of about 0.05. In addition, after the MSE's stable value reduces with the sampling number, these training details indicate that the PLP-All method has better classification ability, as shown in Figure 6. All of these results emphasize that the PLP-All method can effectively distinguish the CD algae with large differences in the microstructure. In addition, the PLP-All method can still distinguish the microalgal samples (DS, CP, PG) with little differences in appearance and structure. This is related to the effect of polarization information on size, shape, microstructure, and morphology, in order to extract high-quality data from the polarization pulses.

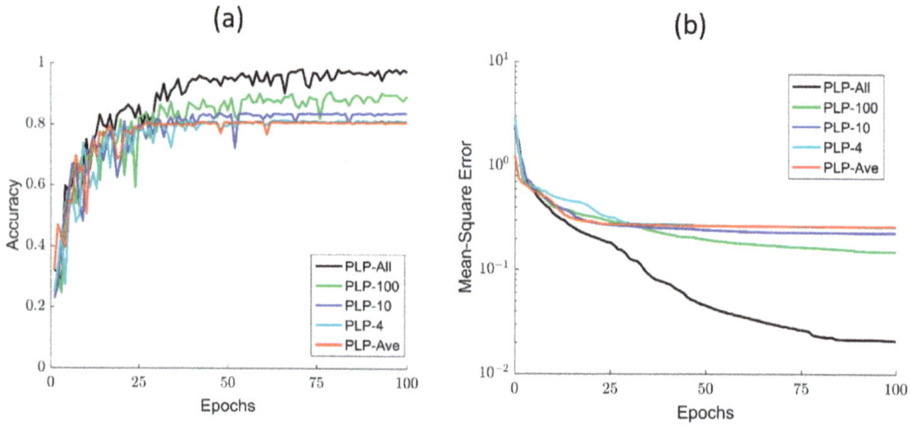

Figure 6. (**a**) The accuracy curve of different processing methods; (**b**) the MSE of different processing methods.

4.2. *Accuracy of Four Microalgal Samples*

Figure 4c shows the average accuracy of five models. Here, as shown in Figure 6, we provide the accuracy of five models for four microalgal samples. From Figure 7, the accuracy of the three types of microalgae (DS, CP, PG) increases with the sampling number of the polarized light pulse, which is consistent with the average accuracy in Figure 4c. However, the CD is a little different, and its accuracy increases when the sampling number grows from 4 to 100, and finally reaches the maximum in the PLP-All model. In addition, the accuracy of PLP-Ave is larger than PLP-4 and PLP-10, which is quite different from the other models. Therefore, we notice that the accuracy of CD samples is larger than the other microalgal samples. In addition, the accuracy changes in a different manner than the models for the four samples.

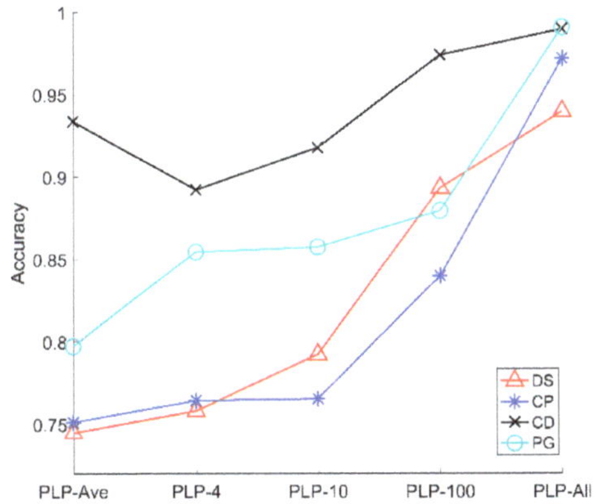

Figure 7. The average accuracy of five polarized light pulses processing methods for signal particles.

Essentially, the accuracy of models for the microalgal samples is subjected to the sensitivity of the polarization parameters to the physical properties of the microalgae cells. The CD cells have flagella and their shape are long oval, which is quite different from the other microalgae. This specific microstructure contributes to the higher accuracy of CD at all the models than the other microalgae.

In addition, we should recall that the averaging in the polarized light pulse will reduce the noise and then, lead to the accurate measurement of the polarization parameters. However, the samplings of the polarized light pulse will increase the information amount, which are both positive for the classification results. When we divide the polarized light pulse in 4, 10, and 100 times or take all the sampling points of the polarized light pulse into account, we suffer more from the noise but increase the information amount of the microalgae cells. As a result, the classification accuracy would be the tradeoff between these two factors. For CD, averaging plays a more important role at first, thus the accuracy of PLP-Ave is larger than PLP-4 and PLP-10, but then the information amount increases and dominates the classification. Therefore, the accuracy of PLP-100 and PLP-All is larger than PLP-Ave. Moreover, the inherent difference between the microalgal types and the tradeoff between the averaging and information amount determine the classification accuracy of the models.

4.3. *Origin of the Performance of PLP-All Method*

Of note, the PLP-All method suffers the most from the noise of the data, but the classification accuracy is the best. Evidently, the increase of the data amount is one origin for the best performance. In addition, in Figure 4c, it can be seen that increasing the data amount will promote the accuracy. However, we would like to emphasize another origin, which is possibly more essential than the amount of data. Figure 8 shows part of the data, $[I, q, u, v]$, of a polarized light pulse processed by the PLP-Ave and PLP-All methods, respectively and their original signal. The red lines are the average values given by the PLP-Ave method and the blue lines are the temporal values given by the PLP-All method. The temporal values change in the polarized light pulse. Here, we can imagine that if we use the average values given by the PLP-Ave method to replace those of the PLP-All model, the same data amount is ensured, but the classification effect would be no better than the PLP-Ave model. Therefore, the key origin of the best performance of PLP-All method is implied by the time-changing values of polarization parameters of the individual particle, as shown in Figure 8.

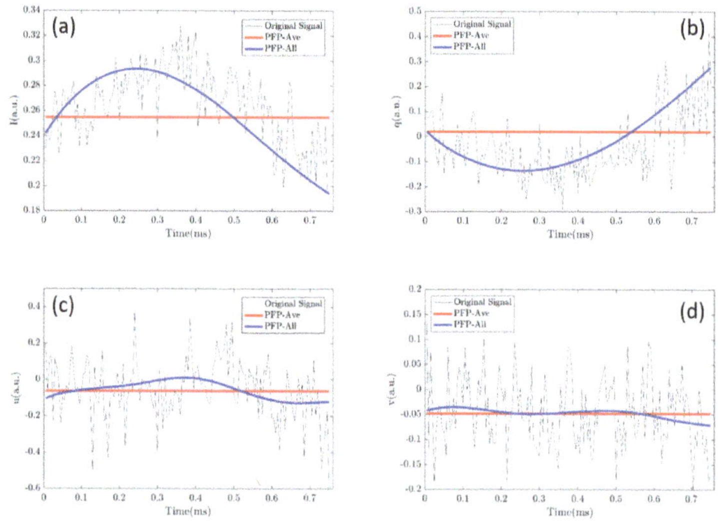

Figure 8. Temporal values of a polarized light pulse processed by the PLP-All method and the dash line is the value given by the PLP-Ave method: (**a**) Intensity values (I); (**b**) polarization parameter (q); (**c**) polarization parameter (u); (**d**) polarization parameter (v).

Moreover, Figure 8 shows the original signals in the pulse, which are noisy and not filtered by the low-pass filter. If we use these original signals to feed the classifier, the data amount is the same, but the final classification effect is the worst. Although the entire information of the particle is in the original signal, the noises destroy the classification ability. The envelope of the pulse gained by the low-pass filtering in Figures 2 and 8 possibly loses some particle information, but it reduces the noise, which indicates that the PLP-All method takes advantage of the denoised and high-quality data.

For each suspended particle crossing the scattering volume, the transient location and orientation change with the time, according to the setup's optical system, which leads to the changes of their scattered polarization states and parameters. Note that the polarization parameters are sensitive to the structures and orientation of the particles. Therefore, they can well characterize this essential information, and finally provide high-quality data. The dense sampling of each pulse records these transient states of the individual particle, which enhances the information collection. In summary, the PLP-All method takes advantage of the polarization parameters and dense sampling, and then achieves the best performance.

In addition, compared with the existing classification methods of suspended particles, the method proposed in this paper has the advantage of not requiring a sample pretreatment and the ability of detecting the rich polarization information of the individual particles. However, the time efficiency needs to be enhanced in the next step. Moreover, beyond the current Stokes vector measurement, the Muller matrix measurement of the individually suspended particles may be added to the future worklist.

4.4. Comparative Different Machine Learning Algorithms

To further confirm that the different polarized light pulse methods for particle classification depend little on the machine learning algorithm, we additionally built the support vector machine (SVM) algorithm. SVM was used to classify the four types of microalgae, DS, CP, CD, and PG for different polarized light pulse process methods and the results are shown in Table 4. A comparison of Tables 1 and 4 showed that the classification accuracy of the PLP-Ave model using the BPNN algorithm was larger by about 6% than the SVM algorithm. Similarly, the accuracy of PLP-All model was larger by about 3% than the SVM algorithm. This indicates that the BPNN algorithm was more powerful in extracting the polarization parameter features for achieving higher accuracy than the SVM algorithm. In addition, we would like to emphasize that the different machine learning algorithms led to the similarly excellent performance of PLP-All method for particle classification.

Table 4. Accuracy of PLP-Ave and PLP-All models using the SVM algorithm.

Dataset	Training Set	Test Set
PLP-Ave model	74.39%	74.38%
PLP-All model	94.71%	94.57%

5. Conclusions

In this paper, we proposed an optimization method for the classification of suspended particles in seawater by dense sampling of polarized light pulses. We built an experimental setup to measure the suspended particles and collect the polarized light pulses. Then, we investigated the classification results of the four types of microalgae using different dense sampling methods. For each method, we sampled the pulse with a certain number and then built the specific model to classify the four types of microalgae. The results showed that the classification accuracy increased with the sampling numbers. In addition, the PLP-All model achieved the best classification performance. Moreover, we classified the four types of microalgae cells in the mixed suspensions and the results indicated that the PLP-All model was feasible. Furthermore, we conducted eight types of suspended particles including microalgae, microplastics, and sediment, and the classification results showed that the PLP-All model had a good generalization ability. In the discussion part, the classification accuracy increased with the sampling number, but the MSE decreased. In addition, for each type of microalgae, the PLP-All model was still the best and the dense sampling improved the classification performance. Finally, the best performance of the PLP-All model can be attributed to taking advantage of the high-quality polarization parameters and dense sampling. In summary, the method based on dense sampling of polarized light pulses had an excellent ability of classifying the suspended particles. It can be expected that the underwater polarization scattering instrument equipped with dense sampling can effectively and accurately help in obtaining the information of particle compositions in seawater. Furthermore, the dense sampling idea can be used in the future development of Muller matrix polarimetry of the suspended particles.

Author Contributions: Writing—original draft preparation, Z.G.; writing—review and editing, R.L.; visualization, H.D. and J.L.; project administration, H.M. R.L. conceived the project aim. All authors have read and agreed to the published version of the manuscript.

Funding: This research was funded by the National Key Research and Development Program of China (2018YFC1406600); the National Natural Science Foundation of China (NSFC) (41527901, 61975088); and the Key-Area Research and Development Program of Guangdong Province (2020B1111 040001).

Institutional Review Board Statement: Not applicable.

Informed Consent Statement: Not applicable.

Data Availability Statement: We did not report any data.

Conflicts of Interest: The authors declare no conflict of interest.

References

1. Rao, J. Coordination degree of marine biological ecological resources based on multi-source monitoring data. *Arab. J. Geosci.* **2021**, *14*, 654. [CrossRef]
2. Ledoux, L.; Turner, R.K. Valuing ocean and coastal resources: A review of practical examples and issues for further action. *Ocean Coast. Manag.* **2002**, *45*, 583–616. [CrossRef]
3. Borja, A.; Ranasinghe, A.; Weisberg, S.B. Assessing ecological integrity in marine waters, using multiple indices and ecosystem components: Challenges for the future. *Mar. Pollut. Bull.* **2009**, *59*, 1–4. [CrossRef] [PubMed]

4. Haward, M. Plastic pollution of the world's seas and oceans as a contemporary challenge in ocean governance. *Nat. Commun.* **2018**, *9*, 667. [CrossRef] [PubMed]
5. Mao, C.; Chen, J.; Yuan, X.; Yang, Z.; Balsam, W.; Ji, J. Seasonal variation in the mineralogy of the suspended particulate matter of the lower Changjiang River at Nanjing, China. *Clay Clay Miner.* **2010**, *58*, 691–706. [CrossRef]
6. Shang, D.; Qin, R.; Xu, H.; Xu, C.; Sun, K.; Zhou, Y. Variation of Suspended Particles in the Bottom Layer of the East China Sea with Data from Seafloor Observatory. *Sensors* **2019**, *19*, 5156. [CrossRef]
7. Hallegraeff, G.M. Ocean climate change, phytoplankton community responses, and harmful algal blooms: A formidable predictive challenge. *J. Phycol.* **2010**, *46*, 220–235. [CrossRef]
8. Griffith, A.W.; Gobler, C.J. Harmful algal blooms: A climate change co-stressor in marine and freshwater ecosystems. *Harmful Algae* **2020**, *91*, 101590. [CrossRef] [PubMed]
9. Hoagland, P.; Kirkpatrick, B.; Jin, D.; Kirkpatrick, G.; Fleming, L.E.; Ullmann, S.G.; Beet, A.; Hitchcock, G.; Harrison, K.K.; Li, Z.C.; et al. Lessening the Hazards of Florida Red Tides: A Common Sense Approach. *Front. Mar. Sci.* **2020**, *7*, 538. [CrossRef]
10. Lebreton, L.; Joost, V.; Damsteeg, J.W.; Slat, B.; Andrady, A.; Reisser, J. River plastic emissions to the world's oceans. *Nat. Commun.* **2017**, *8*, 15611. [CrossRef] [PubMed]
11. Wright, S.L.; Thompson, R.C.; Galloway, T.S. The physical impacts of microplastics on marine organisms: A review. *Environ. Pollut.* **2013**, *178*, 483–492. [CrossRef] [PubMed]
12. Sul, J.; Costa, M.F. The present and future of microplastic pollution in the marine environment. *Environ. Pollut.* **2014**, *185*, 352–364.
13. Pang, C.; Wei, Y.U.; Yang, Y.; Han, D. An improved method for evaluating the seasonal variability of total suspended sediment flux field in the Yellow and East China Seas. *Int. J. Sediment Res.* **2011**, *26*, 1–14. [CrossRef]
14. Bi, N.; Yang, Z.; Wang, H.; Hu, B.; Ji, Y. Sediment dispersion pattern off the present Huanghe (Yellow River) subdelta and its dynamic mechanism during normal river discharge period. *Estuar. Coast. Shelf Sci.* **2010**, *86*, 352–362. [CrossRef]
15. Zhang, T.; Qin, M.; Wei, C.; Li, D.; Lu, X.; Zhang, L. Suspended particles phoD alkaline phosphatase gene diversity in large shallow eutrophic Lake Taihu. *Sci. Total. Environ.* **2020**, *728*, 138615. [CrossRef] [PubMed]
16. Greenstein, K.E.; Wert, E.C. Using rapid quantification of adenosine triphosphate (ATP) as an indicator for early detection and treatment of cyanobacterial blooms. *Water Res.* **2019**, *154*, 171–179. [CrossRef] [PubMed]
17. Gilerson, A.; Zhou, J.; Hlaing, S.; Ioannou, I.; Schalles, J.; Gross, B.; Moshary, F.; Ahmed1, S. Fluorescence component in the reflectance spectra from coastal waters. Dependence on water composition. *Opt. Express* **2007**, *15*, 15702–15721. [CrossRef]
18. Qi, L.; Hu, C.; Duan, H.; Zhang, Y.; Ma, R. Influence of Particle Composition on Remote Sensing Reflectance and MERIS Maximum Chlorophyll Index Algorithm: Examples from Taihu Lake and Chaohu Lake. *IEEE Geosci. Remote. Sens. Lett.* **2017**, *12*, 1170–1174.
19. Zhang, F.; Su, R.; Wang, X.; Wang, L.; He, J.; Cai, M.; Luo, W.; Zheng, Z. A fluorometric method for the discrimination of harmful algal bloom species developed by wavelet analysis. *J. Exp. Mar. Biol. Ecol.* **2009**, *368*, 37–43. [CrossRef]
20. Li, J.; Liao, R.; Tao, Y.; Zhuo, Z.; Liu, Z.; Deng, H.; Ma, H. Probing the Cyanobacterial Microcystis Gas Vesicles after Static Pressure Treatment: A Potential In Situ Rapid Method. *Sensors* **2020**, *20*, 4170. [CrossRef]
21. Polle, J.E.W.; Benemann, J.R.; Tanaka, A.; Melis, A. Photosynthetic apparatus organization and function in the wild type and a chlorophyll b-less mutant of *Chlamydomonas reinhardtii*. Dependence on carbon source. *Planta* **2000**, *211*, 335–344. [CrossRef] [PubMed]
22. Khanal, R.; Uk, S.; Kodikara, D.; Siev, S.; Yoshimura, C. Impact of water level fluctuation on sediment and phosphorous dynamics in tonle sap lake, cambodia. *Water Air Soil Poll.* **2021**, *232*, 139. [CrossRef]
23. Okullo, W.; Ssenyonga, T.; Hamre, B.; Frette, O.; Sorensen, K.; Stamnes, J.J.; Steigen, A.; Stamnes, K. Parameterization of the inherent optical properties of Murchison Bay, Lake Victoria. *Appl. Opt.* **2007**, *46*, 8553–8561. [CrossRef]
24. Zhou, J.; Gilerson, A.; Ioannou, I.; Schalles, J.; Gross, B.; Moshary, F.; Ahmed, S.A. Retrieving quantum yield of sun-induced chlorophyll fluorescence near surface from hyperspectral in-situ measurement in productive water. *Opt. Express* **2008**, *16*, 17468–17483. [CrossRef] [PubMed]
25. Sandven, H.; Kristoffersen, A.S.; Chen, Y.C.; Hamre, B. In situ measurements of the volume scattering function with LISST-VSF and LISST-200X in extreme environments: Evaluation of instrument calibration and validity. *Opt. Express* **2020**, *28*, 37373–37396. [CrossRef]
26. Olson, R.J.; Sosik, H.M. A submersible imaging-in-flow instrument to analyze nano-and microplankton: Imaging FlowCytobot. *Limnol. Oceanogr. Meth.* **2007**, *5*, 195–203. [CrossRef]
27. Olson, R.J.; Shalapyonok, A.; Kalb, D.J.; Graves, S.W.; Sosik, H. Imaging FlowCytobot modified for high throughput by in-line acoustic focusing of sample particles. *Limnol. Oceanogr. Meth.* **2017**, *15*, 867–874. [CrossRef]
28. He, H.; He, C.; Chang, J.; Lv, D.; Wu, J.; Duan, C.; Zhou, Q.; Zeng, N.; He, Y.; Ma, H. Monitoring microstructural variations of fresh skeletal muscle tissues by *Mueller matrix* imaging. *J. Biophotonics* **2017**, *10*, 664–673. [CrossRef] [PubMed]
29. Feng, S.; Chen, R.; Lin, J.; Pan, J.; Wu, Y.; Li, Y.; Chen, J.; Zeng, H. Gastric cancer detection based on blood plasma surface-enhanced *Raman spectroscopy* excited by polarized laser light. *Biosens. Bioelectron.* **2011**, *26*, 3167–3174. [CrossRef]
30. Shen, Y.; Huang, R.; He, H.; Liu, S.; Dong, Y.; Wu, J.; Ma, H. Comparative study of the influence of imaging resolution on linear retardance parameters derived from the *Mueller matrix*. *Biomed. Opt. Express* **2020**, *12*, 211–225. [CrossRef]
31. Chen, Y.; Zeng, N.; Chen, S.; Zhan, D.; He, Y.; Ma, H. Study on morphological analysis of suspended particles using single angle polarization scattering measurements. *J. Quant. Spectrosc. Radiat. Transf.* **2019**, *224*, 556–565. [CrossRef]

32. Sun, M.; He, H.; Zeng, N.; Du, E.; Guo, Y.; Liu, S.; Wu, J.; He, Y.; Ma, H. Characterizing the microstructures of biological tissues using Mueller matrix and transformed polarization parameters. *Biomed. Opt. Express* **2014**, *5*, 4223–4234. [CrossRef] [PubMed]
33. Wang, Y.; Liao, R.; Dai, J.; Liu, Z.; Xiong, Z.; Zhang, T.; Chen, H.; Ma, H. Differentiation of suspended particles by polarized light scattering at 120°. *Opt. Express* **2018**, *26*, 22419–22431. [CrossRef] [PubMed]
34. Liao, R.; Li, Q.; Mao, X. A prototype for detection of particles in sea water by using polarize-light scattering. In Proceedings of the OCEANS 2019, Marseille, France, 17–20 June 2019.
35. Wang, Y.; Dai, J.; Liao, R.; Zhou, J.; Meng, F.; Yao, Y.; Chen, H.; Ma, H. Characterization of physiological states of thesuspended marine microalgae using polarized light scattering. *Appl. Opt.* **2020**, *59*, 1307–1312. [CrossRef] [PubMed]
36. Zhou, W.; Wang, G.; Li, C.; Xu, Z.; Gao, W.; Shen, F. Retrieval of phytoplankton cell size from chlorophyll a specific absorption and scattering spectra of phytoplankton. *Appl. Opt.* **2017**, *56*, 8362. [CrossRef] [PubMed]
37. Coesel, S.N.; Baumgartner, A.C.; Teles, L.M.; Ramos, A.A.; Henriques, N.M.; Cancela, L. Nutrient limitation is the main regulatory factor for carotenoid accumulation and for psy and pds steady state transcript levels in *Dunaliella salina* (chlorophyta) exposed to high light and salt stress. *Mar. Biotechnol.* **2008**, *10*, 602–611. [CrossRef] [PubMed]
38. Xu, X.J.; Lundholm, N.; Li, Y. A study of *Chaetoceros debilis* sensu lato species (bacillariophyceae), with ememdation of *C. debilis* and description of *C. galeatus* sp. nov. 1. *J. Phycol.* **2020**, *56*, 784–797. [CrossRef]
39. Song, H.; Chen, Y.; Kg, E.; Liu, S.; Yu, Z.; Chen, N. High genetic diversity of the harmful algal bloom species *Phaeocystis globosa* revealed using the molecular marker cox1. *Harmful Algae* **2021**, *107*, 102065. [CrossRef]
40. Weng, H.X.; Qin, Y.C.; Sun, X.W.; Dong, H.; Chen, X.H. Iron and phosphorus effects on the growth of *Cryptomonas* sp. (cryptophyceae) and their availability in sediments from the pearl river estuary, china. *Estuar. Coast. Shelf Sci.* **2007**, *73*, 501–509. [CrossRef]
41. Li, W.; Yang, Y.; Li, Z.; Xu, J.; Gao, K. Effects of seawater acidification on the growth rates of the diatom *Thalassiosira* (conticribra) weissflogii under different nutrient, light, and UV radiation regimes. *J. Appl. Phycol.* **2017**, *29*, 133–142. [CrossRef]
42. He, C.; He, H.; Chang, J.; Chen, B.; Ma, H.; Booth, M.J. Polarisation optics for biomedical and clinical applications: A review. *Light-Sci. Appl.* **2021**, *10*, 194. [CrossRef] [PubMed]
43. Lu, S.Y.; Chipman, R.A. Mueller matrices and the degree of polarization. *Opt. Commun.* **1998**, *146*, 11–14. [CrossRef]
44. Parker, J.R. Rank and response combination from confusion matrix data. *Inform. Fusion* **2000**, *2*, 113–120. [CrossRef]

sensors

MDPI

Article

Stokes Dynamic Polarimeter for Non-Organic and Organic Samples Characterization

Dora-Luz Almanza-Ojeda [1], Daniela Rodriguez-Sotelo [1], Rogelio Castro-Sanchez [1,*], Rene Martinez-Celorio [2] and Mario-Alberto Ibarra-Manzano [1]

[1] Department of Electronics Engineering, Universidad de Guanajuato, Salamanca 36885, Mexico; dora.almanza@ugto.mx (D.-L.A.-O.); daniela.rodriguez@ugto.mx (D.R.-S.); ibarram@ugto.mx (M.-A.I.-M.)
[2] Physics Department, Miami Dade College, Miami, FL 33132, USA; rmarti23@mdc.edu
* Correspondence: castro@ugto.mx; Tel.: +1-464-647-9940

Abstract: The light polarization properties provide relevant information about linear–optical media quality and condition. The Stokes–Mueller formalism is commonly used to represent the polarization properties of the incident light over sample tests. Currently, different Stokes Polarimeters are mainly defined by resolution, acquisition rate, and light to carry out accurate and fast measurements. This work presents the implementation of an automatic Stokes dynamic polarimeter to characterize non-biological and biological material samples. The proposed system is configured to work in the He-Ne laser beam's reflection or transmission mode to calculate the Mueller matrix. The instrumentation stage includes two asynchronous photoelastic modulators, two nano-stepper motors, and an acquisition data card at 2% of accuracy. The Mueller matrix is numerically calculated by software using the 36 measures method without requiring image processing. Experiments show the efficiency of the proposed optical array to calculate the Mueller matrix in reflection and transmission mode for different samples. The mean squared error is calculated for each element of the obtained matrix using referenced values of the air and a mirror. A comparison with similar works in the literature validates the proposed optical array.

Keywords: light polarization; Mueller matrix; photoelastic modulator; synchronization; surface fruit

Citation: Almanza-Ojeda, D.-L.; Rodriguez-Sotelo, D.; Castro-Sanchez, R.; Martinez-Celorio, R.; Ibarra-Manzano, M.-A. Stokes Dynamic Polarimeter for Non-Organic and Organic Samples Characterization. *Sensors* **2022**, 22, 2155. https://doi.org/10.3390/s22062155

Academic Editor: Maria Lepore

Received: 28 January 2022
Accepted: 6 March 2022
Published: 10 March 2022

1. Introduction

The design of biosensors has recently shown a considerable advance in human diagnosis through proteins and enzymes detection that characterize bacteria and virus diseases [1]. In detecting specific viruses such as SARS-CoV 2 [2] and more biomedical and clinical diagnoses [3], some of these biosensors use the polarization of light as a detection technique. The study of polarized light and methods of polarization measurements have increased due to their variety and vast applications. Polarimetry techniques allow for the identification of properties of surfaces by analyzing the changes in the polarized light [4]. Some applications of polarized light include the characterization of microstructures in biological tissues [5], the prevention and detection of illnesses like cancer or cirrhosis [6,7], materials classification [8–10], and analysis of components on industrial products, among many other biomedical and industrial areas. The Mueller matrix is used to describe the interaction between light and different optical means to develop potential methods of diagnosis in biomedical science [11,12]. The Mueller matrix is the numerical representation of the polarization states of incident light on the sample surface. Thus, it plays a vital role in characterizing samples' structural properties associated with their Stokes Vectors.

The Stokes–Mueller theory [13] includes 36 measurements representing the different states of the polarization of light. The conventional optical array comprises devices such as polarizers and retarder waveplates. Usually, instrument calibration based on polarized light [14] is carried out by calculating the Mueller matrix for one or several optical elements, considered as reference, and shown in the state-of-the-art [15–17]. Once the polarimeter is calibrated, the samples under test are measured.

A polarimeter is considered dynamic when its design includes photoelastic modulators. These modulators generate output signals based on the birefringence properties [18] produced by the SiO2 piece located at the electronic circuit's output. The birefringence properties change in function of the input voltage of the modulator, generating linear and circular polarization states in the output signal. Thus, the dynamic modulation range of the polarimeter works with a random and polarized light source that passes through the optical array composed of polarizers and photoelastic modulators. Usually, this assembly can be used in transmission or reflection mode for different material samples.

This paper presents the design and implementation of a dynamic polarimeter in transmission and reflection mode for calculating the Mueller matrix of different samples. Electronic components are chosen to provide high accuracy and speed during dynamic measurements. The advantage of the proposed system is to compute the Mueller matrix using the 36 measurements method without using image processing. The instrumented array automatically acquired the measurements to calculate the numerical coefficients of the matrix. The matrices are obtained using a Graphical Unit Interface based on LabView of National Instruments®. The proposed polarimeter is calibrated by calculating the Mueller matrices of the air and a mirror and comparing with reference values presented in state-of-the-art. Then, preliminary apple and banana matrices are proposed for establishing a quantitative reference of the polarization properties of fruit tissues. To the best of our knowledge, Mueller matrices have not been submitted to describe the behavior of the polarized light through the fruit surfaces. However, the main contribution of our system is the instrumentation and control device to automatically perform the polarization states set up, which allows an economical, accurate, and easy-to-use polarimeter. The proposed electronic components can be seen as modules that can be included in the whole optical array and the PC for matrices calculation by software. Overall system performance is validated experimentally, achieving high values of repeatability and precision.

The rest of the document explains the design and experimental setup of the proposed system. The theoretical concepts of polarization properties of light are defined in Section 2. Section 3 describes the optical array to implement the transmission and reflection modes of the system. Section 4 shows the experimental results using different types of medium and compares them with similar findings. Conclusions and perspectives of the proposed work are presented at the end of this document.

2. Background

The Mueller matrix plays an essential role in characterizing the material samples' microstructural and physical properties. According to the polarization states, different interactions occur between the incident light and the material sample. The individual elements of the matrix register the incident beam changes during the light-sample interaction. Hence, calculating such numerical elements in the matrix requires the study of essential concepts about the polarized light generation and the Stokes formalism, which will be briefly presented in the following subsections.

2.1. Polarized Light Generator

A Photoelastic Modulator (PEM) and a polarizer could be time-adjusted by software to generate different polarization states of a reflected beam of light. The variations on the reflected beam show a periodicity on the time domain that allows proposing an array to measure the polarization states dynamically. The basic idea of the PEM is to modulate polarized light [18] that passes through its optical header when the amplitude of the applied periodic voltage varies. This amplitude variation modifies the property of optical birefringence. The optical header of PEM is a device that generates a programmed phase-retarded between the orthogonal components of the electric field. Thus, being the light a transversal electromagnetic wave, the beam's polarization state allows modulating periodically in the time domain the polarization state of the light beam. PEM is used in multiple experimental methods [19,20], showing essential features such as high sensibility, broad spectral range, and high precision in phase modulation under the Stokes-Mueller formalism.

2.2. Stokes-Mueller Formalism

Recently, light polarimetry contributed to analyzing and diagnosing biological tissues, revealing essential information about healthy or pathological tissue status [21]. The structural features in organic or non-organic samples alter the incident light in the sample surface. The changes produced on the incident light after interacting with a material sample can be described by the elements of the Mueller matrix. These elements represent the coefficients of the incident and output Stokes vector. Figure 1 depicts the Stokes–Mueller formalism. An input Stokes vector $\mathbf{S}_i$ affected by a linear system with optical features given by M provides the output Stokes vector $\mathbf{S}_o$. M is the Mueller matrix that models the optical properties of any material sample under test.

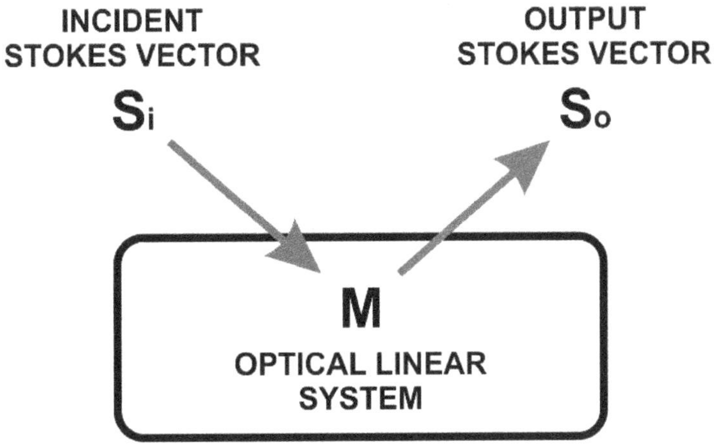

Figure 1. Incident beam light interacting on a lineal optical media.

The mathematical representation of Stokes vectors is defined by Equation (1) [16]:

$$\mathbf{S}_o = \mathbf{M}\mathbf{S}_i \quad (1)$$

where **M** is the Mueller matrix, $\mathbf{S}_x$ represents the lineal polarization states of incident light ($x = i$) and reflected light ($x = o$) into the optical media. Linear polarization states of light are composed of: linear horizontal (h), linear vertical (v), linear to $45°(+)$, linear to $135°(-)$, and circular polarization states, right circular (r) and left circular (l). Combining these states helps determine variations on physical structures of material samples during polarized light incidence.

The Mueller matrix of a lineal optical media is a 4×4 matrix whose elements provide the media's anisotropy information [22]. In this work, the Mueller matrix is calculated using the 36 measurements method [23]. Thus, six polarization states are generated from the incident source of light ($\mathbf{S}_i$); six reflected light ($\mathbf{S}_o$) states are analyzed. This method is mainly used to measure unknown optical properties of samples under tests (arbitrary physical systems). Equation (2) illustrates Mueller matrix elements based on the irradiances I. Each element m_{ab} of the matrix depends on two irradiances I_{xy}. The subscripts at each member of M represent the polarized input states established by the incident Stokes vector $\mathbf{S}_i$ and the polarized output states corresponding to the reflected Stokes vector $\mathbf{S}_o$ [22].

$$\begin{array}{ll} m_{11} = \frac{1}{2}\left(I_{hh} + I_{hv} + I_{vh} + I_{vv}\right) & m_{12} = \frac{1}{2}\left(I_{hh} + I_{hv} - I_{vh} - I_{vv}\right) \\ m_{13} = \frac{1}{2}\left(I_{+h} + I_{+v} - I_{-h} - I_{-v}\right) & m_{14} = \frac{1}{2}\left(I_{rh} + I_{rv} - I_{lh} - I_{lv}\right) \\ m_{21} = \frac{1}{2}\left(I_{hh} - I_{hv} + I_{vh} - I_{vv}\right) & m_{22} = \frac{1}{2}\left(I_{hh} - I_{hv} - I_{vh} + I_{vv}\right) \\ m_{23} = \frac{1}{2}\left(I_{+h} - I_{+v} - I_{-h} + I_{-v}\right) & m_{24} = \frac{1}{2}\left(I_{rh} - I_{rv} - I_{lh} + I_{lv}\right) \\ m_{31} = \frac{1}{2}\left(I_{h+} - I_{h-} + I_{v+} - I_{v-}\right) & m_{32} = \frac{1}{2}\left(I_{h+} - I_{h-} - I_{v+} + I_{v-}\right) \\ m_{33} = \frac{1}{2}\left(I_{++} - I_{+-} - I_{-+} + I_{--}\right) & m_{34} = \frac{1}{2}\left(I_{r+} - I_{r-} - I_{l+} + I_{l-}\right) \\ m_{41} = \frac{1}{2}\left(I_{hr} - I_{hl} - I_{vl} + I_{vr}\right) & m_{42} = \frac{1}{2}\left(I_{hr} - I_{hl} - I_{vr} + I_{vl}\right) \\ m_{43} = \frac{1}{2}\left(I_{+r} - I_{+l} - I_{-r} + I_{-l}\right) & m_{44} = \frac{1}{2}\left(I_{rr} - I_{rl} - I_{lr} + I_{ll}\right) \end{array} \quad (2)$$

3. Materials and Methods

3.1. Experimental Setup

The schematic diagram of the proposed dynamic polarimeter in reflection mode is shown in Figure 2. The dynamic polarimeter is an optical assembly configured in reflection mode that consists of five parts: a light source module (laser), a polarization state generator (PSG), a sample under test, a polarization state analyzer (PSA), and a photodetector (PD).

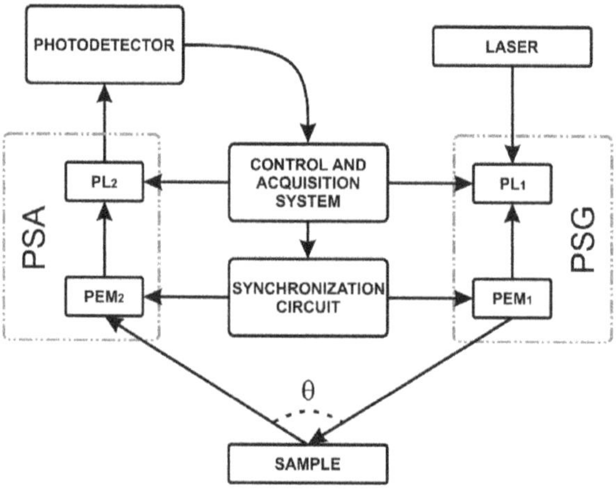

Figure 2. Block diagram of the dynamic polarimeter in reflection mode.

Hence, the first array adopted was configured in reflection mode composed of a 632 nm He-Ne Laser with power output at 17 mW in random polarization. The light beam passes through the polarization state generator (PSG) to generate the six polarization states. The PSG consists of a linear polarizer with a transmission axis fixed at +45° to the horizontal axis; additionally, it includes an optical head PEM100 Photoelastic modulator (Hinds Instruments, OR, USA, [18]) with a transmission axis aligned to 0° with the horizontal axis. After that, polarized light is transmitted or reflected through a sample, which could also be a material substance or the air. Next, the polarization state Analyzer (PSA) consists of an optical array to analyze six polarization states of the light. This PSA includes a Photoelastic modulator (PEM100, Hinds Instruments®, OR, USA), with a transmission axis aligned to 0° with the horizontal axis and a transmission axis oriented to −45° with the horizontal. Both polarizers, PSG and PSA, use two Stepper motors (NR360S, Thorlabs®, Newton, NJ, USA) to generate the combination of circular and linear polarization states. Finally, the proposed assembly includes a switchable gain photodetector PD (PDA36A, Thorlabs®, Newton, NJ, USA) to measure the modulated polarized light.

As mentioned above, the main contribution of the proposed optical assembly is the generation of the circular and linear polarization states autonomously, combining static and dynamic methods and employing an embedded synchronization circuit. The six polarization states are generated in the PSG block and six more in the PSA block. The combination among these polarization states generates the 36 measurements in the photodetector to determine the Mueller matrix. From these 36 measurements of the polarization states, 16 correspond to linear states, and 20 to linear and circular states. The overall methodology consists of four stages: (1) alignment, (2) generation of linear polarization states, (3) generation of circular polarization states, and (4) synchronization. These four stages were also instrumented in transmission mode for this study.

3.2. Alignment

The calibration of the proposed polarimeter requires that each component of the experimental assembly be aligned. Initially, the alignment is performed for transmission mode by setting the light source, the PSG, the sample, the PSA, and the photodetector in the same line as illustrated in Figure 3. A sequence of pulses emitted by software allows extracting minimum, mean, maximal, and normalized values for the air. After that, the 36 measures are automatically computed on the software interface (LabVIEW) to obtain the Mueller matrix of the air as a reference. The matrix of the air shows a positive "1" value on its diagonal elements [24]. Hence, the PSG or PSA must be realigned when diagonal values differ to "1".

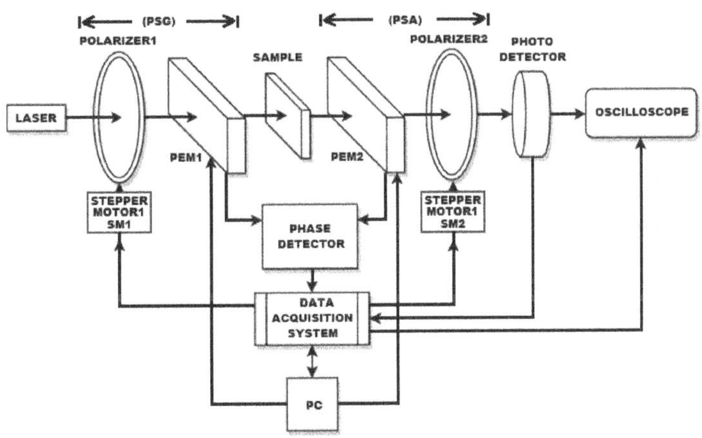

Figure 3. Block diagram of the dynamic polarimeter alignment in transmission mode.

As illustrated in Figure 2, the light source module is oriented at 30° in the PSA and photodetector directions in reflection mode. This configuration setup uses a front surface mirror as a reference sample. Like transmission mode, 36 measurements are automatically calculated using the graphical user interface on LabVIEW to determine the Mueller matrix parameters. Ideally, the diagonal elements m_{11} and m_{22} in the Mueller matrix of the mirror are close to 1. Unlike, m_{33} and m_{44} are close to -1. The diagonal elements with low similarity in the ideal values indicate that the PSG must be realigned.

3.3. Combination of Linear Polarization States

The generation and analysis of the linear polarization states require the PSG and PSA stages. Each polarizer is mounted on an N360S motorized rotation stage of Thorlabs®. A control system programs both motors to set the transmission axis for each polarizer in four specific positions. The 16 combinations required for input and output are calculated from these positions. The data card model USB-6259 distributed by National Instruments® acquires the signals of each polarization state to the PC and generates the synchronized signals to control the instrumented array. Thus, the modules are automated using LabVIEW®. Figure 4 shows the flow diagram programmed in a subVI of LabView software. This strategy implements the proposed dynamic polarimeter that generates and controls various position combinations for each stepper motor.

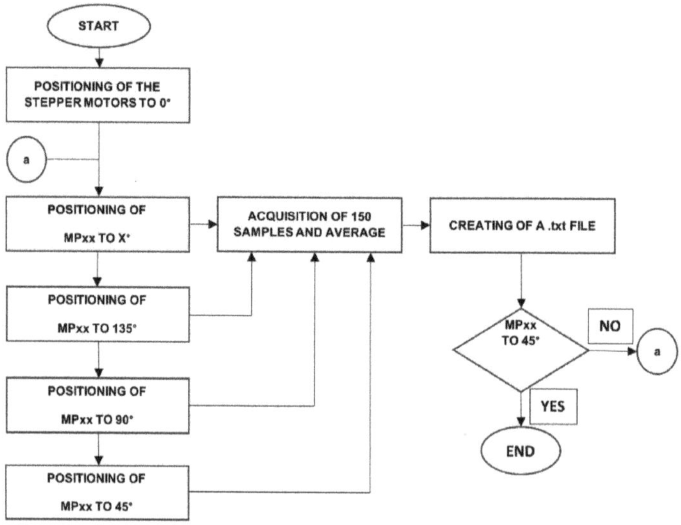

Figure 4. Block diagram of acquisition and states generation of the linear polarimeter.

Linearly polarized light generates polarization by setting off the PEM optical headers but keeping on the data acquisition process. Table 1 shows the combinations performed by our optical array system in transmission mode to determine the Mueller matrix for different linear-optical media. Initially, the reference point of the stepper motors is set to 0°. The first column of Table 1 shows the combination of linear polarization states used for passing a laser beam through the transmission axis of the polarizers. The second column shows each position of the stepper motor. The third and fourth columns illustrate the theoretical and normalized expected values.

Table 1. Setup parameters to calculate linear polarization states and the normalized irradiance.

State	Angular Position PL1/PL2	Theoretical Irradiance	Normalized Irradiance
hh	0°/0°	Maximal	1
$++$	45°/45°		
vv	90°/90°		
$--$	135°/135°		
$h+$	0°/45°	Medium	0.5
$+h$	45°/0°		
h	0°/135°		
$-h$	135°/0°		
$v+$	90°/45°		
$+v$	45°/90°		
v	90°/135°		
$-v$	135°/90°		
hv	0°/90°	Minimal	0
vh	90°/0°		
$l-$	45°/135°		
$-+$	135°/45°		

Several measurements had been acquired to increase the number of samples taken from the polarization states of Table 1. These measurements are averaged for each pair of polarization states during a fixed time interval. The measurements are stored in a file to calculate the Mueller matrix elements shown in Equation (2).

Linear and Circular Polarization States Combinations

Figures 2 and 3 show the schematic representation of the instrumented optical array used to generate linear and circular polarization states. Table 2 illustrates the 20 linear and 16 circular polarizations states to determine the Mueller matrix. The first column of Table 2 shows the states generated and analyzed. Linear polarizers PL1 and PL2 are presented in the second and fifth columns. The third and sixth columns illustrate the combined on/off conditions for the PEM1 and PEM2. The fourth and seventh columns show retarder waveplates values (λ) introduced to generate some circular polarization states.

Table 2. Parameters to generate linear and circular polarization states.

State	Angle of PL_1 (PSG)	PEM_1 Status (PSG)	Λ	Angle of PL_2 (PSA)	PEM_2 Status (PSA)	λ
$-l, -r$	135°	Off	-	135°	On	$\lambda/4$
$l-, r-$	45°	On	$\lambda/4$	135°	Off	-
$l+, r+$	45°	On	$\lambda/4$	45°	Off	-
$+l, +r$	45°	Off	-	135°	On	$\lambda/4$
hr, hl	0°	Off	-	135°	On	$\lambda/4$
rh, lh	45°	On	$\lambda/4$	0°	Off	-
vr, vl	90°	Off	-	135°	On	$\lambda/4$
rv, lv	45°	On	$\lambda/4$	90°	Off	-
rr, ll, rl, lr	45°	On	$\lambda/4$	135°	On	$\lambda/4$

Using Table 2, linear and circular polarization states are generated by combining different angles and states of the polarimeters. This experiment is similar to the states generated using the PSG and PSA blocks in reflection mode shown in Figure 2. For instance, in the four first states ($-l, -r$), the state "$-$" (represented by a minus sign) is performed by setting to 135° the PL1 polarizer in the PSG block and setting the PEM1 to off. A similar combination is performed for the "l" state, but in this case, the PL2 polarizer in the PSA block is set to 135°. The PEM2 is also on and delayed by $\lambda/4$. The two states ($-r$) are generated and analyzed using similar values for PSG and PSA. The states l and r are dynamically generated at different times, while PEM2 is set to *on* and delayed by $\lambda/4$ programmed on LabVIEW. Similarly, the rest of the states proposed in Table 2 are generated. In the last row of Table 2, for the states (rr, ll, rl, lr), both PEM1 and PEM2 are turned on and synchronized [25]; furthermore, PL1 and PL2 polarizers are set to 45° and $-45°$ to the horizontal, respectively [26].

3.4. Synchronization

The dynamic polarimeter proposed uses two optical heads (Hinds Instruments) to integrate the PEM modules. In practice, we found a difference of around 6 Hz in the oscillation frequencies in both devices. The measurements registered from PEM modules must be synchronized to provide correct data values. A phase detector is included to synchronize both PEM modules with a reference point. To set this reference point, we measure the phase difference between both oscillation frequencies oriented at 0° and 180°. For that purpose, we take advantage of each PEM that provides a TTL output signal equivalent to the input signal frequency that passes through the PEMs. Then, a digital phase detector composed of a type II phase comparator is used to detect TTL digital output and the frequency differences to achieve the synchronization PEMs [27].

Figure 5 shows the graphical output signal of the phase-cero detector circuit monitored by an oscilloscope to demonstrate the functionality of the synchronization stage. This circuit provides an output digital pulse ④ when the phase difference is 0° between ① and ② input signals. This pulse is used as a reference if periodic polarization states in the PEMs are detected. Using PEMs synchronization parameters, it is possible to obtain 20 measurements in specific points related to the combination of linear and circular polarization states. In our case, the reference point is the transition from low to high level at the output of the phase detector (see channel ④ in the oscilloscope).

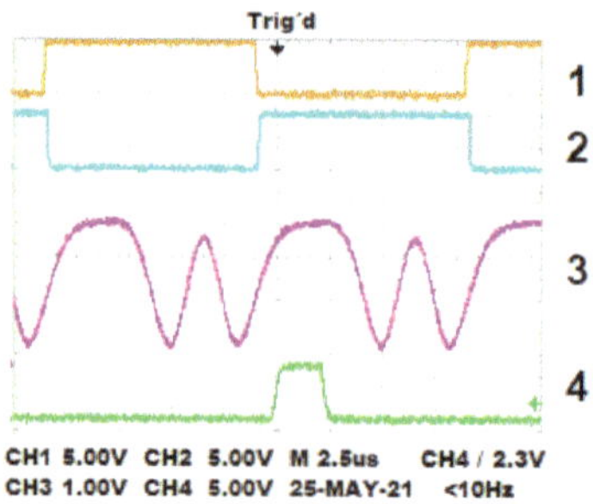

Figure 5. Measured signals: (1) TTL output of PEM1, (2) TTL output of PEM2, (3) output signal of the photodetector, (4) output signal of the phase detector.

Figure 6 illustrates the use of the phase detector to synchronize the NI-USB acquisition card and the oscilloscope during the detection of the polarization states.

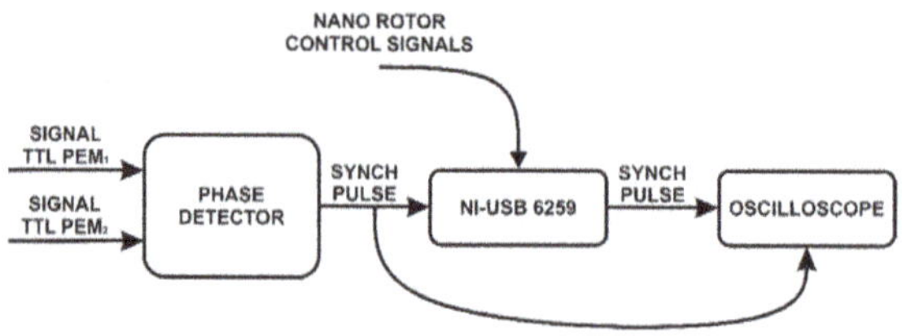

Figure 6. Block diagram of acquisition and states generation of circular polarization.

The polarization states *r* and *l* in the signal propagated through the optical header of each PEM are specified in the datasheet device [18]. Figure 7 illustrates the PEMs monitored by an oscilloscope. Two dynamic polarization states are measured using the acquisition data card at the output of the zero-phase detector. Figure 7a displays the oscilloscope screen for *r* and *l* polarization states, and Figure 7b shows the *l*,*l* states.

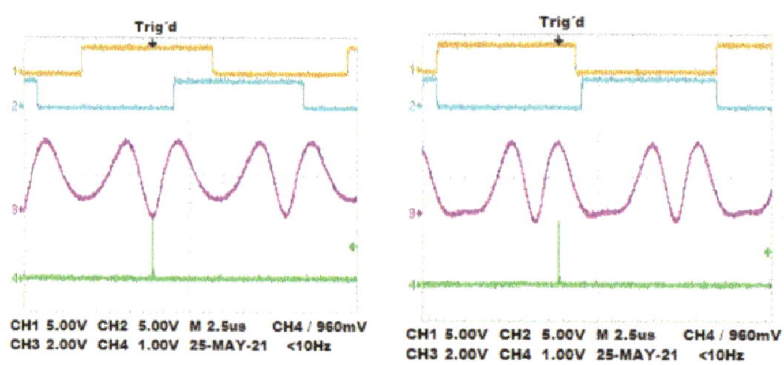

(a) (b)

Figure 7. The plot of the polarization states generated. (**a**) Combinations for the *r*,*l* states. (**b**) Combinations for the *l*,*l* states.

Once the system is referenced to a specific point in time established for the synchronization process, we acquire various measurements automatically at this reference point and calculate the mean values of the instantaneous measurements for obtaining a reliable value.

4. Results

Experimental tests validate the performance and accuracy of our dynamic polarimeter assembly. In the experiments, the room temperature was between 20 °C and 25 °C, and the laser was pre-heated for 30 min before use. For each set of measurements, the overall system is calibrated by software using frequency differences between the PEMs. To obtain the differences, we average the location of maximum and minimum values measured at the output of the photodetector. These values could vary depending on the selected gain in the photodetector; in these experimental tests, we used a fixed gain of 20 db. The retarded output signal pulse measurements in the phase detector showed a periodicity of 100 ns. The air and a mirror are used as standard samples in transmission and reflection mode, respectively, to verify the system calibration. Figure 8 shows the proposed optical system implemented in reflection mode for two different degrees of orientation, 30° and 165°.

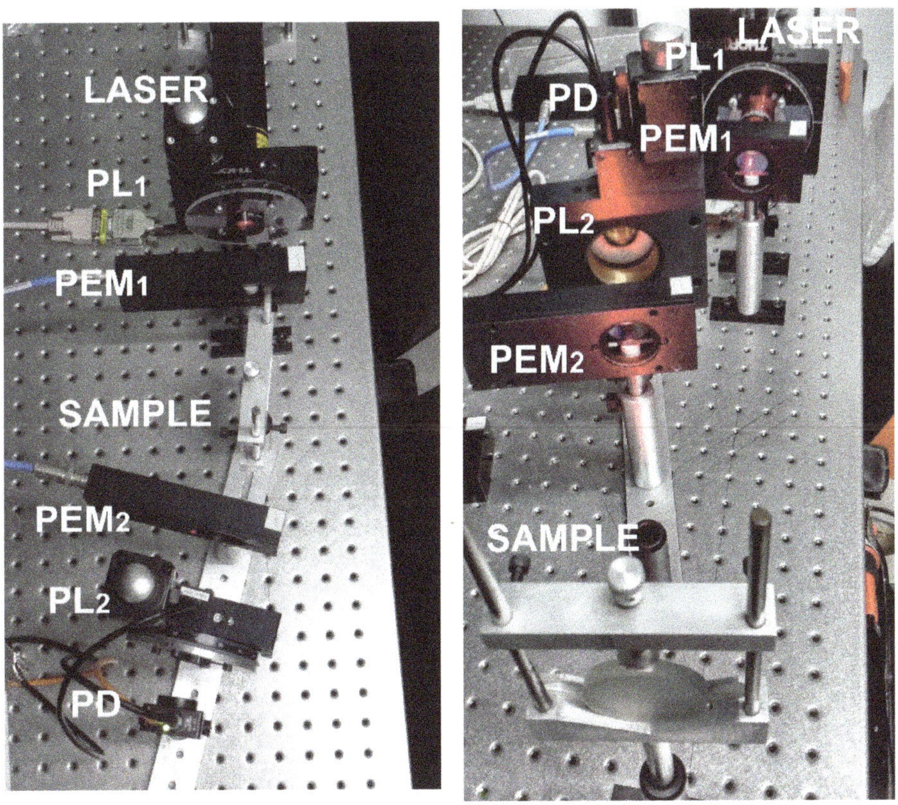

(a) (b)

Figure 8. Dynamic Stokes polarimeter in reflection mode: (**a**) oriented at 30°; (**b**) oriented at 165°.

Each Mueller matrix is compared with its corresponding ground-truth matrix proposed in the literature [22]. The difference between the proposed matrix and the ideal matrix established for standard samples is calculated using the mean square error metric (see Equation (3)). The mean error obtained indicates the deviation of matrix elements to their ideal values but in general. The further numerical analysis allows the determination of the specific details in the matrix with maximal differences.

$$MSE = \frac{1}{16}\sum_{i=1}^{4}\sum_{j=1}^{4}\left|M_{ij}^{ideal} - M_{ij}\right| \quad (3)$$

The following subsections describe the Mueller matrix measured by our proposed dynamic polarimeter and compare the matrix elements that show the higher accuracy error. The comparison results demonstrate the reliability of the instrumented system capable of measuring linear and circular polarization effects of some organic and non-organic samples without using CCD sensors. Additional methods, like Mueller matrix transformation or decomposition, provide a set of microstructural properties of most biological tissues; however, the comparison and computation of such properties are reserved for future works.

4.1. Mueller Matrix in Transmission Mode

The Mueller matrix in transmission mode measures two samples: the air and a linear polarizer with the transmission axis perpendicular to the horizontal. The first row of Table 3 shows our proposed Mueller matrix for the air [16]. According to this reference matrix, the diagonal elements are one or close to one, indicating that our system is calibrated correctly. The mean error indicates 2.12%, being m_{13}, m_{23}, and m_{32} the elements with maximal error. The remaining elements of the matrix show a minimal accuracy error regarding ideal values. This comparison indicates that such elements must be considered the most critical during matrix calibration due to the higher accuracy errors.

Table 3. Mueller matrix for air and a polarizer with the transmission axis set to 90°.

Sample	Experimental Mueller Matrix	Mean Error (e)	Elements with Higher Error
Air	$\begin{bmatrix} 1.0000 & -0.0203 & -0.0407 & -0.0032 \\ -0.0204 & 0.9936 & 0.0677 & -0.0007 \\ 0.0819 & -0.0673 & 1.0165 & -0.0009 \\ -0.0013 & 0.0009 & -0.0052 & 0.9947 \end{bmatrix}$	0.0212	m_{13} m_{23} m_{32}
Polarizer 90°	$\begin{bmatrix} 1.0000 & -0.9782 & -0.0540 & 0.0014 \\ -0.9786 & 0.9767 & 0.0536 & -0.0012 \\ 0.0727 & -0.0725 & -0.0242 & -0.0004 \\ -0.0010 & 0.0015 & -0.0042 & 0.0040 \end{bmatrix}$	0.0223	m_{13} m_{23} m_{31} m_{32}

The second row in Table 3 shows the Mueller matrix of a polarizer with the transmission axis set to 90° to the horizontal. The elements m_{11}, m_{12}, m_{21}, and m_{22} in this matrix are close to 1, similar to ideal values proposed in the literature for a linear polarizer [28]. The mean error obtained for this matrix is 2.23%, being m_{13}, m_{23}, m_{31}, and m_{32} the elements with maximal error.

4.2. Mueller Matrix in Reflection Mode

A typical material used for calculating Mueller matrices is a front mirror surface. The surface of this mirror perfectly reflects a beam of light like a half-wave plate [4,25,26]. As a reference, we calculate the matrix for a front mirror surface in reflection mode, yielding an error of 0.31% compared with previous work (see Table 4). For a mirror, m_{33} and m_{44} diagonal elements show a minus sign [29], validating the assumption that a front mirror surface is perfectly reflected, like a retarder half-wave plate.

Table 4. Mueller matrix of a mirror in reflection mode.

Sample	Experimental Mueller Matrix	Mean Error (e)
Mirror	$\begin{bmatrix} 1.0000 & 0 & 0.0013 & 0.0037 \\ 0 & 1.0000 & -0.0009 & 0.0028 \\ 0.0005 & 0.0001 & -1.0000 & -0.0003 \\ 0.0065 & 0.0018 & 0.0012 & -0.9702 \end{bmatrix}$	0.0031

Before testing with organic samples, similar experimental tests were performed with the dynamic polarimeter in reflection mode using different reflected surfaces. Thus, we replaced the front mirror surface with a polarizer with the transmission axis oriented at 90° and a rough aluminum surface. The experimental results are shown in Table 5. The Mueller matrices for both elements are not provided in the literature to the best of our knowledge. Hence, we have included both matrices without an MSE score. Note that diagonal elements in both matrices show similar behavior to matrix elements of a mirror; the first two elements are positive and the last ones negative. The experimental array mounted for measuring rough aluminum surface is illustrated in Figure 9.

Table 5. Mueller matrix of a polarizer with the transmission axis oriented at 90° and aluminum surface in reflection mode.

Sample	Experimental Mueller Matrix
Polarizer 90°	$\begin{bmatrix} 0.9644 & -0.0228 & 0.1286 & 0.0008 \\ -0.0119 & 0.9529 & -0.2915 & 0.0910 \\ -0.1615 & 0.0972 & -0.9749 & 0.0136 \\ 0.0260 & 0.0650 & 0.0502 & -0.9463 \end{bmatrix}$
Aluminum surface	$\begin{bmatrix} 1.0079 & 0.0073 & 0.1743 & 0.0052 \\ -0.0082 & 0.9808 & -0.3409 & -0.0019 \\ 0.1497 & 0.1023 & -0.9799 & -0.0011 \\ -0.0022 & -0.0018 & 0.0014 & -0.9736 \end{bmatrix}$

Figure 9. Dynamic Stokes polarimeter in reflection mode through the aluminum surface.

4.3. Mueller Matrix for Fruit Surfaces

The dynamic polarimeter is validated using different organic samples in reflection modes, such as an apple and a banana. Table 6 illustrates the corresponding Mueller matrices calculated in reflection mode.

Table 6. Reflection Mueller matrices in fruit surfaces.

Sample	Experimental Mueller Matrix
Apple surface	$\begin{bmatrix} 0.9912 & -0.0076 & -0.0106 & 0.0012 \\ -0.0391 & 0.9301 & -0.6605 & -0.0020 \\ 0.1022 & 0.0972 & -0.9749 & 0.0136 \\ -0.0020 & 0.0002 & 0.0030 & -0.3438 \end{bmatrix}$
Banana surface	$\begin{bmatrix} 0.9451 & -0.0418 & 0.0025 & -0.0008 \\ -0.0691 & 0.8942 & -0.6369 & 0.0025 \\ -0.1848 & 0.0212 & -0.6108 & 0.0003 \\ -0.0001 & -0.0002 & 0.0030 & -0.1874 \end{bmatrix}$

Figures 10 and 11 show the physical experimental array implemented to measure the Mueller matrices of an apple and a banana, respectively. The m_{11}, m_{22}, m_{23}, and m_{33} elements of the matrix are sensitive to linear polarizing light beams. The m_{44} element represents the reaction to circular polarization. We note that this element was significantly reduced to the mirror surface, which is -1, indicating circular depolarization of the surface.

Figure 10. Dynamic Stokes polarimeter in reflection mode through the apple surface.

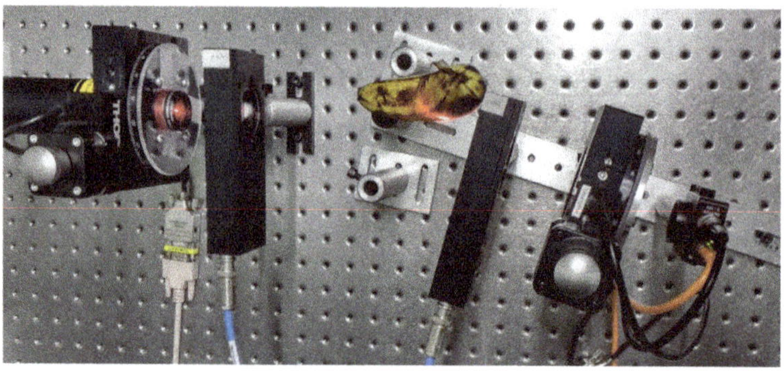

Figure 11. Dynamic Stokes polarimeter in reflection mode through the banana surface.

The Mueller matrices illustrated in Table 6 are general approximation matrices obtained from an instrumented dynamic polarizer. This is the first time that Mueller matrices for fruits are proposed to the best of our knowledge. We can reduce some errors in the matrix elements associated with the number of tests performed by measuring different apples and bananas from a batch of fruit. These matrices provide much information on the sample surface based on the fruits' maturity degree.

5. Discussion

This work proposes a Stokes polarimeter system that calculates the Mueller matrix of organic and non-organic samples by transmission or reflection. Instrumentation of polarizers and stepper motors allow the 36 measurements method to achieve the same precision for every experimental sample. The main contribution of this method is to provide an instrumented array for calculating Mueller matrices by measuring polarized light behavior on the sample surfaces. This method proposes an alternative to the typical image of the Mueller matrix; instead, we propose numerically calculating each element of the matrix by acquiring measured signals and using the 36 measurements method. Mueller matrix of fundamental elements such as air, polarimeters, and mirror are used to calibrate and validate the proposed system. Specifically, Mueller matrices of different organic samples are calculated (fruit surfaces) and shown on the main screen of the designed software. The mode can be selected by reflection or transmission. Moreover, even if some matrices cannot be compared to similar experimental values, we achieve a minimal MSE score for common elements like air or polarizers.

Overall, the Mueller matrices obtained for different elements provide a reliable calibration for ideal reference samples and show that our instrumented array can be configured and adapted to measure lighting properties of additional organic and non-organic materials. Related works proposed Mueller matrices measured using a CCD camera with higher error accuracy values than our proposed optical array. For instance, in [30], the Mueller matrices of the air or polarizers using a CCD camera show around 4% of error and 2% in our case. In contrast, reference [31] also measures the matrix for the air achieving an error of 0.2% and even lower for some individual elements of the matrix. Both works measure the optical properties of the air in transmission mode but using two different models of CCD cameras, the camera with a higher dynamic range provides notably more accurate results. In our case, we have proposed a photodetector instead of a CCD camera because we can vary the acquisition rate of the photodetector and acquire a high number of samples to calculate the Mueller matrix. Additionally, the photodetector measurements require fewer saving storage and memory resources. Moreover, this device provides high bandwidth for measuring the polarization properties of the samples under tests. So far, the photodetector offers an alternative to the CCD camera commonly used for Mueller matrix determination.

Some additional aspects for improvement in future works are listed: (1) the design of a compact and portable prototype for measuring different kinds of samples under the same conditions using reflection mode; (2) Extend the current wavelength of the incident light to a broader spectrum range for measuring microstructural properties that characterize biological tissues; (3) Compute other Mueller matrix properties for a suitable characterization of material samples in industry or clinical applications.

Author Contributions: Conceptualization, R.C.-S.; Formal analysis, M.-A.I.-M.; Methodology, R.A.M.-C.; Software, D.R.-S.; Validation, D.-L.A.-O. All authors have read and agreed to the published version of the manuscript.

Funding: The APC was funded by Universidad de Guanajuato under POA 2022 and by DAIP, the research office of the Universidad de Guanajuato.

Conflicts of Interest: The authors declare no conflict of interest. Also, the funders had no role in the design of the study; in the collection, analyses, or interpretation of data; in the writing of the manuscript, or in the decision to publish the results.

References

1. Ahmadivand, A.; Gerislioglu, B. Photonic and Plasmonic Metasensors. *Laser Photonics Rev.* **2022**, *16*, 2100328. [CrossRef]
2. Behrouzi, K.; Lin, L. Gold nanoparticle based plasmonic sensing for the detection of SARS-CoV-2 nucleocapsid proteins. *Biosens. Bioelectron.* **2022**, *19*, 113669. [CrossRef] [PubMed]
3. He, C.; He, H.; Chang, J.; Chen, B.; Ma, H.; Booth, M.J. Polarisation optics for biomedical and clinical applications: A review. *Light: Sci. Appl.* **2021**, *10*, 194. [CrossRef] [PubMed]
4. Goldstein, D.; Goldstein, D.H. *Polarized Light, Revised and Expanded*, 2nd ed.; CRC Press: Boca Raton, FL, USA, 2003. [CrossRef]

5. Qi, J.; Elson, D. A high definition Mueller polarimetric endoscope for tissue characterisation. *Sci Rep.* **2016**, *6*, 25953. [CrossRef] [PubMed]
6. Hui, M.; Honghui, H.; Dong, Y.; Liu, T. Quantitative detection of breast ductal carcinoma tissues at different progression stages using Mueller matrix microscope. *Int. Soc. Opt. Photonics* **2018**, *10493*, 104930O. [CrossRef]
7. Wang, Y.; He, H.; Chang, J.; Zeng, N.; Li, M.; Ma, H. Differentiating characteristic microstructural features of cancerous tissues using Mueller matrix microscope. *Micron* **2015**, *79*, 8–15. [CrossRef] [PubMed]
8. Li, W.; Jiang, H.; Zhang, C.; Chen, X.; Gu, H.; Liu, S. Characterization of curved surface layer by Mueller matrix ellipsometry. *J. Vac. Sci. Technol. B* **2016**, *34*, 020602. [CrossRef]
9. Dong, Y.; He, H.; He, C.; Zhou, J.; Zeng, N.; Ma, H. Characterizing the effects of washing by different detergents on the wavelength-scale microstructures of silk samples using Mueller matrix polarimetry. *Int. J. Mol. Sci.* **2016**, *17*, 1301. [CrossRef]
10. Hall, S.A.; Hoyle, M.A.; Post, J.S.; Hore, D.K. Combined Stokes vector and Mueller matrix polarimetry for materials characterization. *Anal. Chem.* **2013**, *85*, 7613–7619. [CrossRef]
11. Nguyen, A.L.T.; Nguyen, K.H.D.; Pham, H.T.T. A novel method for extracting characteristics of bio-materials utilizing optical polarizing system. In *Procceeding of the 6th International Conference on the Development of Biomedical Engineering in Vietnam (BME6)*; Vo Van, T., Nguyen Le, T.A., Nguyen Duc, T., Eds.; Springer: Singapore, 2018; Volume 63, pp. 553–558. [CrossRef]
12. Phan, Q.-H.; Lo, Y.-L. Stokes-Mueller matrix polarimetry system for glucose sensing. *Opt. Lasers Eng.* **2017**, *92*, 120–128. [CrossRef]
13. Ghosh, N.; Vitkin, I.A. Tissue polarimetry: Concepts, challenges, applications, and outlook. *J Biomed. Opt.* **2011**, *16*, 110801. [CrossRef] [PubMed]
14. Gonzalez, M.; Montejo, K.; Krupp, K.; Srinivas, V.; DeHoog, E.; Chue-Sang, J.; Sevilla, N.; Madhivanan, P.; Ramella-Roman, J. Calibration of low cost, portable Mueller matrix polarimeter. In *Biophotonics Congress: Biomedical Optics Congress, Proceedings of the Microscopy Histopathology and Analytics 2018, Hollywood, FL, USA, 3–6 April 2018*; Optical Society of America: Washington, DC, USA, 2018. [CrossRef]
15. Azzam, R.M.; Bashara, N.M. *Ellipsometry and Polarized Light*, 3rd ed.; North-Holland Publishing Co.: Amsterdam, The Netherlands, 1987; ISSN 0925-5818. [CrossRef]
16. Bickel, W.S.; Bailey, W.M. Stokes vectors, Mueller matrices, and polarized scattered light. *Am. J. Phys.* **1985**, *53*, 468–478. [CrossRef]
17. Savenkov, S.N. Principles of the Mueller matrix measurements. In *Light Scattering Reviews 9*; Kokhanovsky, A.A., Ed.; Springer & Praxis Publisher: Berlin/Heidelberg, Germany, 2015; pp. 213–255. [CrossRef]
18. Hinds Instruments. *PEMTM 100 Photoelastic Modulator User Manual*; Hinds Instruments: Hillsboro, OR, USA, 2006.
19. Hipps, K.W.; Crosby, G.A. Applications of the photoelastic modulator to polarization spectroscopy. *J. Phys. Chem.* **1979**, *83*, 555–562. [CrossRef]
20. Kuldkepp, M.; Hawkes, N.C.; Rachlew, E.; Schunke, B. Accurate polarization measurements with a dual photoelastic modulator. *Appl. Opt.* **2005**, *44*, 5899–5904. [CrossRef] [PubMed]
21. He, H.; Liao, R.; Zeng, N.; Li, P.; Chen, Z.; Liu, X.; Ma, H. Mueller Matrix Polarimetry—An Emerging New Tool for Characterizing the Microstructural Feature of Complex Biological Specimen. *J. Lightwave Technol.* **2019**, *37*, 2534–2548. [CrossRef]
22. Hecht, E. *Optics*, 5th ed.; Pearson Education: New York, NY, USA, 2017; p. 720.
23. Liu, Y.; Jones, G.A.; Peng, Y.; Shen, T.H. Generalized theory and application of stokes parameter measurements made with a single photoelastic modulator. *Int. J. Appl. Phys.* **2006**, *100*, 063537. [CrossRef]
24. Martínez-Celorio, R.A.; Espinosa-Luna, R.; Rivera-Vázquez, J.O.; Castro-Sánchez, R. Metodología para el estudio de los tejidos biológicos usando matrices de Mueller. *Rev. Mex. Ing. Biomed.* **2005**, *26*, 106–116.
25. Castro-Sanchez, R.; Martinez-Celorio, R.; Cibrían, R.M.; Salvador, R.; Hernandez-Fusilier, D.; Hurtado-Ramos, J.B. Synchronization of two photoelastic light modulators to obtain müeller matrix. *IEEE Trans. Instrum. Meas.* **2013**, *62*, 2050–2057. [CrossRef]
26. Rodriguez-Herrera, O.G.; Bruce, N.C. Mueller matrix for an ellipsoidal mirror. *Opt. Eng.* **2006**, *45*, 053602. [CrossRef]
27. Franco, S. *Design with Operational Amplifiers and Analog Integrated Circuit*, 3rd ed.; Mc-Graw Hill International Editions: New York, NY, USA, 2002; p. 635.
28. Pedrotti, F.L.; Pedrotti, L.S. *Introduction to Optics*, 2nd ed.; Prentice Hall International Inc.: Hoboken, NJ, USA, 1993; p. 323.
29. Krijger, J.; Snel, R.; Van Harten, G.; Rietjens, J.; Aben, I. Mirror contamination in space i: Mirror modelling. *Atmos. Meas. Tech.* **2014**, *7*, 3387. [CrossRef]
30. Shukla, P.; Pradhan, A. Mueller decomposition images for cervical tissue: Potential for discriminating normal and dysplastic states. *Opt. Express* **2009**, *17*, 1600–1609. [CrossRef] [PubMed]
31. Wang, J.; Li, Y.; Cao, C.; Zhou, G.; Li, L. High-fidelity and rapid cellular-level Mueller matrix imaging for tissue identification with unstained sections. *Biomed. Opt. Express* **2021**, *12*, 4745–4758. [CrossRef] [PubMed]

sensors

MDPI

Article

Stress Detection of Conical Frustum Windows in Submersibles Based on Polarization Imaging

Hening Li [1,2,3], Ran Liao [1,2,*], Hailong Zhang [4], Guoliang Ma [4], Zhiming Guo [1,2], Haibo Tu [1,5], Yan Chen [1,5] and Hui Ma [2]

[1] Institute for Ocean Engineering, Shenzhen International Graduate School, Tsinghua University, Shenzhen 518055, China; lhn19@mails.tsinghua.edu.cn (H.L.); gzm20@mails.tsinghua.edu.cn (Z.G.); 202071210@yangtzeu.edu.cn (H.T.); 201971160@yangtzeu.edu.cn (Y.C.)
[2] Guangdong Research Center of Polarization Imaging and Measurement Engineering Technology, Shenzhen International Graduate School, Tsinghua University, Shenzhen 518055, China; mahui@tsinghua.edu.cn
[3] Department of Biomedical Engineering, Tsinghua University, Beijing 100084, China
[4] Institute of Deep-Sea Science and Engineering, Chinese Academy of Sciences, Sanya 572000, China; zhanghailong@idsse.ac.cn (H.Z.); magl@idsse.ac.cn (G.M.)
[5] Department of Physics, Yangtze University, Jingzhou 434100, China
* Correspondence: liao.ran@sz.tsinghua.edu.cn

Abstract: Stress detection of the conical frustum window is a very important issue to ensure the safety of deep manned submersibles. In this paper, we propose a method based on polarization imaging to evaluate the stress accumulation and recovery in the conical frustum window. An experimental setup of Mueller matrix polarimetry is built, and the samples are made by referring to the typical conical frustum windows in submersibles. By pressurizing different pressures on the samples, we can find the changes of their Mueller matrix images and further derived polarization parameters. The results show that the polarization parameters can characterize the stress transfer process and the elastic–plastic transformation process of the window under different pressurization pressures. We also use a two-layered wave plate model to simulate the stress distribution in the window, which reveals different performances of the former and latter layers of the window under pressurization. Finally, we use a finite element model to simulate and understand some of the above experimental results. This proposed method is expected to provide new possibilities for monitoring the window stress and further ensure the safety of deep manned submersibles.

Keywords: conical frustum window; PMMA; polarization imaging; stress detection

Citation: Li, H.; Liao, R.; Zhang, H.; Ma, G.; Guo, Z.; Tu, H.; Chen, Y.; Ma, H. Stress Detection of Conical Frustum Windows in Submersibles Based on Polarization Imaging. *Sensors* **2022**, *22*, 2282. https://doi.org/10.3390/s22062282

Academic Editors: Maria Lepore and Ines Delfino

Received: 29 January 2022
Accepted: 14 March 2022
Published: 16 March 2022

1. Introduction

The pressure hull is an important part of a deep manned submersible, which is mainly composed of a pressure shell made of high-strength titanium alloy and observation windows made of polymethyl methacrylate (PMMA) material [1]. The observation window is a key component as it can provide a channel for scientists to observe the external sea conditions. Meanwhile, it is also an important pressure-bearing component, and stress accumulation will occur in the window due to the high external pressure in the diving process of the submersible [2,3]. The stress distribution inside the observation window affects its pressure resistance. The stress release degree of the observation window after the submersible surfaces is an important factor affecting the use frequency of the submersible [4]. Cracks in the window will threaten human safety inside the submersible. Therefore, it is of great significance for the safety and stability of the submersible to monitor the internal stress accumulation of the observation window and provide an evaluation standard of the window stress release degree in the recovery stage.

Many scientists have simulated the internal stress of the observation window by finite element analysis. Du et al. [5] studied the stress and deformation characteristics of the conical viewport window with the flange. Zhou et al. [6] showed the creep behavior of thick PMMA immersed in a liquid scintillator at eight stress levels. Arnold et al. [7] built predictive models for the creep behavior of PMMA, which match the experimental results. Pranesh et al. [8] showed several models of a viewport window to reduce the internal stress, which can reduce the corner stress by selecting a specific fillet radius. Liu et al. [9] identified unknown viscoelastic parameters and accurately analyzed the deep-water damage by comparing experiments with finite element analysis. Wang et al. [10] analyzed the time-deformation behavior of the observation window using the viscoelastic model. There are also several practical stress measurement methods, each of which has its own advantages and disadvantages. For example, the strain gauge method [11,12] can accurately measure the stress but only characterizes the surface stress and strain, while the ultrasonic method [13,14] can measure the internal stress of the bulk sample but sometimes causes damage to the sample, and the laser speckle method [15,16] can simultaneously measure the large area of the sample surface but requires the window surface to be roughened. Considering that stress accumulation occurs inside the window during diving, an in situ detection method for the submarine is still desired. To our knowledge, the development of a nondestructive, in situ and real-time stress detection method is currently challenging for the scientific community.

Polarization is the inherent property of light. Polarimetric techniques have been demonstrated to provide multidimensional parameters, which are sensitive to the microstructure of the samples [17]. Recently, polarimetric techniques have been used in biomedical therapy [18,19], marine particle probing [20,21], aerosol monitoring [22,23], etc.

Usually, we use the Stokes vector $S = (I, Q, U, V)^T$ to describe the polarization state of light. When a beam of polarized light passes through the sample, the sample's polarization property, always represented as a 4×4 Mueller matrix, will affect the polarization state of the incident light. Mathematically, we can obtain Equation (1) to describe the transformation process, where $(I,\ Q,\ U,\ V)^T$ is the Stokes vector of the incident light and $(I',\ Q',\ U',\ V')^T$ is the Stokes vector of the output light.

$$\begin{pmatrix} I' \\ Q' \\ U' \\ V' \end{pmatrix} = \begin{pmatrix} m_{11} & m_{12} & m_{13} & m_{14} \\ m_{21} & m_{22} & m_{23} & m_{24} \\ m_{31} & m_{32} & m_{33} & m_{34} \\ m_{41} & m_{42} & m_{43} & m_{44} \end{pmatrix} \begin{pmatrix} I \\ Q \\ U \\ V \end{pmatrix} \quad (1)$$

At present, there are many polarization systems to measure the Mueller matrix of samples. The classical methods of rotating the optical polarization components, such as the polarizer and the quarter-wave plate, have been extensively used in measuring the Mueller matrix of biomedical tissues and integrated electronic chips [24]. However, reducing the rotational components can often improve the acquisition speed, measurement accuracy and stability of the polarimetry system. Recently, liquid crystal modulators [25,26] and photoelastic crystal devices [27,28] tend to gradually replace rotational components. In the meantime, new types of polarimeters are emerging to directly measure the polarization states of light in one or two dimensions. For example, the division of a focal plane (DoFP) polarimeter is capable of measuring the linear polarization states in a single shot, which consists of a common CCD sensor with a pixelated micro-polarizer array (MPA) in front of it. Several recent studies have used the DoFP polarimeter to measure the polarization properties of biological tissues [29,30]. Compared with traditional Mueller matrix polarimetry, it effectively improves the measurement speed and enhances the system's stability.

In this study, polarization imaging is used to measure the stress and strain of the conical frustum window during pressurization and recovery stages. We firstly build an experimental setup to measure the Mueller matrix of the samples referring to the typical conical frustum window in submersibles. In the experiment, a controllable jack is used to pressurize the sample step by step, and the Mueller matrix of the sample is measured at each pressure level. The results demonstrate that the polarization parameters derived from the Mueller matrix can characterize the stress accumulation and release process of the sample in both the pressurization and recovery stages and provide effective indicators for the elastic–plastic transformation of the internal structure. Moreover, a two-layered wave plate simulation is proposed to describe the stress distribution in the sample, which is consistent with the results of finite element analysis. The results in this work indicate that polarization imaging can effectively detect the stress of the samples referring to the typical conical frustum window in submersibles, and this implies the possibility of monitoring the stress distribution in the conical frustum window quantitatively in the future.

2. Materials and Methods

2.1. Material

Since Piccard first proposed a conical observation window in marine engineering in the 1950s, the conical observation window has been widely used in deep-sea submersibles. For convenience, we refer to the typical conical observation window of submersibles and designed the samples to be made of PMMA material to carry out the pressure experiment. In order to facilitate a description in this context, the large face of the sample is called the former face and the small face is called the latter face, according to the sequence of light passing through the sample in the experiment. The structure of the sample (Tiemao Glass, China) is shown in Figure 1 with the diameter of the latter face $\Phi_1 = 20$ mm, the diameter of the former face $\Phi_2 = 80.5$ mm (to follow the direction of light passing through the sample), the height $H = 32.3$ mm, the former face chamfer angle $R = 115°$, and the cone Angle $\alpha = 90°$. The mechanical parameters of the sample are shown in Table 1.

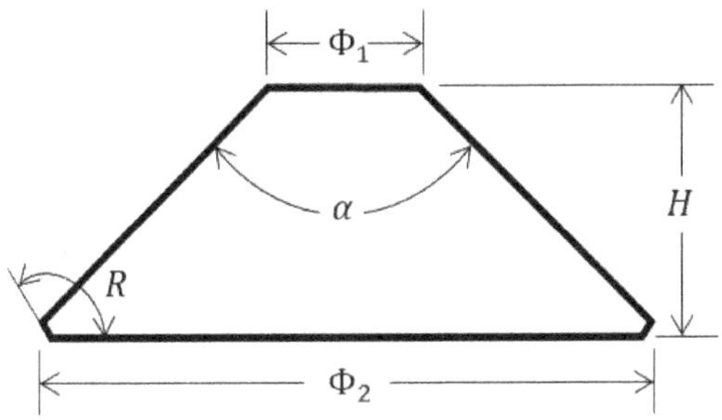

Figure 1. Structural parameters of sample.

Table 1. Mechanical parameters of the PMMA sample.

Properties	Value
Density/(g/cm^3)	1.186
Tensile Modulus/GPa	3.13
Yield Strength/MPa	129
Poisson's ratio	0.37
Refractive index	1.49

2.2. Experiment Setup

The experimental setup, as shown in Figure 2, was built to measure the Mueller matrix images of the sample when pressurizing different pressures on it. The setup consisted of a light collimator, a polarization state generator (PSG), a polarization state analyzer (PSA), and a pressurization device. In the light collimator, the light emitted from an LED lamp with the 630 nm central wavelength and 10 nm bandwidth was collimated by an optical system to finally form a parallel light beam whose transversal homogeneity was larger than 93%. The beam diameter was 20 mm, which can completely cover the latter face of the sample. The light beam successively passed PSG, the sample, and finally PSA, and the light beam's polarization states were modulated by PSG, while its Stokes vector image after passing the sample was detected by PSA. A jack with a maximum working pressure of 63 MPa (RRH-1003, Yuli Electromechanical Equipment Group Co., Taizhou, China) was used to pressurize the sample.

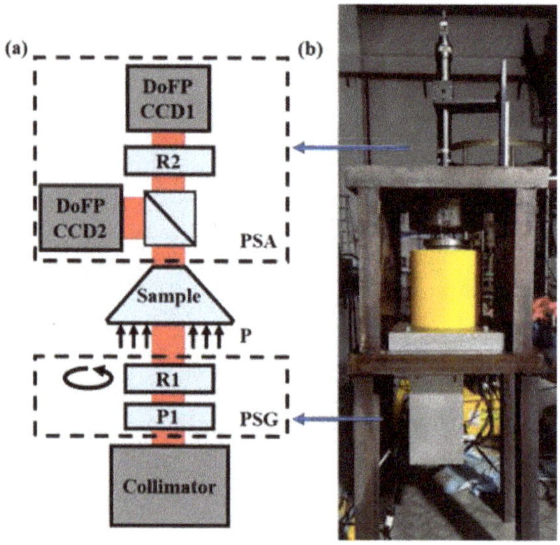

Figure 2. Schematic configuration (**a**) and photograph (**b**) of the experiment setup. P1, polarizer; R1 and R2, achromatic quarter-wave plates.

In the setup shown in Figure 2, PSG includes a fixed linear polarizer P1 (LPNIRB100, Thorlabs Inc., Newton, NJ, USA) and a rotatable zero-order quarter-wave plate R1 (WPQ10E-633, Thorlabs Inc., Newton, NJ, USA). R1 was installed in an electric rotating stage (PRM1/MZ8, Thorlabs Inc., Newton, NJ, USA). PSA consisted of two 16-bit DoFP polarimeters (PHX050S-PC, Lucid Vision Labs Inc., Vancouver, BC, Canada, DoFP-CCD1 and DoFP-CCD2), with 2048 $\times$ 2448 pixels and 21 frames per second. Each DoFP polarimeter is capable of obtaining images in four linear polarization channels of $0°$, $45°$, $90°$ and $135°$ in a single measurement. Two DoFP polarimeters were installed at the transmission and reflection ends of a 50:50 non-polarizing beam splitter prism (CCM1-BS013/M, Thorlabs Inc., Newton, NJ, USA), and a fixed zero-order quarter-wave plate R2 (WPQ10E-633, Thorlabs Inc., Newton, NJ, USA) was installed between the transmission end of the prism and DoFP-CCD1.

To improve the measurement accuracy and reduce the instantaneous field-of-view (IFOV) error, a calibration strategy was used to calibrate the Mueller matrix image measurement of the setup [30]. Firstly, to reduce the error caused by the polarization direction, the extinction ratio and the intensity response of the pixels, the parallel light beam with known polarization states was used to evaluate the DoFP instrument matrix to calibrate the pixels of DoFP polarimeters. The polarization states were generated by PSG and measured by a standard polarimeter (PAX1000VIS, Thorlabs, Newton, NJ, USA). Secondly, PSA can be also calibrated by these well-calibrated DoFP polarimeters by using the same parallel beam with known polarization states as the incident light beam. In addition, a so-called PSA instrument matrix was obtained to calculate the Stokes vector image of the incident light beam from the pixel values of the two DoFP polarimeters. Thirdly, we measured a series of Stokes vector images by rotating the wave plate in PSG at angles in a given angle set, when the sample was not pressurized. Here, we considered the air as the standard, whose Mueller matrix was the unit matrix. Up to this point, we obtained a controllable PSG and a qualified PSA. After the calibration, the error of measured Mueller matrix elements normalized by m_{11} was less than 0.005.

After the sample was loaded in the setup, we illuminated the sample with the parallel light beam with the known polarization states and, accordingly, recorded the Stokes vector image of the beam after the sample. Since the setup can measure the Mueller matrix with very low error after the above calibration by the air, we can calculate the Mueller matrix of the sample using Equation (1). However, the other error, excluding those in the polarization measurement, such as the image distortion from the deformed sample, may do harm to the measurement of the Mueller matrix of the sample. As a result, we focused much effort to correct the image distortion, in order to obtain the accurate Mueller matrix of the sample under different pressure values.

During the measurement, we optimized the angle set by considering both the small condition number to suppress the error accumulation and the fast measurement speed of the system. Currently, four different angles, $-45°$, $45°$, $-19.6°$ and $19.6°$, of the rotating wave plate in PSG were used in a single measurement [31].

In order to ensure that the light beam can pass entirely through the sample, the pressurization area on the sample was a ring whose inner diameter was 22 mm, larger than the latter face of the sample. Ideally, according to the pressure conversion formula, the transfer of force without loss can be expressed as Equation (2),

$$F = P_A \cdot S_{window} = P \cdot S_{jack} \tag{2}$$

where P_A is the ideal equivalent pressure, S_{window} is the pressurization area of the window, P is the pressure displayed on the jack dashboard [32], and S_{jack}(=175.84 cm^2) is the pressurization area of the jack, which was obtained by asking the company. The transfer efficiency η (=0.8) was introduced considering the transfer loss in the actual situation. The actual equivalent pressure, P'_A on the sample can be calculated by Equation (3) from the pressure value read from the jack,

$$P'_A = \eta \cdot \frac{P \cdot S_{jack}}{S_{window}} \approx 3 \cdot P \tag{3}$$

This means that when the jack operated between 0 and 60 MPa, the actual pressure on the window was about 0–180 MPa, and in the following context, we mention the pressure as the actual pressure. In the pressurization experiment, we took 12 MPa as the step pressure to carry out the experiments. Besides, we also investigated the full recovery of the sample after pressurization and monitored the change of polarization parameters in this process, so as to find the parameters that can effectively characterize the stress change of the sample.

2.3. Image Distortion Correction Method

During the experiment, we found that with the increase in pressure, the sample was deformed, which caused the parallel light beam to experience a certain level of distortion. The pixel-level correspondence between the incident and output light beam was destroyed, which decreased the accuracy of the Mueller matrix imaging. Additionally, the correspondence between the measurement images in different pressures was damaged, which caused harm to the stress characterization of the polarization parameters. Here, we used a correction method to correct the distortion of these images after pressurization.

In order to correct the image distortion accurately, the grid auxiliary line was drawn by a dark soft brush on the latter face of the sample, whose correspondence with the covered areas on the latter surface was physically unchanged, regardless of how the distortion occurred. Then, a manual feature point extraction method was used to correct the image distortion under different pressures according to the grid auxiliary line. The key point of the method was to select the feature points in the original image without pressure and the distorted image under certain pressure, and then obtain each feature point's pixel coordinate index. For the example of the i th feature points, these were (x_0^i, y_0^i)

and (x_1^i, y_1^i). If we obtained entirely n feature points, we obtained x_0 as an $n \times 1$ vector formed by x_0^i, (i = 1, 2, ... , n), and similarly, we obtained y_0, x_1, and y_1. After that, we carried out the projection transformation with a 3×3 transformation matrix T, that is, $(x_0, y_0, \mathbf{u}) = (x_1, y_1, \mathbf{u}) * T$, where **u** is an $n \times 1$ vector formed by 1. Three columns in T respectively represent the transformation of the image in the x, y and z directions. For the two-dimensional transformation shown in this case, the data in the third column will be [0, 0, 1]. T is specified by pairs of feature points. By substituting T into the *imwarp* function, we obtained the corrected image [33].

We compared the images before and after pressure, and the raw intensity images from the 0° linear polarization channel of DoFP-CCD1 were considered as examples. Figure 3a,b respectively show the images before and after loading 120 MPa, which are marked by red dots. We can see that the distortion is very serious, and the grids in Figure 3b are slightly larger than those in Figure 3a, which may originate from the bulge of the latter face of the sample in the pressurization process. We merged these two images together with a false-color algorithm and show it in Figure 3c where the green part is the original image and the pink part is the 120 MPa pressure image. One can see that the overlap between pink and green is very severe, and the ghosts of the grid auxiliary lines are rather obvious.

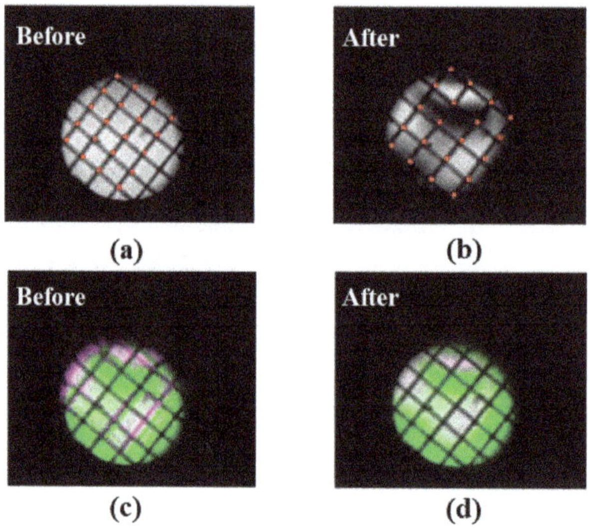

Figure 3. Results of the correction method. (**a**) Original image without pressure; (**b**) distorted images with 120 MPa pressure; (**c**) merged image of (**a**,**b**) with false-color algorithm; (**d**) merged image of original and corrected images with false-color algorithm. Red dots in (**a**,**b**) are the feature points.

We selected the feature points (marked as red points in Figure 3a,b) according to the grid auxiliary line, and then obtained the transformation matrix to correct the distorted image to the corrected image. Figure 3d shows the merged image of the original and corrected images. The pixel correspondence greatly improved after the algorithm, as the pink image was almost completely covered by the green image. The pink part that is not completely covered may have been caused by the inaccuracy of T, due to the insufficiently dense selection of feature points.

To verify the stability of this method, we manually marked the same data six times to test the error introduced by the transformation matrix due to the manual mark of feature points. Each mark process selected more than 20 pairs of feature points, and Table 2 collects and shows the nine transformation matrix elements $t_{11} \sim t_{33}$ calculated in the six marking processes. By calculating the Mean $\pm$ Var values of each element, we evaluated the influence of the manual mark on T. t_{13}, t_{23} were close to 0, t_{33} is 1, which means a change of the image in the z-direction did not occur.

Table 2. Elements of transformation matrix for six independent markings.

Number	t_{11}	t_{12}	t_{13}	t_{21}	t_{22}	t_{23}	t_{31}	t_{32}	t_{33}
1	0.8452	−0.0135	-1.47×10^{-5}	0.0075	0.8699	4.45×10^{-6}	125.6382	150.0923	1
2	0.8322	−0.0158	-2.13×10^{-5}	0.0077	0.8613	5.34×10^{-6}	133.1351	154.9269	1
3	0.8250	−0.0209	-2.19×10^{-5}	0.0074	0.8470	2.20×10^{-6}	130.6332	151.8865	1
4	0.8434	−0.0099	-1.47×10^{-5}	0.0116	0.8678	5.71×10^{-6}	127.6634	150.0143	1
5	0.8436	−0.0126	-1.22×10^{-5}	0.0147	0.8818	1.16×10^{-6}	129.1101	149.3359	1
6	0.8437	−0.0072	-1.60×10^{-5}	−0.0026	0.8559	3.04×10^{-6}	131.3437	152.5151	1
Mean ± var	0.8389 ± 0.0083	−0.0134 ± 0.0048	$-1.44 \times 10^{-5} \pm 7.38 \times 10^{-6}$	0.0077 ± 0.0059	0.8640 ± 0.0121	$5.39 \times 10^{-6} \pm 3.32 \times 10^{-6}$	129.5873 ± 2.6941	151.4619 ± 2.0874	1 ± 0

In order to verify the influence of the standard deviations of t_{31} and t_{32} on the pixel movement, we set all other elements as their own average values but allowed t_{31} and t_{32} to change within their own standard deviations, with which we constructed T accordingly. The corrected image can be calculated for each new setting of t_{31} and t_{32}, and the pixel positions are found to change within 4 pixels. Compared with the 2048 × 2448 image, the relative error of the pixel position is less than 0.2%. Generally, from Table 2, the difference between six independent markings is so small that it indicates the above correction method is not influenced by the manual point marking.

3. Results

3.1. Characterization of Sample Changes during Pressurization and Recovery Stages by Polarization Parameters

The results of several polarization parameters on the sample at pressurization and recovery stages are shown in Figure 4, which are b, t_3, α_r in the Mueller matrix transformation (MMT) technique [34] and parameter δ in the Mueller matrix polar decomposition (MMPD) method [35]. Their calculation formulas are shown in Equations (4)–(12), where R is the total retardance, ε_{ijk} is the Levi–Cività permutation symbol, δ_{ij} is the Kronecker delta, φ is the optical rotation of magnitude. The previous literature shows that b is reversely related to depolarization, t_3 and δ are related to linear retardation, while α_r is related to the medium anisotropy [36].

$$b = \frac{1}{2}(m_{22} + m_{33}), \tag{4}$$

$$t_3 = \sqrt{m_{42}^2 + m_{43}^2}, \tag{5}$$

$$\alpha_r = \frac{1}{2}\arctan\left(\frac{m_{24}}{-m_{34}}\right), \tag{6}$$

$$\alpha_q = \frac{1}{2}\arctan\left(\frac{-m_{24}}{m_{34}}\right), \tag{7}$$

$$M_R = \begin{bmatrix} 1 & 0 \; 0 \; 0 \\ 0 & \\ 0 & m_R \\ 0 & \end{bmatrix} = \begin{bmatrix} 1 & 0 \; 0 \; 0 \\ 0 & \\ 0 & m_{LR} \\ 0 & \end{bmatrix} \begin{pmatrix} 1 & 0 & 0 & 0 \\ 0 & \cos 2\varphi & \sin 2\varphi & 0 \\ 0 & -\sin 2\varphi & \cos 2\varphi & 0 \\ 0 & 0 & 0 & 1 \end{pmatrix} = M_{LR} M_{CR}, \tag{8}$$

$$(m_R)_{ij} = \delta_{ij} \cos \mathbf{R} + a_i a_j (1 - \cos \mathbf{R}) + \sum_{k=1}^{3} \varepsilon_{ijk} a_k \sin \mathbf{R}, i, j = 1, 2, 3 \tag{9}$$

$$\delta = \arccos\left(\sqrt{(m_{R22} + m_{R33})^2 + (m_{R32} - m_{R23})^2} - 1\right), \tag{10}$$

$$\mathbf{R} = \arccos\left[\frac{tr(M_R)}{2} - 1\right] \tag{11}$$

$$a_i = \frac{1}{2\sin R} \sum_{i,j=1}^{3} \varepsilon_{ijk}(m_R)_{jk} \tag{12}$$

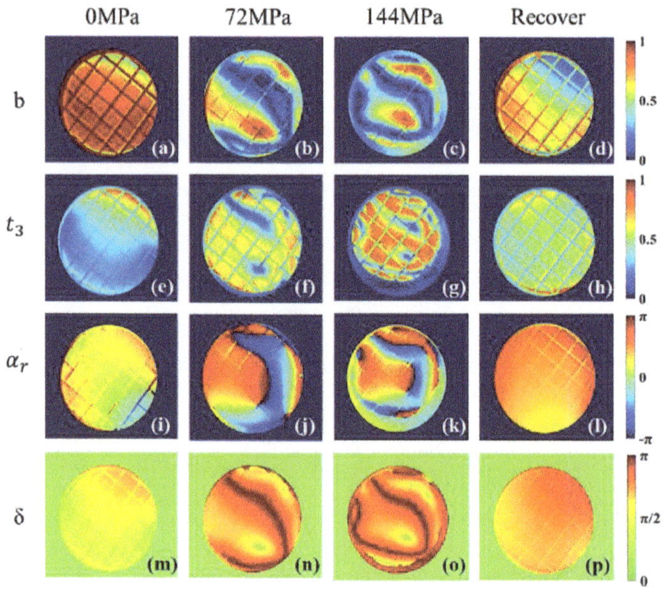

Figure 4. b, t_3, α_r, δ images of sample after 0 MPa, 72 MPa and 144 MPa pressure and those after 24 h recovery. **(a–d)**: b images; **(e–h)**: t_3 images; **(i–l)**: α_r images; **(m–p)**: δ images.

We selected the above four polarization parameters to compare the sample's images after 0 MPa, 72 MPa and 144 MPa and images after the 24 h recovery. In the horizontal direction, the polarization parameters can similarly characterize the changes in pressurization and recovery stages. When the pressure is 0 MPa, the first column images were relatively uniform and b in Figure 4a is close to 1 uniformly, which indicates that the sample's depolarization was weak initially. t_3 in Figure 4e, together with δ in Figure 4m, reflect that the sample without pressure has relatively small variation in the linear retardation. Figure 4i shows the sample has an almost uniform anisotropic structure, except for its peripheries when the pressure is 0 MPa.

The jack exerts ring-shaped pressure on the sample, and the center of the sample is the clear aperture, so there is a stress transfer from the periphery to the center of the sample. When the pressure reaches 72 MPa, for the second column of Figure 4, the central areas become heterogeneous, and there are also annular structures close to the periphery. It can be inferred that at this annular area, there is a strong depolarization effect because of the small b, and a large retardation difference because of the dramatic changes of t_3, α_r and δ. When the sample is pressurized to 144 MPa, a clear and complete ring-shaped area is observed in the third column of Figure 4, where the ring contracts more toward the central area than what happens at 72 MPa. Note that the values of α_r and δ obtained by Equations (6) and (10) may be wrapped, and the retardance of the sample may be larger than those values shown in the third and fourth row of Figure 4 [37]. However, we can see α_r and δ images become more heterogeneous due to the larger pressures. This indicates that the polarization parameters can characterize the process of stress transfer inside the sample during the pressurization stage. When the sample is recovered after 24 h, the fourth column of Figure 4 returns to a certain homogeneity, which is similar to that in the first column, but the values of the parameters are quite different from the initial ones. This indicates that after pressurizing to 144 MPa, the sample experiences changes in its internal structure that cannot be fully recovered in 24 h, and the proposed polarization parameters are sensitive to these changes.

3.2. Characterization of Elastic-Plastic Transformation of Samples Described by Polarization Parameters

In the previous experiments, we found the polarization parameters could not return to the initial state after pressurizing 144 MPa. In order to investigate the elastic–plastic transformation process in the sample, we design the experiment and chose the MMPD parameter γ to characterize the elastic–plastic transformation of the sample. Previous work shows that γ can perfectly reveal the sample's fast axis orientation [38]. The calculation formula of γ is shown by Equations (13) and (14), with a magnitude of linear retardance β:

$$\gamma = \frac{1}{2}\arctan\left(\frac{r2}{r1}\right), \tag{13}$$

$$r_i = \frac{1}{2\sin\beta} \times \sum_{i,j=1}^{3} \varepsilon_{ijk}(m_{LR})_{jk} \tag{14}$$

We carried out the experiments under different pressures and recovery times, and the results are presented in Figure 5, which shows the ability of polarization parameter to characterize the elastic–plastic transformation of the sample. Figure 5a shows the sample's initial image of γ, whose homogeneity indicates that the sample is an anisotropic material with a homogeneous orientation. Figure 5b–d shows the γ images of the sample after 12 h recovery from being pressurized to 36 MPa, 60 MPa and 72 MPa, respectively. With the increase in pressurization pressure, γ gradually changes and the overall homogeneity decreases. Furthermore, since γ is an angle parameter with a cycle of 180 degrees, the cross-cycle variation first appears at the top right corner of Figure 5e, which means that γ changed so dramatically that it skipped the current cycle. Additionally, the area of the cross-cycle parts in Figure 5f,g increases continuously with the increasing pressurization pressure. Combined with the structural characteristics of the sample, the images in Figure 5 indicate that γ's distributions are rather different after 12 h recovery from different pressures, and especially at the top right corners, γ experiences entirely different changes from the other parts. These factors imply that γ is sensitive to the mechanical structure behavior of the sample under pressurization.

In addition, the experimental results reveal that the sample's recovery is closely related to the recovery time. When the maximum pressure rises to 180 MPa, the sample obviously cannot fully recover within 12 h, so we extended the recovery time. Figure 5h,i shows γ images after 24 h and 36 h recovery, respectively. If γ no longer changes significantly, the sample is considered to have reached a stable state, and the degree of unevenness in γ may be used as an indicator for the recovery degree of the sample.

In order to quantitatively describe this process, we define the circular standard deviation of the γ images as V. The V values after 12 h recovery from different pressurization values are collected and shown in Figure 6a. From Figure 6a, V continues to increase with the increase in pressurization pressure, and the slope increases sharply between 96 MPa and 120 MPa, which may be the critical pressure range where the elastic–plastic transition occurs.

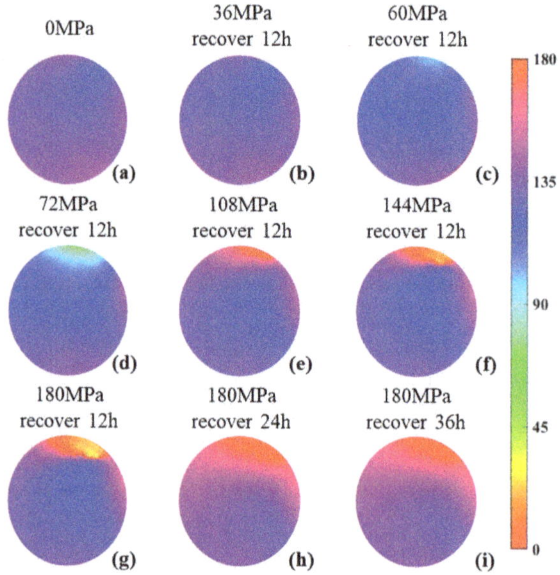

Figure 5. γ images under different pressures and recovery times. (**a–f**): γ images under different pressures; (**g–i**): γ images under different recovery times.

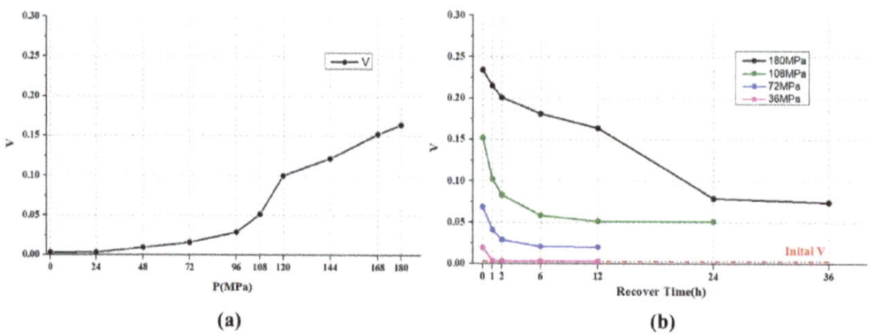

Figure 6. (**a**) V for different pressurization pressures; (**b**) V for different recovery times.

Meanwhile, V values' temporal changes during the sample's recovery process under different pressurization pressures are shown in Figure 6b. Note that the initial V value is estimated when there is no pressure applied to the sample. Under low pressures, V quickly drops to the initial V values. With the increase in pressure, the time for V to reach the stable value increases, and the stable V value gradually deviates from the initial V. Under pressure of 180 MPa, V decreases continuously within 24 h after depressurization to 0 MPa and reaches its stable value after 36 h. This means that the sample may recover to its stable state after 36 h recovery, but it cannot return to the initial state, which indicates that the sample has irreversible plastic deformation during the pressurization process. In this sense, we are able to draw the conclusion that V can effectively characterize the elastic–plastic transformation of the sample under pressure.

3.3. *Two-Layered Wave Plate Simulation*

In the loading experiment, we find a difference between the values of α_q and α_r calculated by Equations (6) and (7), indicating that the sample may have a multilayer structure [39]. In order to describe the structural stratification of this sample more accurately, we build a two-layered wave plate model to describe the polarization properties of the sample and carry out the analysis according to the existing experimental results. In the model, the polarization property of the sample can be approximated to the successiveness of two optical retardation components because the dichroism and depolarization properties of the sample can be neglected under an unpressurized state.

The Mueller matrix of one wave plate can be expressed as Equation (15), where θ is the orientation of the fast axis of this wave plate and δ is the retardance [33]. The Mueller matrix of a two-layered wave plate model can be expressed as Equation (16), as the light passes through M_{LR1} and then through M_{LR2}. θ_1 and δ_1 are the fast axis orientation and the retardance of the first wave plate, respectively, and θ_2 and δ_2 are those of the second wave plate.

$$M_{LR} = \begin{pmatrix} 1 & 0 & 0 & 0 \\ 0 & \cos^2 2\theta + \sin^2 2\theta \cos\delta & \sin 2\theta \cos 2\theta(1-\cos\delta) & -\sin 2\theta \sin\delta \\ 0 & \sin 2\theta \cos 2\theta(1-\cos\delta) & \sin^2 2\theta + \cos^2 2\theta \cos\delta & \cos 2\theta \sin\delta \\ 0 & \sin 2\theta \sin\delta & -\cos 2\theta \sin\delta & \cos\delta \end{pmatrix}, \quad (15)$$

$$M_{LR2}M_{LR1} = \begin{pmatrix} 1 & 0 & 0 & 0 \\ 0 & \cos 2\theta_2 & -\sin 2\theta_2 & 0 \\ 0 & \sin 2\theta_2 & \cos 2\theta_2 & 0 \\ 0 & 0 & 0 & 1 \end{pmatrix} \begin{pmatrix} 1 & 0 & 0 & 0 \\ 0 & 1 & 0 & 0 \\ 0 & 0 & \cos(\delta_1+\delta_2) & \sin(\delta_1+\delta_2) \\ 0 & 0 & -\sin(\delta_1+\delta_2) & \cos(\delta_1+\delta_2) \end{pmatrix} \begin{pmatrix} 1 & 0 & 0 & 0 \\ 0 & \cos 2\theta_1 & \sin 2\theta_1 & 0 \\ 0 & -\sin 2\theta_1 & \cos 2\theta_1 & 0 \\ 0 & 0 & 0 & 1 \end{pmatrix} \quad (16)$$

Then, the Mueller matrix of the sample is deliberately considered to be equivalent to the Mueller matrix defined as Equation (16). The two-layered wave plate model introduces four parameters, $\theta_1, \theta_2, \delta_1, \delta_2$. We expect to solve these four parameters according to the measured Mueller matrix.

The initial $\theta_{1-1}, \theta_{2-1}, \delta_{1-1}, \delta_{2-1}$ are solved by the initial Mueller matrix of the sample, which is measured before the loading pressure, and the final $\theta_{1-2}, \theta_{2-2}, \delta_{1-2}, \delta_{2-2}$ are obtained by the Mueller matrix fully recovered after being loaded to 180 MPa. Then, $\Delta\theta_1(\equiv \theta_{1-2} - \theta_{1-1})$, $\Delta\theta_2(\equiv \theta_{2-2} - \theta_{2-1})$, $\Delta\delta_1(\equiv \delta_{1-2} - \delta_{1-1})$, $\Delta\delta_2(\equiv \delta_{2-2} - \delta_{2-1})$ are obtained, as shown in Figure 7. Figure 7a,c shows the parameter differences $\Delta\theta_1, \Delta\delta_1$ of the former layer, respectively, and Figure 7b,d shows the parameter differences $\Delta\theta_2, \Delta\delta_2$ of the latter layer, respectively. It is obvious that $\Delta\theta_2$ in the latter layer is significantly greater than $\Delta\theta_1$ in the former layer. Both $\Delta\delta_1$ and $\Delta\delta_2$ distribute unevenly and their values are large, but the difference between them is not quite noticeable. This means that, relatively, θ is sensitive to the layers but δ is sensitive to the pressurization pressure.

The part of the sample near the former face is equivalent to the former layer wave plate, and the part of the latter face is equivalent to the latter layer wave plate. As seen in Figure 7, in the process of pressurization and recovery, the change of the latter layer is more significant than that of the former layer. In addition, the texture structure appears in the upper half of each image of Figure 7, which may be the trail of stress transfer during the stage of pressurization.

In order to explore the different reactions of the two-layered model before and after pressurization, we draw θ images in Figure 8. Figure 8a,b shows the θ images of the former and latter layers of the sample before pressurization, and Figure 8c,d represents those after 24 h recovery. In order to be more intuitive, we calculate the correlation coefficients between the matrices respectively, which are indicated on the two-way arrows.

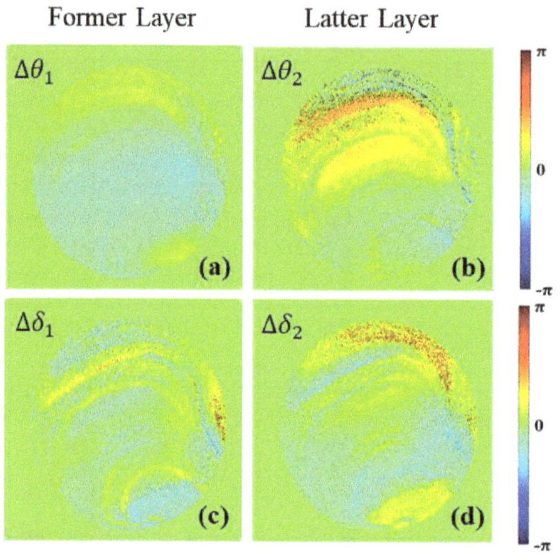

Figure 7. Two-layered difference of retardance parameters $\Delta\theta_1, \Delta\theta_2, \Delta\delta_1, \Delta\delta_2$. (**a**) $\Delta\theta_1$ image; (**b**) $\Delta\theta_2$ image; (**c**) $\Delta\delta_1$ image; (**d**) $\Delta\delta_2$ image.

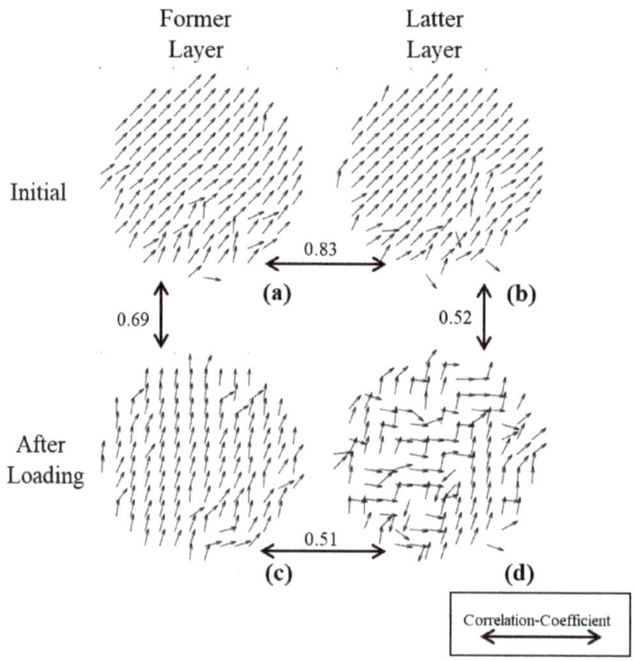

Figure 8. Fast axis orientation flow diagram of two-layered simulation. (**a**) initial image of former layer; (**b**) initial image of latter layer; (**c**) after loading image of former layer; (**d**) after loading image of latter layer.

From the horizontal direction, the value of θ for the two layers before pressurization is highly consistent, and the correlation coefficient is 0.83. After recovery, the value of θ of the two layers change in a disorderly manner, and the correlation coefficient decreased to 0.51.

This indicates that the loading-pressure stage causes irreversible damage to the sample, and the internal structure changes, which echoes the conclusions drawn in Figure 4.

From the vertical direction, Figure 8c shows that the former layer still maintains good θ homogeneity after being pressurized, and the correlation coefficient with the initial image in Figure 8a is 0.69. Figure 8d shows that θ of the latter layer is disordered after being pressurized, and the correlation coefficient decreases to 0.52 with the initial image Figure 8b. This shows that in the pressurization, the latter layer undergoes a greater change than the former layer.

4. Discussion

4.1. Finite Element Analysis of Sample after Pressurization

In order to evaluate the accuracy and effectiveness of the polarization parameters, the finite element simulation of the stress inside the sample is carried out according to the actual experimental situation by Solidworks (2020, educational trial version, Dassault Systèmes SolidWorks Corporation, Concord, CA, USA). The model is built according to the actual sample size, and the interaction mode between the model and the base is set as sliding friction while the finite element parameters of the sample and base are collected and shown in Table 3.

Table 3. Parameters for finite element simulation.

Properties	Value
Sample's Density/(g/cm^3)	1.186
Sample's Tensile Modulus/MPa	3130
Sample's Yield Strength/MPa	121
Sample's Poisson's ratio	0.37
Base's Density/(g/cm^3)	7.85
Base's Tensile Modulus/MPa	200,000
Base's Yield Strength/MPa	551
Base's Poisson's ratio	0.3
Friction coefficient	0.05

There is some mismatch between the central ring-shaped pressurization surface and the clear aperture of the sample due to the machining errors of the mounting bracket, as shown in Figure 9. Note that the hollow area of the pressurization surface entirely covers the clear aperture of the sample. When pressure is applied to the sample, the mismatched offset of the pressure center is set to be 2 mm, and the boundary constraint distribution and mesh element division of the finite element model are shown in Figure 10.

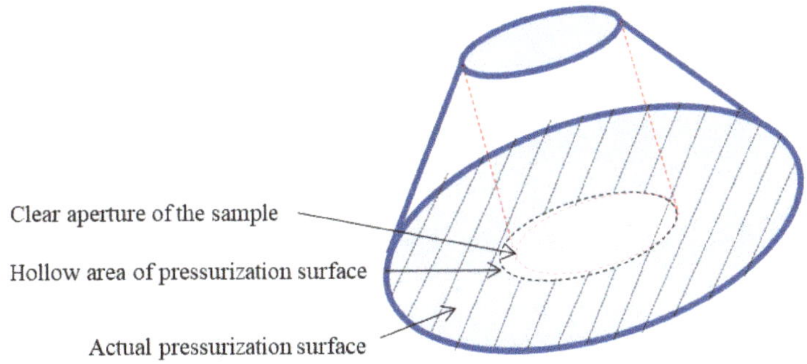

Figure 9. Relationship of the clear aperture of the sample with the pressurization surface.

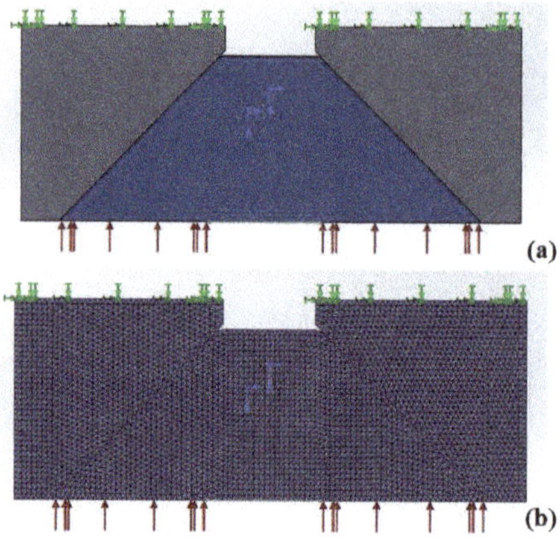

Figure 10. Boundary constraint and mesh division of finite element model. (**a**): boundary constraint image; (**b**): mesh division image.

Figure 11 shows the Mises stress of the sectional sample under 108 MPa. Due to the offset of the annular pressure area, the internal stress of the sample is not symmetrical, and the stress near the latter periphery on the offset direction is larger than that on the opposite side. With the pressure increasing, the stress inside the sample tends to accumulate on this side. This conclusion can correspond to the phenomenon of Figure 5, explaining why the experimental results change unevenly.

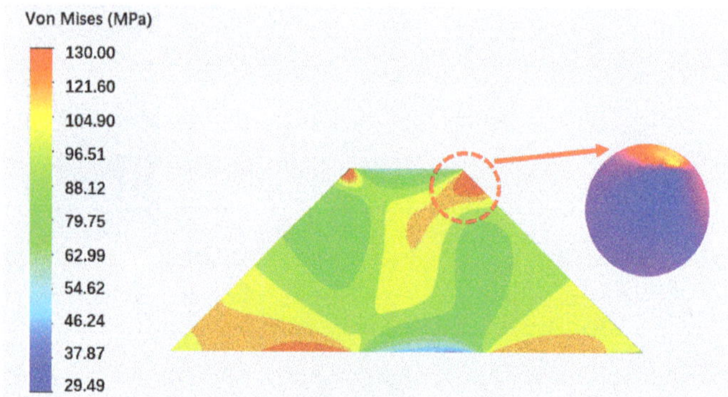

Figure 11. Mises stress diagram of sample.

The analytical results of the two-layered wave plate in Section 3.3 show that after pressurization, the sample's polarization properties can be described as a two-layered model. In order to further understand this result, we especially plot several stress curves inside the sample according to Figure 11. Figure 12 shows the Mises stress changes along the x-axis direction (light propagating direction) when a pressure of 108 MPa is applied to the sample, while the y-axis direction is also determined. We define a value, *r*, as the distance to the central line along the y-axis, and specifically, take the side with strong stress accumulation as the example. For the plot axes in Figure 12, the top horizontal axis represents the thickness of the sample and the left vertical axis represents the Mises stress. The five curves in Figure 12 represent the stress varying with different *r*.

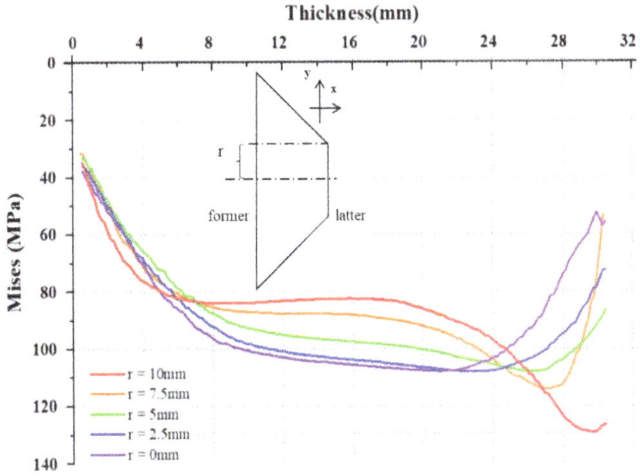

Figure 12. The stress along x-axis of sample as different distances to the central line.

In Figure 12, the orange-shadow areas are accordingly marked in both the model and the plot to show the possible range of the latter layer, and the other parts are considered as the former layer. From the five selected curves in Figure 12, the stress consistency of the latter layer of the sample is far less than that of the former layer, and the maximum stress appears in the curve of the latter layer with r = 10 mm, which corresponds to the position of the sample near the periphery in the actual structure. The maximum stress value is around 130 MPa, exceeding the yield limit of the material, which means that plastic deformation has been produced at the periphery of the sample. These prove that, firstly, with different thicknesses, the stress accumulation inside the sample is unevenly distributed, and the largest value appears at the periphery part of the sample. Meanwhile, one can see that in the latter layer, the values of curves are rather different, which means that the stress change at this layer is striking at both x and y directions when pressurizing a sample. Secondly, both the experimental and simulation results reveal the critical pressure of the elastic–plastic transformation. In Figure 6, we see that the possible critical pressure for the elastic–plastic transition of the sample appears between 96 and 120 MPa, as the slope of V increases significantly in this zone. As this happens, the finite element simulation, with a 108 MPa pressurization surface, results in stress accumulation in the periphery of the sample and reaches the yield limit, which triggers the plastic deformation in this area. So, the results of simulation and experiment can be understood as corresponding. Besides that, we prove that V, the degree of unevenness in γ, is able to characterize the elastic–plastic transformation of the material.

According to Figure 12, it seems as though a multilayer model would be more reasonable. However, it has been proved in the literature that multilayer wave plates can be equivalently represented in the form of two-layered wave plates [40], such that the two-layered model in the work is the simplest but most effective for the sample. However, to accurately describe and analyze the internal stress structure of the sample, the detailed multilayer model or tomographic method should be introduced to interpret or measure the stress properties of this kind of sample.

4.2. Characterization Potential of Other Parameters

In addition to the results described above, we also find that many other parameters are also sensitive to the pressurization process of the sample, as shown in Figure 13. The calculation formulas of A and Ψ parameters are given in Equations (17)–(19).

$$A = 2bt_1/\left(b^2 + {t_1}^2\right), \quad (17)$$

$$t_1 = \frac{1}{2}\sqrt{(m_{22} - m_{33})^2 + (m_{23} + m_{32})^2}, \quad (18)$$

$$\Psi = \frac{1}{2}\arctan[(M_{R21} - M_{R12})/(M_{R11} + M_{R22})] \quad (19)$$

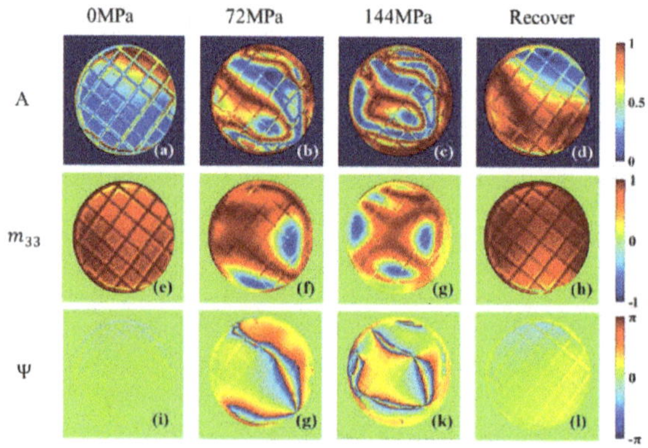

Figure 13. A, m_{33}, Ψ images of sample at 0 MPa, 72 MPa, 144 MPa and images after 24 h recovery. **(a–d)**: A images; **(e–h)**: m_{33} images; **(i–l)**: Ψ images.

Figure 13 shows that A and Ψ, calculated by Equations (17) and (19), can show the ring-shaped area under the loading stage, which is similar to those parameters in Figure 4. However, the boundaries of the ring-shaped area are different due to their different physical meanings, which implies more specific meanings carried by these separate parameters than those in Figure 4. Meanwhile, many other parameters also deserve our attention. They may be related to the individual differences between the samples, and may also contain more mechanical information about the material.

The results in this work encourage us to believe that polarization parameters have great research potential for characterizing the stress change of materials. Note that the annular pressurization method is currently adopted in this study, rather than the uniform water pressure of the observation window in the actual working environment. However, the polarization parameters proven in these experiments are still promising for comprehensively monitoring a working submersible window. Since the polarization parameters describe physical properties such as the orientation angles, retardance, and the anisotropy of the window under pressure, they have little relationship to the manner of pressurization. In the future, a real water pressure environment should be considered to characterize the internal stress changes of the observation window accordingly, and powerful polarization parameters are promising for comprehensively monitoring the working submersible's window in the future.

On the other side, more possibilities can be imagined based on the results of this work. For example, one can directly measure the polarization parameters and does not need to measure the whole Mueller matrix. Two-dimensional imaging is not necessary and can be replaced by specified dot measurement. Additionally, according to the requirement of in situ monitoring of the submersible window, other methods can be developed to overcome the difficulties in this work such as image distortion, polarization calibration, etc.

5. Conclusions

Stress detection of observation windows is a key issue in the process of ensuring the safety of deep manned submersible. In this paper, we present a method based on Mueller matrix imaging and build the experimental setup to measure the stress accumulation and recovery situation inside the window. Experimental results support the idea that some polarization parameters can effectively characterize the internal stress transfer and elastic–plastic transformation of the window. Furthermore, we also use the two-layered wave plate model to analyze the internal stress structure of the window under pressurization, and the results show the difference of the former and latter layers of the window in the pressurization process. In addition, we present a finite element simulation explaining the stress accumulation at different thicknesses of the window, which verifies the effectiveness and innovation of our method in stress evaluation. Finally, more possibilities are discussed amid the target of this work by using other polarization parameters derived from Mueller matrix imaging. It is promising that this method can provide a new method of monitoring window stress, which helps to further ensure the safety of deep manned submersibles.

Author Contributions: Writing—original draft preparation, H.L.; writing—review and editing, R.L.; visualization, Z.G., H.T. and Y.C.; project administration, H.Z., G.M. and H.M. All authors have read and agreed to the published version of the manuscript.

Funding: This research was funded by the National Natural Science Foundation of China (61975088, 41527901), the Key-Area Research and Development Program of Guangdong Province (2020B11110400 01), the National Key Research and Development Program of China (2018YFC1406600) and the Shenzhen-Hong Kong Joint Project (SGDX20201103095403017).

Institutional Review Board Statement: Not applicable.

Informed Consent Statement: Not applicable.

Data Availability Statement: Data sharing not applicable.

Conflicts of Interest: The authors declare no conflict of interest.

References

1. Stachiw, J.D. *Handbook of Acrylics for Submersibles, Hyperbaric Chambers, and Aquaria*; Best Publishing Company: Flagstaff, AZ, USA, 2003.
2. Busby, R.F. *Manned Submersibles*; Office of the Oceanographer of the Navy: Washington, DC, USA, 1976; pp. 55–63.
3. Stachiw, J.D. Conical acrylic windows under long-term hydrostatic pressure of 10,000 psi. *ASME J. Eng. Ind.* **1972**, *94*, 1053–1059. [CrossRef]
4. ASME PVHO-1. *Safety Standard for Pressure Vessels for Human Occupancy*; American Society of Mechanical Engineers: New York, NY, USA, 2012.
5. Du, Q.H.; Hu, Y.; Cui, W.C. Safety assessment of the acrylic conical frustum viewport structure for a deep-sea manned submersible. *Ships Offshore Struct.* **2017**, *12*, 221–229. [CrossRef]
6. Zhou, F.; Hou, S.; Qian, X.; Chen, Z.; Zheng, C.; Xu, F. Creep behavior and lifetime prediction of PMMA immersed in liquid scintillator. *Polym. Test.* **2016**, *53*, 323–328. [CrossRef]
7. Arnold, J.C.; White, V.E. Predictive models for the creep behaviour of PMMA. *Mater. Sci. Eng. A* **1995**, *197*, 251–260. [CrossRef]
8. Pranesh, S.B.; Kumar, D. Numerical and experimental study on the safety of viewport window in a deep sea manned submersible. *Ships Offshore Struct.* **2020**, *15*, 769–779. [CrossRef]
9. Liu, P.F.; Li, J.X.; Wang, S.B.; Leng, J.X. Finite element analysis of viscoelastic creep behaviors of deep-sea manned submersible viewport windows. *Int. J. Press. Vessel. Pip.* **2020**, *188*, 104218. [CrossRef]
10. Wang, F.; Wang, M.; Wang, W.; Yang, L.; Zhang, X. Time-dependent axial displacement of PMMA frustums designed for deep-sea manned cabin based on finite element analysis. *Ships Offshore Struct.* **2021**, *16*, 827–837. [CrossRef]
11. Stachiw, J.D. Deep Submergence Spherical Shell Window Assembly with Glass or Transparent Ceramic Windows for Abyssal Depth Service. *ASME J. Eng. Ind.* **1975**, *97*, 1020–1034. [CrossRef]
12. Du, Q.; Xu, D.; Liu, D. Experimental and numerical analysis on pressure effect of polymethyl methacrylate (PMMA) structure in electric-resistance strain measurement. *Ships Offshore Struct.* **2021**, *15*, 1–11. [CrossRef]

13. Zhu, Y.; Liang, W.; Zhao, X.; Wang, X.; Xia, J. Strength and stability of spherical pressure hulls with different viewport structures. *Int. J. Press. Vessel. Pip.* **2019**, *176*, 103951. [CrossRef]
14. Pranesh, S.B.; Sathianarayanan, D.; Ramadass, G.A.; Chandrasekaran, E.; Murugesan, M.; Rajput, N.S. Design and construction of shallow water spherical pressure hull for a manned cabin. *Ships Offshore Struct.* **2020**, *16*, 1–16. [CrossRef]
15. Anwander, M.; Zagar, B.G.; Weiss, B.; Weiss, H. Noncontacting strain measurements at high temperatures by the digital laser speckle technique. *Exp. Mech.* **2000**, *40*, 98–105. [CrossRef]
16. Huntley, J.M. Laser Speckle and Its Application to Strength Measurement and Crack Propagation. Ph.D. Thesis, University of Cambridge, Cambridge, UK, 1986.
17. Tuchin, V.V. Polarized light interaction with tissues. *J. Biomed. Opt.* **2016**, *21*, 071114. [CrossRef] [PubMed]
18. He, H.; He, C.; Chang, J.; Lv, D.; Wu, J.; Duan, C.; Ma, H. Monitoring microstructural variations of fresh skeletal muscle tissues by Mueller matrix imaging. *J. Biophotonics* **2017**, *10*, 664–673. [CrossRef] [PubMed]
19. Shen, Y.; Sheng, W.; He, H.; Li, W.; Ma, H. Assessing distribution features of fibrous structures using Mueller matrix derived parameters: A quantitative method for breast carcinoma tissues detection and staging. In Proceedings of the Dynamics and Fluctuations in Biomedical Photonics XVII, San Francisco, CA, USA, 1–6 February 2020; Volume 11239, p. 112390F.
20. Li, J.; Liao, R.; Tao, Y.; Liu, Z.; Wang, Y.; Ma, H. Evaluation for gas vesicles of sonicated cyanobacteria using polarized light scattering. *Optik* **2020**, *216*, 164835. [CrossRef]
21. Wang, Y.; Liao, R.; Dai, J.; Liu, Z.; Xiong, Z.; Zhang, T.; Ma, H. Differentiation of suspended particles by polarized light scattering at 120°. *Opt. Express* **2018**, *26*, 22419–22431. [CrossRef]
22. Chen, Y.; Zeng, N.; Chen, S.; Zhan, D.; He, Y.; Ma, H. Study on morphological analysis of suspended particles using single angle polarization scattering measurements. *J. Quant. Spectrosc. Radiat. Transf.* **2019**, *224*, 556–565. [CrossRef]
23. Xu, Q.; Zeng, N.; Guo, W.; Guo, J.; He, Y.; Ma, H. Real time and online aerosol identification based on deep learning of multi-angle synchronous polarization scattering indexes. *Opt. Express* **2021**, *29*, 18540–18564. [CrossRef]
24. Alali, S.; Vitkin, A. Polarized light imaging in biomedicine: Emerging Mueller matrix methodologies for bulk tissue assessment. *J. Biomed. Opt.* **2015**, *20*, 61104. [CrossRef]
25. De Martino, A.; Kim, Y.K.; Garcia-Caurel, E.; Laude, B.; Drévillon, B. Optimized Mueller polarimeter with liquid crystals. *Opt. Lett.* **2003**, *28*, 616–618. [CrossRef] [PubMed]
26. Pust, N.J.; Shaw, J.A. Dual-field imaging polarimeter using liquid crystal variable retarders. *Appl. Opt.* **2006**, *45*, 5470–5478. [CrossRef] [PubMed]
27. Arteaga, O.; Freudenthal, J.; Wang, B.; Kahr, B. Mueller matrix polarimetry with four photoelastic modulators: Theory and calibration. *Appl. Opt.* **2012**, *51*, 6805–6817. [CrossRef]
28. Alali, S.; Yang, T.; Vitkin, I.A. Rapid time-gated polarimetric Stokes imaging using photoelastic modulators. *Opt. Lett.* **2013**, *38*, 2997–3000. [CrossRef]
29. Chang, J.; He, H.; Wang, Y.; Huang, Y.; Li, X.; He, C.; Ma, H. Division of focal plane polarimeter-based 3 × 4 Mueller matrix microscope: A potential tool for quick diagnosis of human carcinoma tissues. *J. Biomed. Opt.* **2016**, *21*, 056002. [CrossRef]
30. Huang, T.; Meng, R.; Qi, J.; Liu, Y.; Wang, X.; Chen, Y.; Ma, H. Fast Mueller matrix microscope based on dual DoFP polarimeters. *Opt. Lett.* **2021**, *46*, 1676–1679. [CrossRef] [PubMed]
31. Chang, J.; He, H.; He, C.; Ma, H. DofP polarimeter based polarization microscope for biomedical applications. In Proceedings of the Dynamics and Fluctuations in Biomedical Photonics XIII, San Francisco, CA, USA, 13–18 February 2016; Volume 9707, p. 97070W.
32. Product Parameters of Jacks. Available online: http://www.yl1988.com/index.php?id=144 (accessed on 25 January 2022).
33. Documentation of Imwarp. Available online: https://ww2.mathworks.cn/help/images/ref/imwarp.html?lang=en (accessed on 25 January 2022).
34. He, H.; Zeng, N.; Li, D.; Liao, R.; Ma, H. Quantitative Mueller matrix polarimetry techniques for biological tissues. *J. Innov. Opt. Health Sci.* **2012**, *5*, 1250017. [CrossRef]
35. Lu, S.Y.; Chipman, R.A. Interpretation of Mueller matrices based on polar decomposition. *JOSA A* **1996**, *13*, 1106–1113. [CrossRef]
36. Du, E.; He, H.; Zeng, N.; Sun, M.; Guo, Y.; Wu, J.; Ma, H. Mueller matrix polarimetry for differentiating characteristic features of cancerous tissues. *J. Biomed. Opt.* **2014**, *19*, 076013. [CrossRef]
37. Arteaga, O. Conversion of a polarization microscope into a mueller matrix microscope. Application to the measurement of textile fibers. *Opt. Pura Apl.* **2015**, *48*, 309.
38. Tao, S.; Liu, T.; He, H.; Wu, J.; Ma, H. Distinguishing anisotropy orientations originated from scattering and birefringence of turbid media using mueller matrix derived parameters. *Opt. Lett.* **2018**, *43*, 4092.
39. Li, P.; Lv, D.; He, H.; Ma, H. Separating azimuthal orientation dependence in polarization measurements of anisotropic media. *Opt. Express* **2018**, *26*, 3791–3800. [CrossRef] [PubMed]
40. Chen, B.; Li, W.; He, H.; He, C.; Ma, H. Analysis and calibration of linear birefringence orientation parameters derived from Mueller matrix for multi-layered tissues. *Opt. Lasers Eng.* **2021**, *146*, 106690. [CrossRef]

sensors

MDPI

Article

Study of Intra-Chamber Processes in Solid Rocket Motors by Fiber Optic Sensors

Andrey A. Zhirnov [1,2,†], Konstantin V. Stepanov [1,*,†], Stanislav G. Sazonkin [1], Tatyana V. Choban [1], Kirill I. Koshelev [1], Anton O. Chernutsky [1], Alexey B. Pnev [1,2], Alexey O. Novikov [1] and Dmitriy A. Yagodnikov [1]

1 Bauman Moscow State Technical University, 2-nd Baumanskaya 5-1, 105005 Moscow, Russia; a.zh@bmstu.ru (A.A.Z.); sazstas@bmstu.ru (S.G.S.); chobantv@yandex.ru (T.V.C.); koshelev-k@yandex.ru (K.I.K.); chernutsky.a@bmstu.ru (A.O.C.); pniov@bmstu.ru (A.B.P.); novikovao@bmstu.ru (A.O.N.); daj@bmstu.ru (D.A.Y.)

2 Kotelnikov Institute of Radioengineering and Electronics of RAS, Mokhovaya 11-7, 125009 Moscow, Russia

* Correspondence: stkv@bmstu.ru

† These authors contributed equally to this work.

Abstract: In this study, an experimental study of the burning rate of solid fuel in a model solid propellant rocket motor (SRM) E-5-0 was conducted using a non-invasive control method with fiber-optic sensors (FOSs). Three sensors based on the Mach–Zehnder interferometer (MZI), fixed on the SRM E-5-0, recorded the vibration signal during the entire cycle of solid fuel burning. The results showed that, when using MZI sensors, the non-invasive control of solid fuel burnout is made possible both by recording the time of arrival of the combustion front to the sensor and by analyzing the peaks on the spectrogram of the recorded FOS signal. The main mode of acoustic vibrations of the chamber of the model SRM is longitudinal, and it changes with time, depending on the chamber length. Longitudinal modes of the combustion chamber were detected by MZI only after the combustion front passed its fixing point, and the microphone was unable to register them at all. The results showed that the combustion rate was practically constant after the first second, which was confirmed by the graph of the pressure versus time at the nozzle exit.

Keywords: Mach–Zehnder interferometer; fiber optic sensor; solid fuel; solid rocket motors

Citation: Zhirnov, A.A.; Stepanov, K.V.; Sazonkin, S.G.; Choban, T.V.; Koshelev, K.I.; Chernutsky, A.O.; Pnev, A.B.; Novikov, A.O.; Yagodnikov, D.A. Study of Intra-Chamber Processes in Solid Rocket Motors by Fiber Optic Sensors. *Sensors* **2021**, *21*, 7836. https://doi.org/10.3390/s21237836

Academic Editors: Maria Lepore and Ines Delfino

Received: 2 November 2021
Accepted: 23 November 2021
Published: 25 November 2021

Publisher's Note: MDPI stays neutral with regard to jurisdictional claims in published maps and institutional affiliations.

1. Introduction

In modern engines and, particularly, in a solid propellant rocket motor (SRM), energy conversion processes are characterized by extreme temperatures and released power. For example, the thermodynamic temperature is around 3600 K in the combustion chamber of the Ariane-V launch vehicle's accelerator, EAP P241, which produces a thrust of 7.08 MN [1]. Under such conditions, the SRM operation parameters change frequently due to rates of the main intra-chamber processes—for example, fuel combustion. Therefore, we can describe the working process parameters in SRM as fast-flowing. They include vibrations, pressure in the combustion chamber, and acousto-optical and electrophysical characteristics [2,3].

There are various ways to monitor these parameters. Some methods include rather complex installations using, for example, X-ray analysis [4–6] for intra-chamber process control. This method allows for the observation of the fuel mass combustion patterns through the walls of the housing [7]. However, the sampling frequency is low, and the complexity of the setup makes this method inaccessible for common usage. There are also methods for the invasive monitoring of such processes, for example, by inserting thermocouples into test holes. This is simpler, but violates the integrity of the engine and probably changes the operation's parameters. Additionally, the connection wires require a lot of space, add mass to the setup, and can produce sparks. If we place thermocouples on the motor casing in a non-invasive way, the measurement obtained will be incorrect due to low thermal conductivity in the casing in the outer direction, sensor inertia, and the ambient temperature. In such conditions, fiber-optic sensors (FOSs) present a very promising solution for verifying the simulation data of engine processes [8]. Nowadays, many tasks can be solved in science and technics with the help of FOSs. They are useful for measuring deformation [9], temperature [10–12], vibration [13], concentration of substances [14], rotation speed [15], refractive index [16], pressure [17], liquid level [18], acceleration [19], acoustics [20–22], and other parameters [2,23–25]. Additionally, new types of fiber promise to open new opportunities for such monitoring devices [26,27]. The main FOS related advantage for SRM monitoring is its insensitivity to any kind of electromagnetic interference. Optical fibers, unlike wired connection sensors, cannot originate ignition or explosion, since sparking is excluded during the operation. Optical fibers are produced from dielectric materials, which allow measurements of objects with high electrical voltage to be obtained, and allow for their use in liquid media and environments with high humidity. The chemical inertness of FOS materials allows them to be used under the influence of aggressive, gaseous, and dusty environments, which are realized during tests of SRM. In addition, FOSs that are made of special radiation-resistant fibers can be used when background radiation is high [28]. These performance advantages and their constant evolution predict the rapid deployment of fiber sensors in various aerospace applications [29].

Nowadays, FOSs can be directly integrated into the composite materials of aircraft construction [30]. Fiber Bragg gratings (FBG) are the main sensors used in motors. They detect delamination in the composite housings of rocket engines [31,32]. Polymer fibers [33] and fixed-in-fuel FBGs [34–36] have been used to control the state of solid fuel, but this research was conducted during storage and cannot provide information about the working process. Fiber sensors have been used to study rapidly occurring combustion, deflagration, and detonation in SRM [37,38]. Additionally, a spectroscopic analysis of the combustion process was performed using FOSs [39]. Despite the wide range of FOS applications, areas of SRM ignition and combustion processes have not been investigated as of yet. In comparison with X-ray, a simple installation for non-invasive control can be created, with the help of FOSs to control the working process.

In this paper, we propose the concept and operating principle of an FOS based on an Mach–Zehnder interferometer (MZI) for monitoring the combustion of solid fuel. It allows for the registration of combustion front displacement via two principles. The first is based on the housing deformation measurement, and the second uses the registration of acoustic longitudinal modes to measure the combustion chamber parameters. These investigations use the same MZI interference signal, but with different modes of processing.

2. Theory

Pressure, temperature, deformation, and vibration constitute the most important parameters of the working process in a model SRM chamber. Currently, only a limited number of FOSs are able detect these values [28,40,41]. The available data suggests the following requirement: an expected total measurement time of less than 10 seconds; typical oscillation frequencies of intra-chamber processes starting from tens of Hz, increasing to hundreds of kHz. In this case, optical time-domain reflectometry (OTDR) systems and devices based on FBG are insufficient due to the low sampling frequency of 30 kHz for phi-OTDR [42–44] and 10 kHz for FBG in configurations where there are a few sensors in one line [45–47]. In turn, the small SRM length and the requirement of less sensing points allows for the application of sensors based on the MZI. A phase-unwrapping technique was used for dynamic range improvement, based on the 3×3 output splitter providing phase-shifted signals.

A fiber MZI registers a phase difference between two arms, one of which is a reference and the other a sensing arm, as shown in Figure 1. No load is applied to the reference arm during the experiment. The sensing arm is fixed on the controlled object, and its length changes depending on the object's deformation. This leads to a signal intensity fluctuation, which is proportional to the cosine function of deformation. A 3 × 3 output splitter produces a $2\pi/3$ phase shift between neighboring fibers. Thus, the recorded intensity on each photoreceiver can be determined by the following expression [48]:

$$\begin{bmatrix} I_{PD1}(t) = I_1 + I_2 + 2\sqrt{I_1 \cdot I_2} \cdot \cos^2\left(\frac{\pi}{\lambda}\Delta(t) + \varphi_0\right), \\ I_{PD2}(t) = I_1 + I_2 + 2\sqrt{I_1 \cdot I_2} \cdot \cos^2\left(\frac{\pi}{\lambda}\Delta(t) + \varphi_0 + 2\pi/3\right), \\ I_{PD3}(t) = I_1 + I_2 + 2\sqrt{I_1 \cdot I_2} \cdot \cos^2\left(\frac{\pi}{\lambda}\Delta(t) + \varphi_0 - 2\pi/3\right), \end{bmatrix} \quad (1)$$

where I_1, I_2 are the intensities from the reference and sensing arms, respectively; λ is the wavelength of laser radiation, m; $\Delta(t)$ is the optical path difference between the MZI arms, m and φ_0 is the initial phase difference, rad.

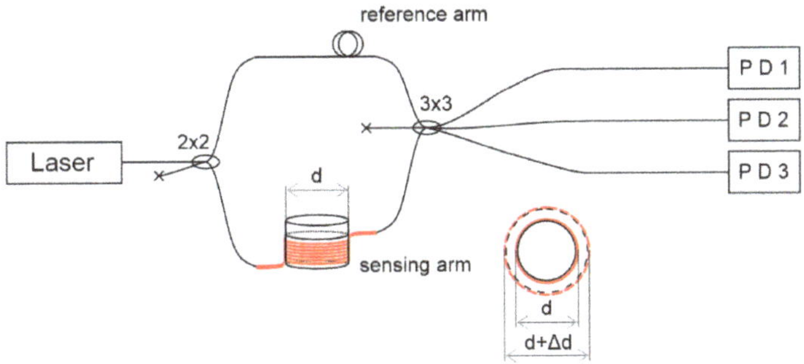

Figure 1. Scheme for the deformation measurement using an MZI.

The presence of two or more signals with a $2\pi/3$ phase shift allows for the restoration of the phase φ of the deformation without uncertainty caused by a cosine function. We implemented a scheme with two photoreceivers for each MZI, reduced the number of photoreceivers, and simplified the measurement scheme. The deformation phase change $\Delta\phi$ was obtained by the algorithm described in [49]. In this case, at each time t, it was calculated by the formula:

$$(t) = \int_0^t \left[S_1(t) \cdot S_2'(t) - S_2(t) \cdot S_1'(t)\right] dt, \quad (2)$$

where $S_1(t) = I_{PD1}(t) - I_{PD2}(t)$, $S_2(t) = I_{PD1}(t) + I_{PD2}(t)$.

The phase change depends on the deformation of the fiber, which was influenced by thermal, mechanical, and acoustic effects. In the experiment, the fiber was coiled in certain places on the outer surface of a model SRM, as shown in Figures 1 and 3. The total sensitive fiber length was $L_{sens} = N\pi d$, where d = 19 mm is the initial outer diameter of the SRM housing and N = 10 is the number of turns. The SRM diameter increase, and phase change are linked through the fluctuations of the sensing arm length ΔL_{sens} by the formula:

$$\Delta\varphi = \frac{2\pi n \Delta L_{sens}}{\lambda} = \frac{2\pi n (N\pi\Delta d)}{\lambda} = \frac{2\pi^2 n N \Delta d}{\lambda},$$

where n is the effective refractive index of the fiber core, and consequently,

$$\Delta d = \frac{\Delta\varphi\lambda}{2n\pi^2 N}. \quad (3)$$

A frequency analysis of the phase change provides additional information about the combustion process. The SRM chamber's acoustic vibration modes depend on its size. The main types are the first longitudinal f_{lon}, tangential f_{tan}, and radial f_{rad}, determined by the formula [50–52]:

$$\begin{bmatrix} f_{lon} = \dfrac{a}{2l(t)} \\ f_{tan} = 0.586\dfrac{a}{d_i} \\ f_{rad} = 1.22\dfrac{a}{d_i} \end{bmatrix}, \quad (4)$$

where a is the speed of sound, m/s, and $l(t)$ is the combustion chamber length, m, at time t, s, as shown in Figure 2. Length $l(t)$ varied from 8 to 109 mm during the experiments; d_i = 15 mm is the chamber internal diameter.

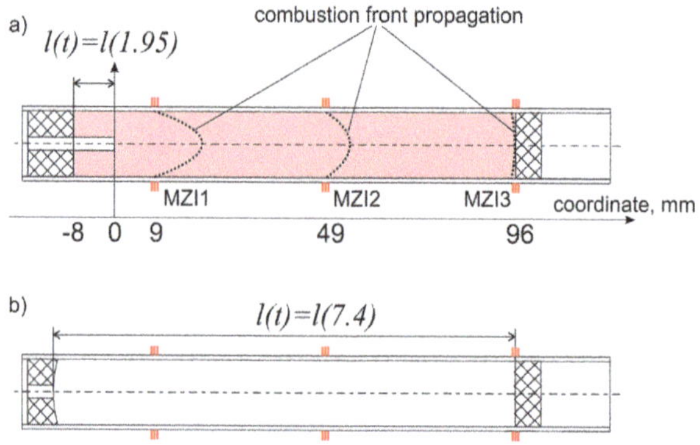

Figure 2. Combustion chamber scheme before (**a**) and after (**b**) work.

These frequencies contribute to phase change fluctuations and can be observed in the spectra. This observation method can precisely determine the chamber length.

3. Experiment

3.1. Description of the Experimental Setup and Methods for Recording the Characteristics of Intra-Chamber Processes

In this study, a model SRM E-5-0 is the research object. It operates via a black powder [53], that was pressed into a cylindrical body made of cardboard. A graphite nozzle block was installed on the bottom with a critical section diameter of 3.4 mm. Fuel ignition was performed using a 0.5 g black powder sample via a combustible wire. The model SRM characteristics are shown in Table 1. Figure 3 shows the SRM photograph (a), a diagram with dimensions between the main components and the MZIs (b), and a section of the SRM after the study was conducted (c). The experimental setup included three MZIs; its scheme is shown in Figure 4. The MZI sensing arms were fixed equidistantly along the entire fuel length.

Table 1. Characteristics of the model SRM E-5-0 used in the work.

Parameter	Value
External diameter, mm	19.0
Internal diameter, mm	15.0
Full length, mm	141.2
Burning time of a solid fuel charge, s	5.5 ± 0.1
Solid fuel charge length, mm	105
Fuel	black powder ($C_{11.362}H_{6.493}O_{22.768}N_{7.319}S_{3.243}K_{7.319}$)

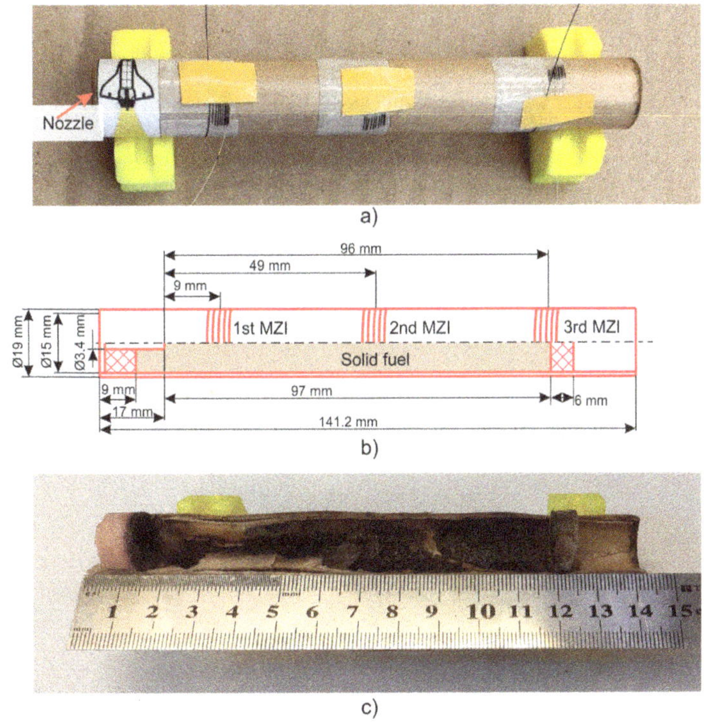

Figure 3. View of the model SRM E-5-0 with installed MZI (**a**), a diagram with dimensions between the main SRM components and the MZI (**b**), and the SRM section after work (**c**).

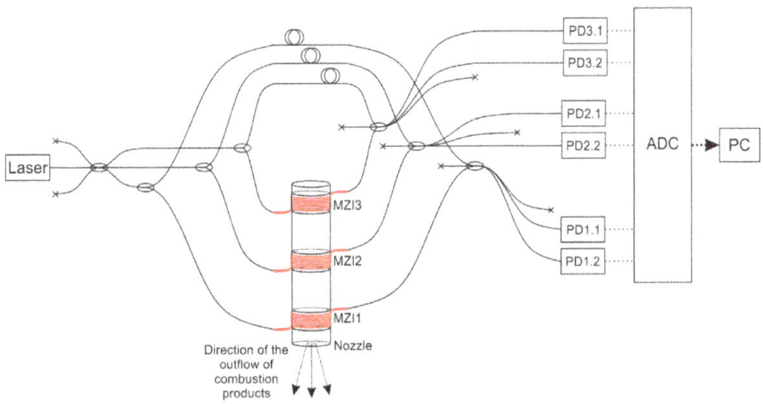

Figure 4. Scheme of the experimental setup. Laser: radiation source; PD: photodiode; ADC: analog-digital converter (L-Card E20-10); PC: personal computer.

A narrow-band NKT BASIK MIKRO fiber laser with a central wavelength of 1550 nm and a bandwidth of less than 0.1 kHz was used. Its radiation emitted through a 3×3 splitter to three independent equal-arm MZIs. The supporting arms were at rest, and the sensing arms were coiled on the SRM housing (see Figure 1). The measuring arm of each MZI consisted of $L_{\text{sens}} = N\pi d = 10 \cdot \pi \cdot 19$ mm ≈ 0.6 m of SMF-28. We glued this fiber loop-to-loop using one layer of double-sided tape. This method of construction increased the MZI sensitivity to fluctuations of housing diameter [54]. The SRM was fixed on the metal table by clamps. Such mounting proved adequate to complete the measurements; the motor shifted slightly at the start as a result of the highest pressure, and the sensing fiber remained connected to the housing at all times. The process was recorded, and is provided in the attached Video S1. Two fibers of each MZI 3 × 3 splitter outputs were transferred to photodiodes (PD). The signals were digitized on an ADC with a sampling rate of 2.5 MHz. This value determined the maximum detectable vibration frequency (1.25 MHz), according to the Nyquist theorem. An image of a laboratory setup with the measurement and registration systems is shown in Figure 5. Before the experiment, we checked the setup's integrity and its ability of deformation registration.

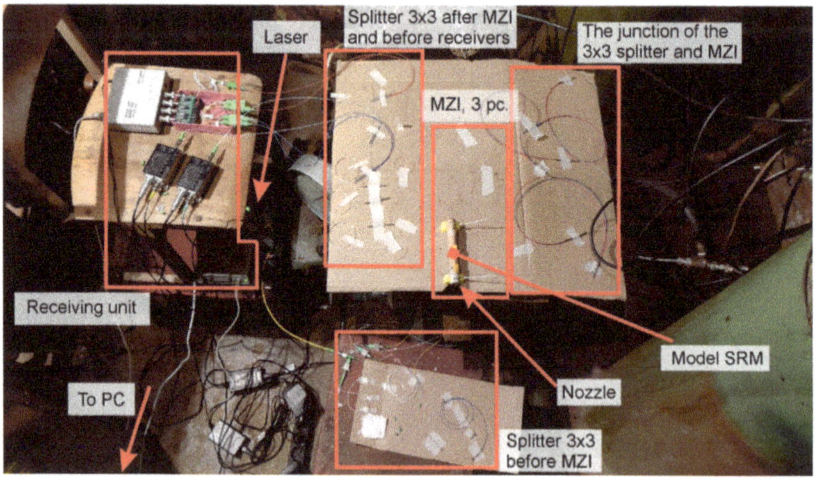

Figure 5. Photo of the assembled laboratory setup.

3.2. Analysis of the Investigation Results

Images of the SRM stages are shown in Figure 6, including start-up (a,b), operation in nominal mode (c,d), and shutdown (e,f). It is worth noting that the tracks of the condensed phase particles flowed out of the engine nozzle, which is common for the combustion products of powder and metal-containing fuels. The total operating time was about 5.5 s, during which the optical fiber did not undergo any damage or changes due to the effect of high-temperature combustion products.

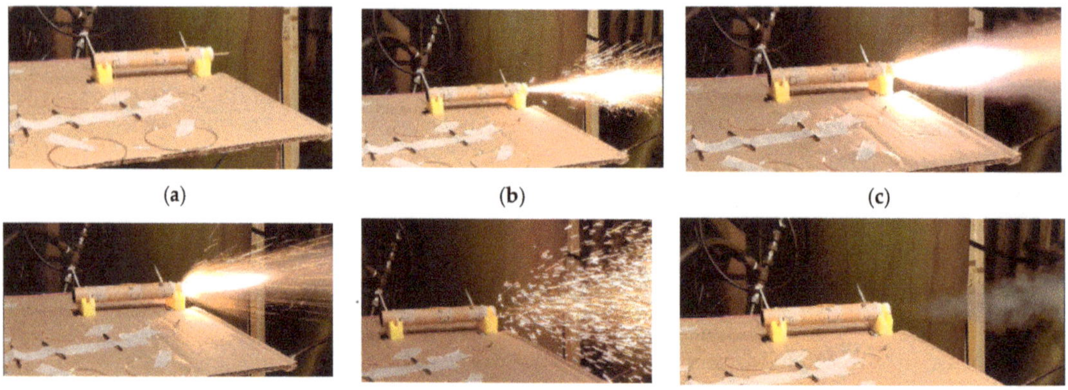

(a) (b) (c) (d) (e) (f)

Figure 6. Photos of the experiment: before the beginning (**a**); launch (**b**); nominal mode (**c**,**d**); engine shutdown and combustion termination (**e**,**f**).

The recorded data from each MZI were processed in the time and frequency domains. The data ranges from 2 seconds before fuel ignition through the combustion process to around 2 seconds after its completion. Until the engine was turned on, the signal at each PD changed with a small amplitude. The high-frequency component occurred due to the PD and the laser phase noise, and the low-frequency fluctuation was a result of the installation temperature drift and the laser wavelength drift. At launch, the amplitude increased in signal oscillations on all PDs. An example of the initial data from one channel of each MZI is presented in Figure 7—the oscillation amplitude increased on all interferometers from the moment the engine was launched, but it only reached the maximum contrast when the combustion surface of the solid fuel reached the MZI sensing arm on the SRM housing.

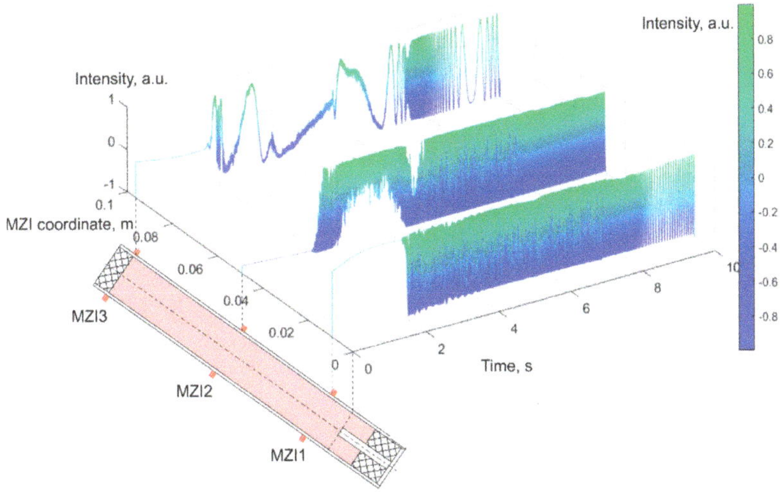

Figure 7. Plots of one data channel for each MZI.

For each sensor, the phase-unwrapping procedure was carried out according to Formula (2). An absolute value of the optical signal phase change, from the initial state (before the engine launch), was obtained and was found to be proportional to the change of the fiber length on the model SRM according to Equation (3). The results highlight that the closer the MZI to the nozzle, the more changes it experienced. The plots for the housing diameter increase are presented in Figure 8.

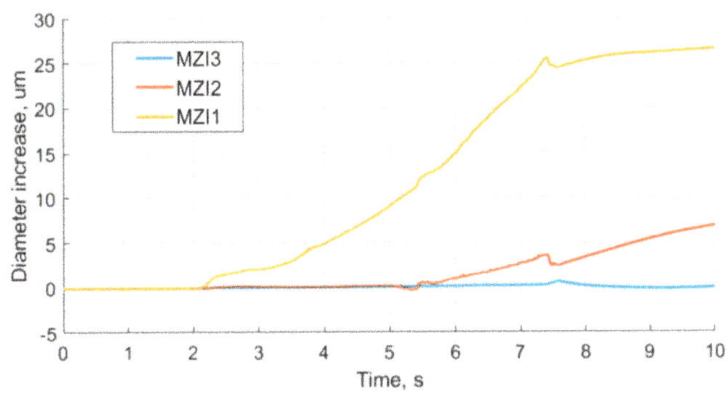

Figure 8. Plots of housing diameter increase.

The derivatives of the housing diameter expansion graphs were calculated with a 60 ms window, allowing for the exclusion of high-frequency oscillations and their influence on the derivative stability. All of the sensors had a moment of initial expansion at the engine start, after which the diameter value became relatively stable, without a noticeable trend of expansion. Graphs illustrating the derivatives from each sensor are shown in Figure 9. A sharp increase in the derivative was observed when the combustion surface coordinate reached the sensor fixing point. These points are marked with circles in Figure 9.

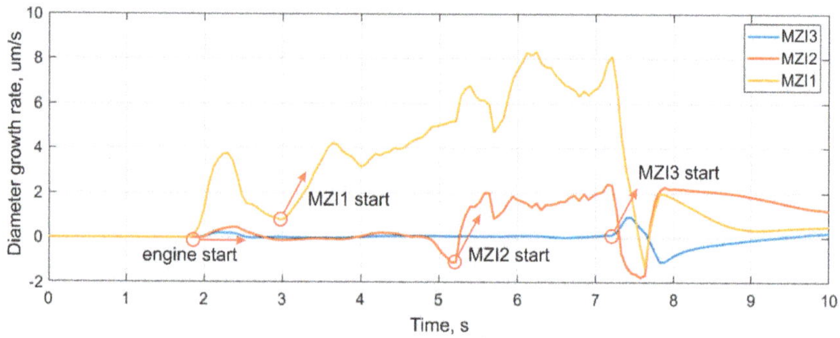

Figure 9. Derivative plots of increasing diameter.

The coordinates of the combustion-front propagation were determined by the time of derivative sharp growth for the sensors, and are shown in Table 2. Based on these values, we graphed the combustion surface movement, as presented in Figure 10.

Table 2. Time and coordinates of burning front propagation.

Point	Fuel Bottom Edge	MZI1	MZI2	MZI3
Coordinate, mm	0	9	49	96
Time of the derivative growth beginning	1.95	2.97	5.20	7.21

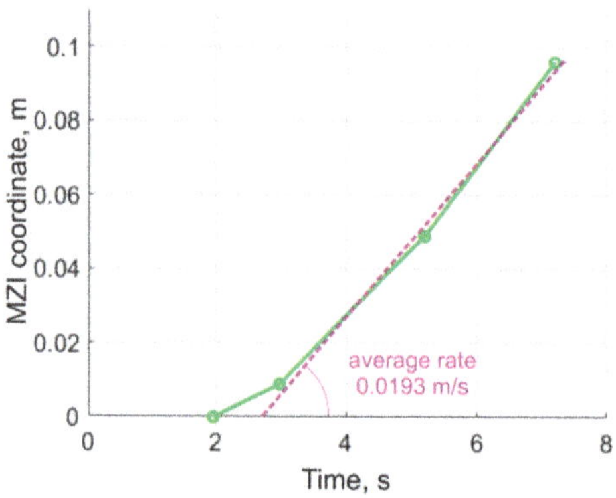

Figure 10. Graph of the combustion surface movement.

The burning rate is non-linear in the first and smaller section of the graph due to the uneven combustion front, caused by the presence of a groove at the end of the solid fuel, as well as the combustion of the igniter sample, also made of black powder. After the end of the ignition period, the time dependence of the combustion surface movement was found to be close to linear with an average linear displacement velocity of 0.0193 m/s.

This dependence (nonlinear during the ~1 second and then linear) is consistent with the results of a similar SRM test, showing that the pressure in the combustion chamber after around 1 second, following the engine start, became almost constant, as shown in Figure 11. The pressure in the combustion chamber during the experiment was measured using a special setup. The SRM was installed in a stainless steel external chamber with a pressure sensor. This setup was the only method by which to fix the pressure sensor to SRM. The described modification slightly increased the combustion time to 6.5 s. However, in general, the pressure change during the investigation remained unchanged for all SRMs of such a model. The fuel and housing construction provide a constant combustion surface area for when the fuel burns, therefore, a constant pressure in the chamber after ~1 second after start becomes apparent even in the presence of deviations in the initial temperature, solid fuel composition, critical section diameter, etc.

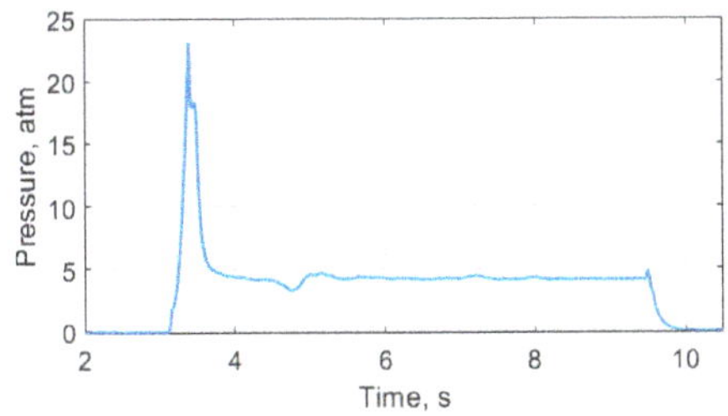

Figure 11. Pressure plot for the similar SRM.

Spectrograms of the unwrapped signal were calculated for each sensor to complete the frequency analysis. They are shown in Figure 12. Some peaks in the characteristic frequencies can be expected. The values of the first longitudinal f_{lon}, tangential f_{tan}, and radial f_{rad} modes of chamber sound vibrations, according to Equation (4), are as follows:

$$\left[\begin{array}{c} f_{lon} = \text{from 45 kHz at SRM start to 3.30 kHz at finish} \\ f_{tan} = 28.13\ \text{kHz} \\ f_{rad} = 58.56\ \text{kHz} \end{array}\right.$$

A shifting peak, in the range from 3 to 20 kHz, and its harmonics are visible in the spectrograms, and have been caused by the changing longitudinal modes. They have a lower frequency in comparison to tangential and radial modes, so longitudinal modes were the most probable. It is possible to calculate the speed of sound, which is determined by the used fuel. Based on the boundary conditions—the minimum frequency of the longitudinal mode in Figure 12a is 3.3 kHz, and the length of the combustion chamber, which was 109 mm—the following results are found using Equation (4):

$$a = 2l(t)f_{lon}(t) = 2l(7.4)f_{lon}(7.4) = 2{\cdot}0.109{\cdot}3300 \approx 720\text{m/s}$$

The spectrograms in Figure 12 show that the peak of longitudinal oscillations only appeared in the interferometer signal when the burning front reached the MZI fixing point. We also analyzed the spectrogram of the audio signal, which was recorded by a microphone during the experiment, and is presented in Figure 13. This plot did not reveal any changing peaks during the burning process. This highlights the advantage provided by the FOS, which was able to detect vibrations that have been generated via sound longitudinal modes. Thus, the fiber MZI worked as a small, light, fire-safe, and easily installed sensor for SRM monitoring.

We plotted the graphs of the ideal longitudinal vibration modes with a known speed of sound and the length of the combustion chamber, calculated according to Equation (4). They are shown in Figure 12b,d,f and are in good agreement with the experimental data. The length changes of the combustion chamber account for two aspects. The first is a meniscus of the burning front, among other factors, caused by a deepening in the solid fuel, as shown in Figure 2. The second is a partial burnout of the plug with the nozzle, from 9 to

5 mm in the center, as shown in Figure 2b. It should be noted that the frequencies of the peaks on the MZI1 and MZI2 spectrograms coincide during their occurrence.

The graphs obtained allow us to conclude that the rate of fuel burnout in the model SRM was almost constant, since the burning front reached the coordinate of the MZI1 after 1 s following initiation.

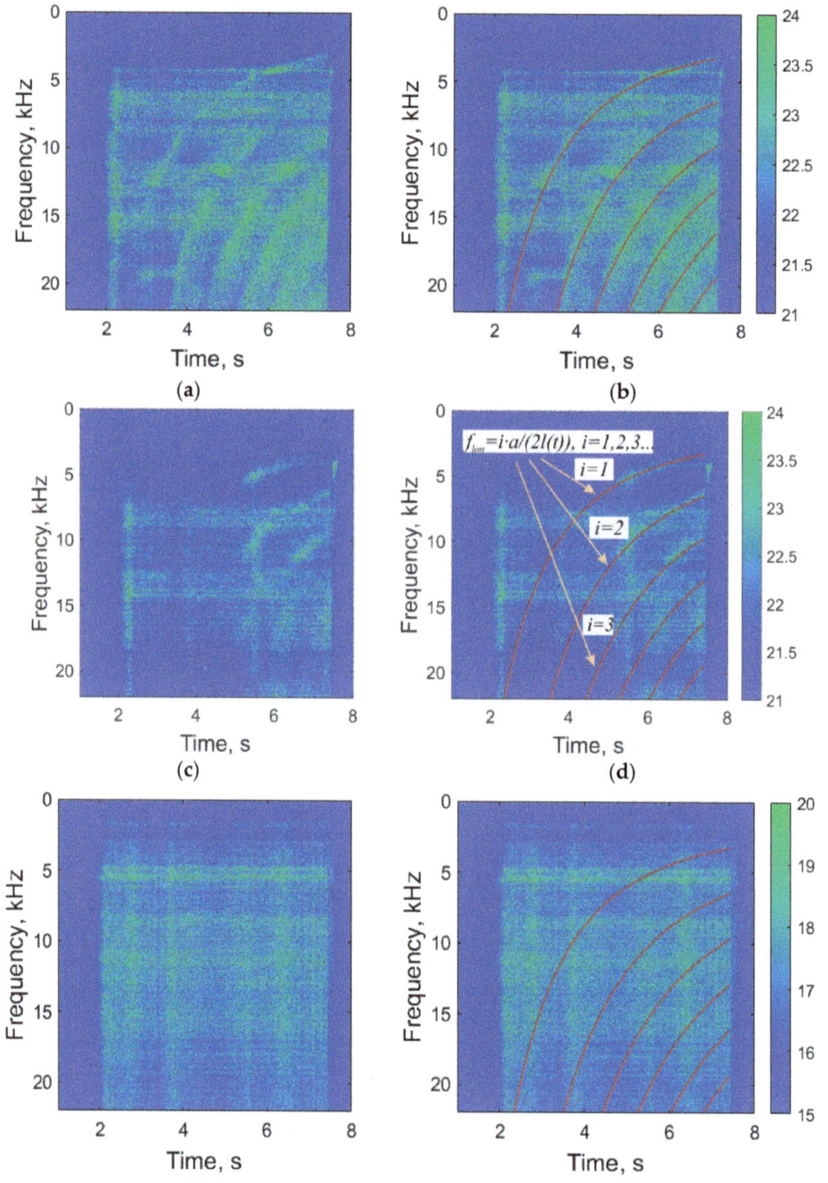

Figure 12. Spectrograms of the housing diameter deformation signals recorded by sensors: pure graphs MZI1 (**a**), MZI2 (**c**), MZI3 (**e**), and the same with marked ideal longitudinal mode and its harmonics for MZI1 (**b**), MZI2 (**d**), and MZI3 (**f**).

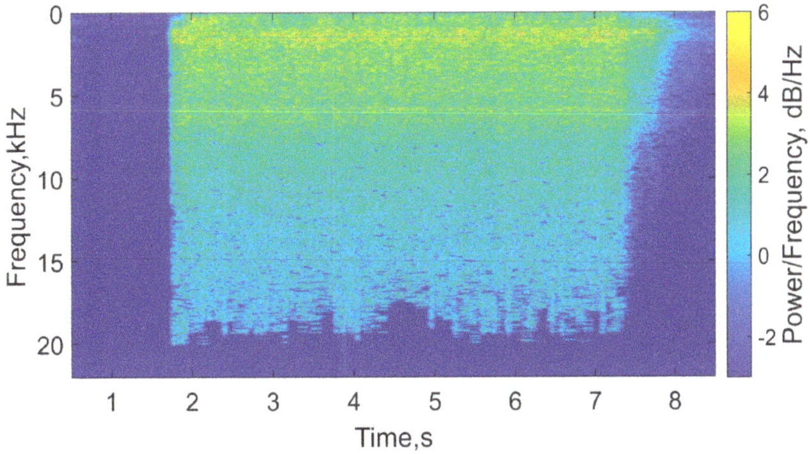

Figure 13. Spectrogram of the audio signal recorded during the experiment.

4. Discussion

A non-invasive diagnostic technique using fiber-optic MZIs as sensors has been developed. This technique makes it possible to determine the characteristics of intra-chamber processes—particularly the burning rate of solid fuel and the length of the combustion chamber—at a given time. The calculations aim to determine when the burning front passes through the MZI fixation points. For each MZI, this can be determined by the derivative growth. Additionally, the resonance frequencies of the acoustic vibration longitudinal modes in the combustion chamber can be determined via the shifting peaks in the spectrogram. As a result, the calculation of the combustion chamber length and the burning rate of solid fuel during the overall SRM worktime can be performed.

For the tested SRM, an uneven combustion of the fuel was detected during the first phase of the work due to deepening occurring at the point at which combustion begins. Then, the burnout rate became almost constant; for our experimental conditions, the burnout rate was approximately 0.0193 m/s.

Supplementary Materials: The following are available online at https://www.mdpi.com/article/10.3390/s21237836/s1, Video S1: Process of SRM work.

Author Contributions: Conceptualization, A.B.P. and D.A.Y.; methodology, K.V.S.; software, A.A.Z.; validation, A.O.C. and K.I.K.; formal analysis, T.V.C. and A.A.Z.; investigation, K.V.S., S.G.S. and T.V.C.; resources, A.O.N.; data curation, A.A.Z.; writing—original draft preparation, K.V.S., S.G.S. and A.A.Z.; writing—review and editing, A.O.C., A.O.N. and D.A.Y.; visualization, A.A.Z. and K.V.S.; supervision, A.B.P.; project administration, D.A.Y. All authors have read and agreed to the published version of the manuscript.

Funding: This research was carried out within the framework of the state task.

Institutional Review Board Statement: Not applicable.

Informed Consent Statement: Not applicable.

Data Availability Statement: The data presented in this study are available on request from the corresponding author.

Conflicts of Interest: The authors declare no conflict of interest. The funders had no role in the design of the study; in the collection, analyses, or interpretation of data; in the writing of the manuscript, or in the decision to publish the results.

References

1. Stephan, S.; Wu, J.; Radespiel, R. Propulsive jet influence on generic launcher base flow. *CEAS Space J.* **2015**, *7*, 453–473. [CrossRef]
2. Mazur, M.M.; Pustovoit, V.I.; Suddenok, Y.A.; Shorin, V.N. Acousto-Optic Monochromator with a Controlled Width of The Instrumental Function. *Phys. Bases Instrum.* **2018**, *7*, 20–37. [CrossRef]
3. Yagodnikov, D.A.; Rudinskiy, A.V. Diagnostics of rocket and jet engines through characteristics of the intrinsic electromagnetic field of combustion products. *High Temp.* **2017**, *55*, 828–845. [CrossRef]
4. Remakanthan, S.; Kk, M.; Gunasekaran, R.; Thomas, C.; Thomas, C.R. Analysis of Defects in Solid Rocket Motors Using X-ray radiography. *E J. Nondestruct. Test.* 2015, *20*. Available online: https://www.ndt.net/search/docs.php3?id=17881 (accessed on 23 November 2021).
5. Gupta, G.; Jawale, L.; Mehilal, D.; Bhattacharya, B. Various methods for the determination of the burning rates of solid propellants: An overview. *Cent. Eur. J. Energ. Mater.* **2015**, *12*, 593–620. Available online: http://yadda.icm.edu.pl/yadda/element/bwmeta1.element.baztech-0c480ad3-921e-47b6-88fe-fc654e0ef5df (accessed on 23 November 2021).
6. Xianggeng, W.; Tao, B.; Pengbo, W.; Xinjian, M.; Yongchun, L.; Jian, C. Burning Rate Enhancement Analysis of End-Burning Solid Propellant Grains Based on X-Ray Real-Time Radiography. *Int. J. Aerosp. Eng.* **2020**, *2020*, 7906804. [CrossRef]
7. Collard, D.N.; McClain, M.S.; Rahman, N.A.; Dorcy, N.H.; Meyer, T.R.; Son, S.F. Dynamic X-Ray Imaging of Additively Manufactured Reactive Components in Solid Propellants. *J. Propuls. Power* **2021**, *37*, 362–368. [CrossRef]
8. Fabignon, Y.; Anthoine, J.; Davidenko, D.; Devillers, R.; Dupays, J.; Gueyffier, D.; Hijlkema, J.; Lupoglazoff, N.; Lamnet, J.M.; Tessé, L.; et al. Recent advances in research on solid rocket propulsion. *J. Aerosp.* **2016**, *11*, 1–15.
9. Taranov, M.A.; Gorshkov, B.G.; Alekseev, A.E.; Potapov, V.T. Distributed strain and temperature sensing over 100 km using tunable-wavelength OTDR based on MEMS filters. *Appl. Opt.* **2021**, *60*, 3049–3054. [CrossRef]
10. Lukashova, T.O.; Nanii, O.E.; Nikitin, S.P.; Treshchikov, V.N. Measurement accuracy and spatial resolution of a distributed temperature sensor based on a two-pulse differential coherent reflectometer. *Quantum Electron.* **2020**, *50*, 882. [CrossRef]
11. Nikitin, S.P.; Kuzmenkov, A.I.; Gorbulenko, V.V.; Nanii, O.E.; Treshchikov, V.N. Distributed temperature sensor based on a phase-sensitive optical time-domain Rayleigh reflectometer. *Laser Phys.* **2018**, *28*, 085107. [CrossRef]
12. Ososkov, Y.Z.; Chernutsky, A.O.; Dvoretskiy, D.A.; Sazonkin, S.G.; Kudelin, I.S.; Orekhov, I.O.; Pnev, A.B.; Karasik, V.E. Fiber optic Raman distributed temperature sensor based on an ultrashort pulse mode-locked fiber laser. *Opt. Spectrosc.* **2019**, *127*, 664–668. [CrossRef]
13. Alekseev, A.E.; Gorshkov, B.G.; Potapov, V.T.; Taranov, M.A.; Simikin, D.E. Dual-pulse phase-OTDR response to propagating longitudinal disturbance. *Laser Phys.* **2020**, *30*, 035107. [CrossRef]
14. Manuylovich, E.; Tomyshev, K.; Butov, O.V. Method for Determining the Plasmon Resonance Wavelength in Fiber Sensors Based on Tilted Fiber Bragg Gratings. *Sensors* **2019**, *19*, 4245. [CrossRef] [PubMed]
15. Shang, K.; Zhang, Y.; Galea, M.; Brusic, V.; Korposh, S. Fibre optic sensors for the monitoring of rotating electric machines: A review. *Opt. Quantum Electron.* **2021**, *53*, 75. [CrossRef]
16. Tomyshev, K.A.; Manuilovich, E.S.; Tazhetdinova, D.K.; Dolzhenko, E.I.; Butov, O.V. High-precision data analysis for TFBG-assisted refractometer. *Sens. Actuator A Phys.* **2020**, *308*, 112016. [CrossRef]
17. Fajkus, M.; Nedoma, J.; Martinek, R.; Fridrich, M.; Bednar, E.; Zabka, S.; Zmij, P. Pressure Membrane FBG Sensor Realized by 3D Technology. *Sensors* **2021**, *21*, 5158. [CrossRef] [PubMed]
18. Liu, Q.; Liu, T.; He, T.; Li, H.; Yan, Z.; Zhang, L.; Sun, Q. High resolution and large sensing range liquid level measurement using phase-sensitive optic distributed sensor. *Opt. Express* **2021**, *29*, 11538–11547. [CrossRef] [PubMed]
19. Xu, H.; Li, F.; Gao, Y.; Wang, W. Simultaneous measurement of tilt and acceleration based on FBG sensor. *IEEE Sens. J.* **2020**, *20*, 14857–14864. [CrossRef]
20. Plotnikov, M.Y.; Lavrov, V.S.; Dmitraschenko, P.Y.; Kulikov, A.V.; Meshkovskiy, I.K. Thin Cable Fiber-Optic Hydrophone Array for Passive Acoustic Surveillance Applications. *IEEE Sens. J.* **2019**, *19*, 3376–3382. [CrossRef]
21. Egorova, D.A.; Kulikov, A.V.; Nikitenko, A.N.; Gribaev, A.I.; Varzhel, S.V. Investigation of bending effects in chirped FBGs array in multicore fiber. *Opt. Quantum Electron.* **2020**, *52*, 130. [CrossRef]
22. Monteiro, C.S.; Raposo, M.; Ribeiro, P.A.; Silva, S.O.; Frazão, O. Acoustic Optical Fiber Sensor Based on Graphene Oxide Membrane. *Sensors* **2021**, *21*, 2336. [CrossRef]
23. Majumder, M.; Gangopadhyay, T.K.; Chakraborty, A.K.; Dasgupta, K.; Bhattacharya, D.K. Fibre Bragg gratings in structural health monitoring-Present status and applications. *Sens. Actuator A Phys.* **2008**, *147*, 150–164. [CrossRef]
24. Glišić, B.; Inaudi, D. *Fibre Optic Methods for Structural Health Monitorin*; John Wiley & Sons: Hoboken, NJ, USA, 2007.
25. Wang, B.; Mao, Y.; Ashry, I.; Al-Fehaid, Y.; Al-Shawaf, A.; Ng, T.K.; Yu, C.; Ooi, B.S. Towards Detecting Red Palm Weevil Using Machine Learning and Fiber Optic Distributed Acoustic Sensing. *Sensors* **2021**, *21*, 1592. [CrossRef] [PubMed]
26. Leonov, S.O.; Yelistratova, E.A.; Demidov, V.V.; Pryamikov, A.D. Birefringence properties of anti-resonant octagonal-core and nodeless hollow-core fibers. *Appl. Opt.* **2020**, *59*, 5013–5019. [CrossRef] [PubMed]
27. Chernutsky, A.O.; Dvoretskiy, D.A.; Orekhov, I.O.; Sazonkin, S.G.; Ososkov, Y.Z.; Denisov, L.K.; Stepanov, L.V.; Zhirnov, A.A.; Pnev, A.B.; Karasik, V.E. High-spatial-resolution distributed temperature sensing system based on a mode-locked fiber laser. In Proceedings of the 2020 International Conference Laser Optics (ICLO), St. Petersburg, Russia, 2–6 November 2020. [CrossRef]
28. Butov, O.V.; Bazakutsa, A.P.; Chamorovskiy, Y.K.; Fedorov, A.N.; Shevtsov, I.A. All-Fiber Highly Sensitive Bragg Grating Bend Sensor. *Sensors* **2019**, *19*, 4228. [CrossRef] [PubMed]

29. Udd, E.; Scheel, I.U. Fiber optic sensors: Technical trends from the mid-1970s to the present. In Proceedings of the Fiber Optic Sensors and Applications XV, Orlando, FL, USA, 14 May 2018. [CrossRef]
30. Pan, X.W.; Liang, D.K.; Li, D. Optical fiber sensor layer embedded in smart composite material and structure. *Smart Mater. Struct.* **2006**, *15*, 1231. [CrossRef]
31. Chang, X.L.; He, X.Y.; Jian, B.; Li, Z.L. The experimental research of delamination damage location based on FBG sensors network in solid rocket motor shell. *Key Eng. Mater.* **2009**, *413*, 47–54. [CrossRef]
32. Chang, X.; He, X.; Hu, J.; Li, J. Experimental research on embedded fiber Bragg grating sensors network for solid rocket motors health monitor. In Proceedings of the 2008 First International Conference on Intelligent Networks and Intelligent Systems, Wuhan, China, 1–3 November 2008. [CrossRef]
33. Riziotis, C.; Eineder, L.; Bancallari, L.; Tussiwand, G. Fiber optic architectures for strain monitoring of solid rocket motors' propellant. *Sens. Lett.* **2013**, *11*, 1403–1407. [CrossRef]
34. Lopatin, C.; Grinstein, D. Active sensing for monitoring the properties of solid rocket motor propellant grains *Propellants Explos. Pyrotech* **2015**, *40*, 295–302. [CrossRef]
35. Zhang, L.; Chang, X.; Zhang, Y.; Liu, W. Response of FBG sensors embedded in SRM interface of combustor when subjected to tri-axial normal loadings. *IEICE Electron. Express* **2017**, 14–20170657. [CrossRef]
36. Zhang, L.; Chang, X.L.; Zhang, Y.H.; Chen, X.D.; Zhang, Q. On-line Monitoring of SRM Grain Structure Based on embedded FBG Sensor. In Proceedings of the 7th International Conference on Power Science and Engineering (ICPSE 2018), Sibiu, Romania, 5–7 June 2019. [CrossRef]
37. Udd, E.; Benterou, J. Improvements to high-speed monitoring of events in extreme environments using fiber Bragg grating sensors. *Proc. SPIE* **2012**, *8370*, 83700L. [CrossRef]
38. Benterou, J.; Bennett, C.V.; Cole, G.; Hare, D.E.; May, C.; Udd, E.; Mihailov, S.J.; Lu, P. Embedded fiber-optic Bragg grating (FBG) detonation velocity sensor. *Proc. SPIE* **2009**, *7316*, 73160E. [CrossRef]
39. Glatkowski, P.J.; Druy, M.A.; Stevenson, W.A. Demonstration of remote fiber optic spectroscopy using mid-IR sensors for in-situ monitoring of solid rocket propellent cure and aging. *Proc. Fiber Opt. Phys. Sens. Manuf. Transp.* **1994**, *2072*, 109–116. [CrossRef]
40. Stepanov, K.V.; Zhirnov, A.A.; Chernutsky, A.O.; Koshelev, K.I.; Pnev, A.B.; Lopunov, A.I.; Butov, O.V. The Sensitivity Improvement Characterization of Distributed Strain Sensors Due to Weak Fiber Bragg Gratings. *Sensors* **2020**, *20*, 6431. [CrossRef]
41. Yatseev, V.A.; Zotov, A.M.; Butov, O.V. Combined frequency and phase domain time-gated reflectometry based on a fiber with reflection points for absolute measurements. *Results Phys.* **2020**, *19*, 103485. [CrossRef]
42. Hu, Y.; Meng, Z.; Zabihi, M.; Shan, Y.; Fu, S.; Wang, F.; Zhang, X.; Zhang, Y.; Zeng, B. Performance Enhancement Methods for the Distributed Acoustic Sensors Based on Frequency Division Multiplexing. *Electronics* **2019**, *8*, 617. [CrossRef]
43. He, Q.; Zhu, T.; Zhou, J.; Diao, D.; Bao, X. Frequency Response Enhancement by Periodical Nonuniform Sampling in Distributed Sensing. *IEEE Photonics Technol. Lett.* **2015**, *27*, 2158–2161. [CrossRef]
44. Xu, N.; Wang, P.; Wang, Y.; Liu, X.; Bai, Q.; Gao, Y.; Zhang, H.; Jin, B. Crosstalk Noise Suppressed for Multi-frequency-OTDR using Compressed Sensing. *J. Lightwave Technol.* **2021**, *39*, 7343–7350. [CrossRef]
45. Brindisi, A.; Vendittozzi, C.; Travascio, L.; Di Palma, L.; Ignarra, M.; Fiorillo, V.; Concilio, A. A Preliminary Assessment of an FBG-Based Hard Landing Monitoring System. *Photonics* **2021**, *8*, 450. [CrossRef]
46. Wang, B.; Sun, W.; Wang, H.; Wan, Y.; Xu, T. Location Determination of Impact on the Wind Turbine Blade Surface Based on the FBG and the Time Difference. *Sensors* **2021**, *21*, 232. [CrossRef]
47. FBGA-IRS Analyzer. Available online: https://www.bayspec.com/telecom-fiber-sensing/fbga-irs-analyzer/ (accessed on 23 November 2021).
48. Hui, R.; O'Sullivan, M. *Fiber Optic Measurement Techniques*; Elsevier: London, UK, 2009.
49. Koo, K.P.; Tveten, A.B.; Dandridge, A. Passive stabilization scheme for fiber interferometers using (3 × 3) fiber directional couplers. *Appl. Phys. Lett.* **1982**, *41*, 616–618. [CrossRef]
50. Васильев, А.П.; Кудрявцев, В.М.; Кузнецов, В.А. Основы Теории и Расчета Жидкостных Ракетных Двигателей; Высшая школа: Moscow, Russia, 1993.
51. Carmicino, C.; Pastrone, D. An Analytical Model to Predict Longitudinal Acoustic Modes Frequency of Hybrid Rockets Combustion Chamber. In Proceedings of the 53rd AIAA/SAE/ASEE Joint Propulsion Conference, Atlanta, GA, USA, 10–12 July 2017; p. 4645. [CrossRef]
52. Carmicino, C.; Di Martino, G.D. Analytical modeling of the one-dimensional acoustic field in hybrid rocket combustion chamber. *J. Sound Vib.* **2018**, *437*, 180–197. [CrossRef]
53. Жуков, Б.П. ЭнергетическиеКонденсированныеСистемы. Краткий ЭнциклопедическийСловарь; Янус-К: Moscow, Russia, 2000.
54. Stepanov, K.V.; Zhirnov, A.A.; Koshelev, K.I.; Chernutsky, A.O.; Khan, R.I.; Pnev, A.B. Sensitivity Improvement of Phi-OTDR by Fiber Cable Coils. *Sensors* **2021**, *21*, 7077. [CrossRef] [PubMed]

precision motion stage, which is available in most of the optical lab). However, for the case of the stitching calibration of long-range linear scale, the sub-aperture topography is one-dimensional, and the calibration error will propagate fully and could result in a second-order accumulative error when applying the stitching algorithm. Figure 1 shows the effect of the calibration error on the final stitching result. From the Figure, it can be deduced that as the measured object size increases, the accumulative second-order stitching error would be enlarged correspondingly and should not be neglected in the stitching calibration.

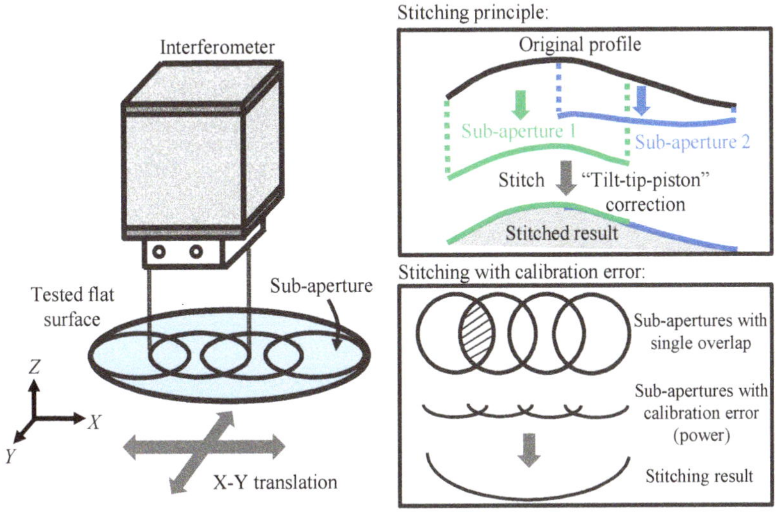

Figure 1. Schematic of the concept of the Z-directional surface form stitching interferometry and the influence of calibration error on stitching result.

2.2. Self-Calibrated Stitching of X-Directional Pitch Deviation of the Long-Range Linear Scale

The stitching calibration method for the long-range linear scale pitch deviation evaluation can be developed by integrating the stitching interferometry for the Z-direction phase error compensation, just as it is used to evaluate the surface form of a large size optical component. On the other hand, since the periodic pattern of the linear scale is arranged along the X-direction, the X-directional stitching for pitch deviation should also be included. The pitch deviation evaluation of the linear scale requires the measurement of the first-order diffracted beams from the linear scale by setting it in the Littrow setup as Figure 2 shows. In the setup, the diffracted beam can be back-reflected directly to the direction of the incident beam and be captured by the interferometer. When using the Fizeau interferometer for the measurement of the diffracted wavefront of the linear scale, the positive and negative first-order phase outputs of the interferometer can be expressed by [16]

$$I_{X+1}(x,y) = 2\pi\frac{e_X(x,y)}{g} + 2\pi\frac{2e_Z(x,y)}{\lambda}\cos\theta - 2\pi\frac{2e_R(x,y)}{\lambda} \tag{1}$$

$$I_{X-1}(x,y) = -2\pi\frac{e_X(x,y)}{g} + 2\pi\frac{2e_Z(x,y)}{\lambda}\cos\theta - 2\pi\frac{2e_R(x,y)}{\lambda} \tag{2}$$

where $I_{X+1}(x, y)$ and $I_{X-1}(x, y)$ are the positive and negative first-order phase outputs, respectively. $e_Z(x, y)$ and $e_R(x, y)$ are Z-directional out-of-flatness of the scale grating and reference flat. λ is the wavelength of the light source of the Fizeau interferometer, $e_X(x, y)$ is the X-directional pitch deviation of the linear scale, g is the nominal pitch of the grating, θ is the Littrow angle. It should be pointed out that in the experiment only the first-order diffracted beams from the grating are utilized for calibration. Although higher-order diffracted beams can also be used to calibrate pitch deviation, the measurement could suffer from low diffraction efficiency and loss of lateral information when applying a larger tilt angle. Consequently, the X-directional pitch deviation of the linear scale can be calculated as [16]

$$e_X(x,y) = \frac{g}{4\pi}[I_{X+1}(x,y) - I_{X-1}(x,y)] \tag{3}$$

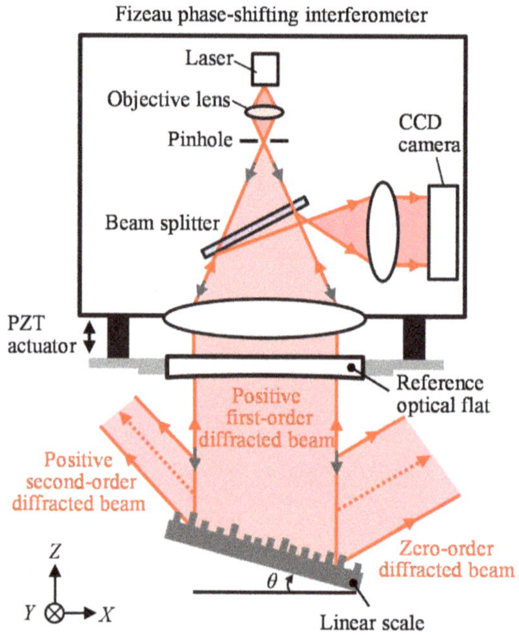

Figure 2. Measurement of the positive first-order diffracted beams from the grating by using the Littrow setup.

In the stitching measurement, multiple measurements are required by shifting the linear scale along the X-direction to obtain several overlapped results and stitching them together. When measuring after shifting the linear scale with a known distance, phase error would be generated by the pitch deviation in different areas. According to the phase shift theory of diffraction grating, the phase error caused by the pitch deviation of the scale grating is accumulated and can be stitched. Figure 3 shows the case that two adjacent areas *A* and *B* with an overlapped area of the scale grating are measured by the Fizeau interferometer. The translation distance is a known value *a*. In the figure, g_0 represents the nominal pitch, g_i (i = 1, 2, 3, . . . , M) represents the actual pitch. More conveniently, the pitch at any coordinate can be expressed as the pitch function $g(x)$. When the linear scale is measured at area *A*, the phase error caused by the pitch deviation at coordinate x can be calculated as

$$\Delta\phi_D(x) \cong \frac{2m\pi}{g_0}\int_0^x \frac{(g_0 - g(x))}{g(x)}dx = \frac{2m\pi}{g_0}\int_0^x \frac{\Delta g(x)}{g(x)}dx \tag{4}$$

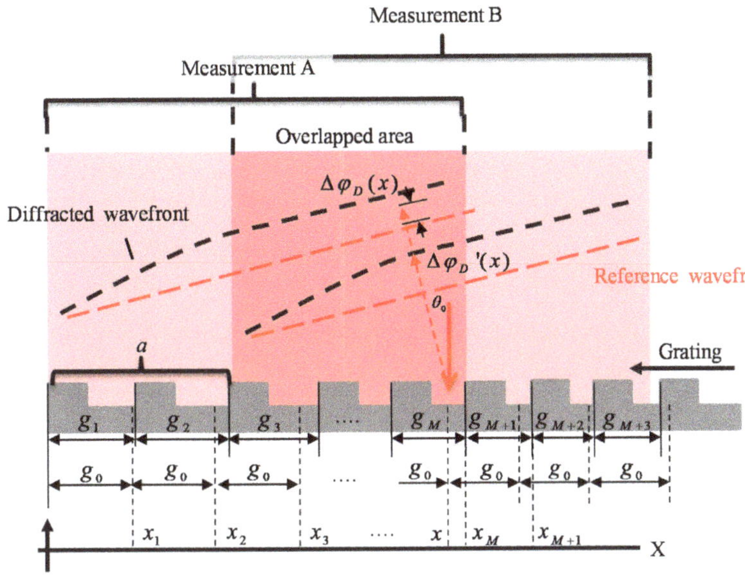

Figure 3. Schematic of the phase error generated by the X-directional pitch deviation in stitching measurement.

After moving the linear scale to the next area B with a distance of value a, the phase error caused by the pitch deviation at the coordinate x would be changed to

$$\Delta\phi'_D(x) \cong \frac{2m\pi}{g_0}\int_a^x \frac{(g_0 - g(x))}{g(x)}dx = \frac{2m\pi}{g_0}\int_a^x \frac{\Delta g(x)}{g(x)}dx \tag{5}$$

Therefore, at the same position x from the field-of-view (FOV) of the interferometer, the phase error is generated due to the change of the evaluated area. The phase difference caused by the shifting process can be expressed as

$$\Delta\varphi(x) = \Delta\phi_D(x) - \Delta\phi'_D(x) = \frac{2m\pi}{g_0}\int_0^a \frac{(g_0 - g(x))}{g(x)}dx = c \tag{6}$$

Equation (6) indicates that the difference of the phase shift caused by the two measurements is only a constant value c, and the additional piston error caused by the pitch deviation should also be corrected. On the other hand, regarding the grating itself as a rigid body in the stitching measurement, the piston/tilts errors would be generated in the shifting process. Combing the additional phase errors caused by the X-directional pitch deviation and Z-directional out-of-flatness error in the stitching measurement, the first-order phase outputs in the overlapped area can be expressed as

$$I_{X\pm1,i}(x,y) = \pm 2\pi\frac{e_{X,i}(x,y)}{g} + \frac{4\pi}{\lambda}[e_{Z,i}(x,y)\cos\theta - e_{R,i}(x,y)] + a_{X\pm1,i}x + b_{X\pm1,i}y + c_{X\pm1,i} \tag{7}$$

where the coefficients $a_{X\pm1,i}$, $b_{X\pm1,i}$, are the tilt coefficients along the X- and Y- directions in i-th sub-aperture corresponding to the positive and negative first-order diffracted beams measurement, respectively, while $c_{X\pm1,i}$ is the coefficient of the piston error along the Z-direction in i-th sub-aperture of the positive and negative first-order diffracted beams measurement. To correct the tilts and piston error and obtain the full aperture of the phase maps, it is desired to minimize the sum of the square differences for all overlapped areas at the same time [32]

$$\min = \sum_{i=1\ldots N} \sum_{j=1\ldots N(j\neq i)}^{j\cap i} [(I_{k,i}(x,y) + a_{k,i}x + b_{k,i}y + c_{k,i}) - (I_{k,j}(x,y) + a_{k,j}x + b_{k,j}y + c_{k,j})]^2 \quad (8)$$

where N represents the total number of the sub-apertures, $k = X \pm 1$, i and j represent different i-th/j-th sub-aperture. The error coefficients $a_{k,i/j}$, $b_{k,i/j}$, and $c_{k,i/j}$ can be calculated by solving Equation (8) with least-square techniques [34]. The equation is differentiated and the error coefficients can be calculated by solving a linear matrix equation. Assuming the overlap area is square ($n \times n$ pixels), the time-complexity of the optimization of the equation would be proportional to n^2N. Equation (8) can be applied to calculate the error coefficients for each sub-aperture in the two diffraction orders. It is noted that the Z-directional rotational error is ignored due to the small effect of the cosine error. In addition, the X-directional displacement error is also not considered since the resolution of the precision stage used to translate the measured optics is usually better than that of the form interferometer. According to Equation (3), by using the stitched positive and negative first-order phase outputs, the pitch deviation of the long-range linear scale over its whole area can be evaluated. Since the reference flat error component contained in the i-th positive and negative first-order diffracted beams are the same. Therefore, the second-order error component caused by the reference flat error component would also be the same in the final stitched phase maps, which can be wiped out by carrying out the differential operation. Based on the self-calibration principle, a more direct approach is to stitch the pitch deviation maps in each sub-aperture together by minimizing the following function

$$\min = \sum_{i=1\ldots N} \sum_{j=1\ldots N(j\neq i)}^{j\cap i} [(e_{X,i}(x,y) + a_{k,i}x + b_{k,i}y + c_{k,i}) - (e_{X,j}(x,y) + a_{k,j}x + b_{k,j}y + c_{k,j})]^2 \quad (9)$$

where $e_{X,i}(x, y)$ is the pitch deviation in i-th aperture, which is calculated by

$$e_{X,i}(x,y) = \frac{g}{4\pi}[I_{X+1,i}(x,y) - I_{X-1,i}(x,y)] \quad (10)$$

From the above analysis, the procedure of the interferometric self-calibrated stitching of the long-range linear scale pitch deviation can thus be summarized, as shown in Figure 4. In the stitching measurement, n phase maps are first collected continuously by moving the tested object or the interferometer. Since one-dimensional stitching is conducted for the linear scale, the arrangement of the position of each sub-aperture should be carefully designed to ensure there is enough overlap area between adjacent sub-apertures. With the existence of the calibration errors and dynamic errors (i.e., thermal, mechanical, etc.), the stitching accuracy could be influenced if only a single-overlap strategy with a small overlap area between adjacent sub-apertures is performed. Although there is no rule-of-thumb of the selection of the best overlap ratio, a double-overlap strategy is preferred so that each overlap could be constrained by an independent sub-aperture and the stitching error can be reduced. The digital filtering method is then used to remove the high-frequency noise components in the phase maps and enhance the performance of the stitching algorithm. The next step is to locate the unified global phase map center and sort the different images according to the moving distance between each phase map. With the arrangement of each image, the overlapped area of each image in the global coordinate can thus be determined. Before applying the stitching algorithm, the reference image should be decided. Usually, the image closest to the global image center could act as the global reference and will remain fixed throughout the whole stitching process. Then the objective is to find a transformation that can describe the misalignment of the sub-image sets with respect to the global reference image and then correct for the misalignment. The n phase maps can, thus, be stitched together to obtain the final phase maps. Finally, the pitch deviation can be calibrated using Equation (3). On the other hand, as demonstrated in the previous analysis, another approach to obtain the pitch deviation of the scale grating is to directly stitch the pitch deviation in each sub-aperture together by using Equations (9) and (10).

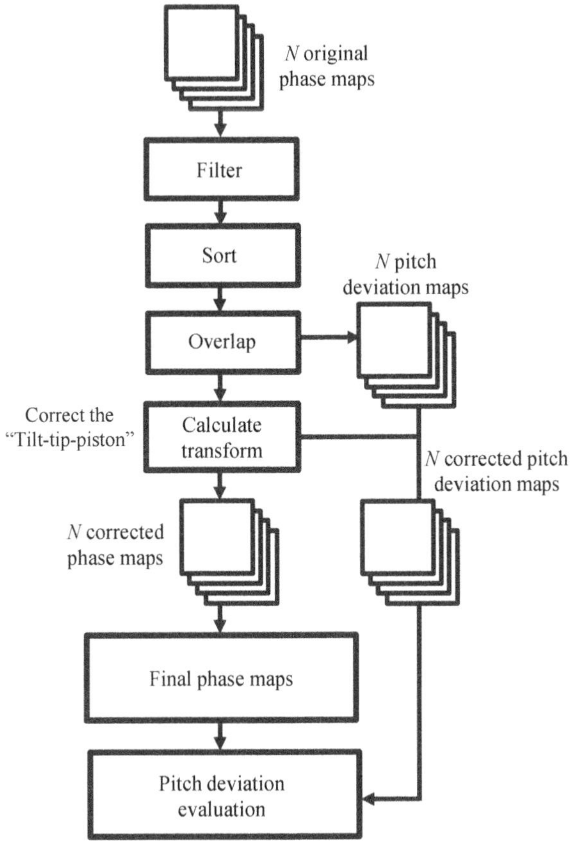

Figure 4. Flow diagram indicating the working principle and the data flow of the stitching algorithm for the calibration of the long-range linear scale pitch deviation.

3. Simulations

Following the theoretical analysis described in the previous sections, numerical calculations are conducted to verify the proposed algorithm. In the simulation, the nominal pitch of the scale grating is set to be 1.6 μm, while the wavelength of the laser source is set as 632.8 nm. First, the form errors of the scale grating $e_Z(x, y)$ and $e_X(x, y)$, as well as the reference flat error $e_R(x, y)$, are simulated with given functions. Note that in the simulation, the X- and Y- coordinate of the linear scale is normalized to $[-1, 1]$ for the sake of simplicity. Then, the phase outputs $I_{X\pm1}(x)$ are simulated by using the previously given form errors of the scale grating and reference flat. Next, the simulated phase outputs are separated into several sub-apertures with known overlap information. Finally, the pitch deviation of the scale grating is reconstructed with the proposed self-calibration method by using both the stitched phase outputs or directly stitching the pitch deviation in each sub-aperture together.

As the first step of the numerical calculation, each of the form errors of the scale grating and the reference flat is given as follows

$$\begin{aligned} e_Z(x,y) &= 0.31 - 26.1x + 3.3y + 147.9x^2 + 18.1y^2 + 20.7x^3 - 0.53x^2y + 2.76xy^2 - 2.9xy^2 \\ &\quad -154.7x^4 + 12.1x^3y + 16.1x^2y^2 + 12.1xy^3 - 27.5y^4 \end{aligned}$$

$$\begin{aligned} e_R(x,y) &= -8.1 - 1.1x - 4.3y - 0.066x^2 - 22.5y^2 + 0.14x^3 + 0.44x^2y - 3.63xy^2 - 45x^4 \\ &\quad -0.55x^3y + 19.1x^2y^2 + 1.02xy^3 - 1.02y^4 \end{aligned}$$

$$e_X(x,y) = -0.84 + 188.7x - 4.2y - 93.4x^2 - 1.8xy - 12.8y^2 - 183.8x^3 + 0.76x^2y + 2.29xy^2 -61x^4 - 1.2x^3y + 7.47x^2y^2 + 2.8xy^3 + 11y^4$$

Figure 5 shows the simulated results of the form errors. After the form errors are simulated, the phase outputs $I_{X\pm1}(x)$ are then calculated according to Equations (1) and (2). To simplify the analysis, three sub-apertures with a rectangular shape are applied totally to separate the simulated phase outputs $I_{X\pm1}(x)$ with known overlap information, as shown in Figure 6.

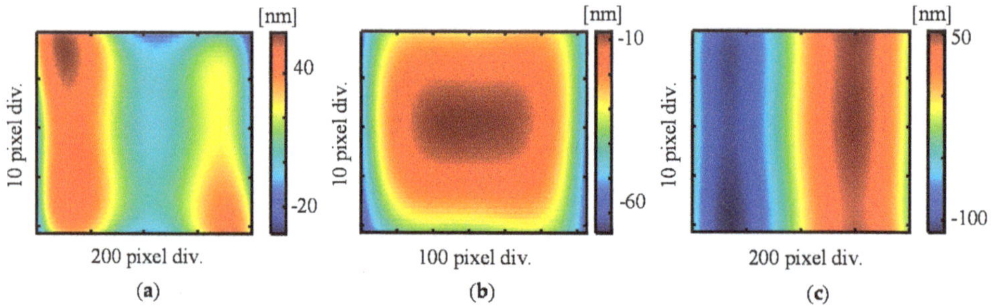

Figure 5. Simulation results of the form errors of a linear scale. (**a**) Out-of-flatness error; (**b**) Reference flat error; (**c**) X-directional pitch deviation.

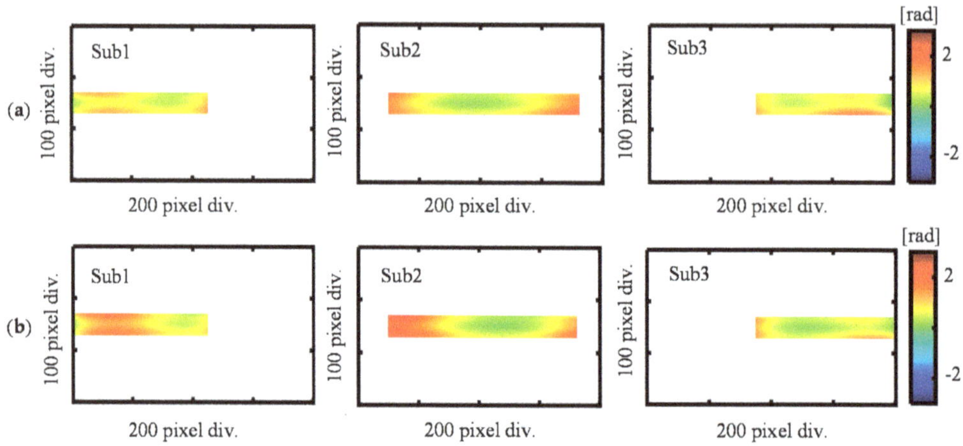

Figure 6. Simulation results of the positive and negative first-order diffracted beams. (**a**) Sub-apertures of the positive first-order diffracted beams; (**b**) Sub-apertures of the negative first-order diffracted beams.

The reference flat error is also considered and added to the phase output in each sub-aperture, as shown in Figure 7. To simulate the possible phase errors caused by the tilt, tip, and piston, the values of the error coefficients a, b, and c are randomly generated for sub-aperture 1 and 3, while these error coefficients are set to be zero for sub-aperture 2 since it is arranged as the reference sub-aperture.

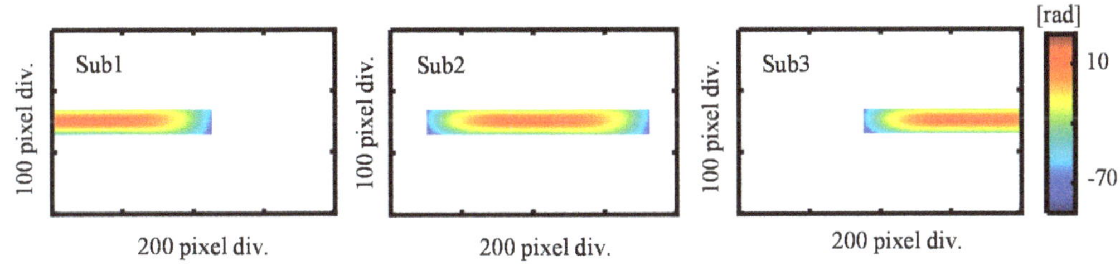

Figure 7. Simulated reference flat error for each sub-aperture.

Figure 8 shows the reconstructed pitch deviation using the proposed self-calibration method. The pitch deviation of the scale evaluated by using the stitched phase outputs $I_{X\pm1}(x)$ is shown in Figure 8a and the evaluation result obtained by directly stitching the pitch deviation is shown in Figure 8b. Figure 8c shows the difference between the two reconstruction results, which indicates that the two reconstruction results are almost the same with a small difference on the level of 10–13 nm. Figure 8d presents the differential result between the reconstructed pitch deviation and simulated pitch deviation. The results verify that the self-calibration stitching method has a stitching accuracy at the level of 10–13 nm.

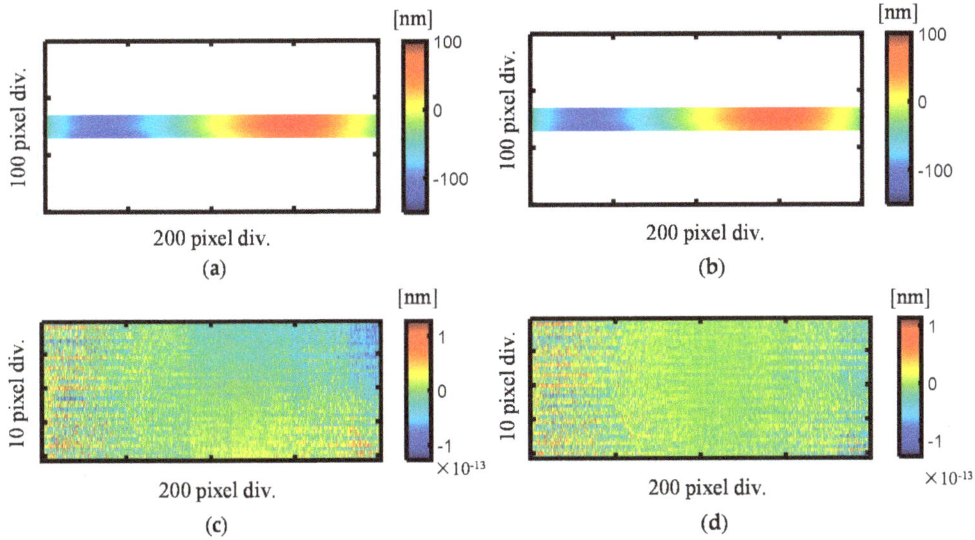

Figure 8. Reconstructed pitch deviation results using the stitched first-order diffracted beams and the pitch deviation maps. (**a**) Evaluated pitch deviation obtained by using the stitched first-order diffracted beams; (**b**) Evaluated pitch deviation obtained by directly stitching the pitch deviation maps; (**c**) Difference obtained from the two results (**a**,**b**); (**d**) Difference between the evaluated pitch deviation and the simulated pitch deviation in Figure 5c.

4. Experiments

4.1. Experimental Setup

A commercial Fizeau interferometer (Verifire^{TM}, Zygo Corp., Middlefield, CT, USA), which has a measurement range of 102 mm in diameter, was used in the experiment. The wavelength of the light source was 632.8 nm. The resolution and accuracy along the *z*-axis were 0.05 nm and $\lambda/20$, respectively. Figure 9 shows the experiment setup with major components. A precision tilt stage (TS-211, Chuo Precision Industrial Co., Ltd., Tokyo, Japan) was employed in the experiments to set the linear scale in the Littrow configuration.

A precision two-axis positioning stage with a resolution of 10 μm and a manual rotary stage was employed to adjust the in-plane position of the linear scale. In addition, a precision manual linear stage with a resolution of 100 μm and a travel range of 100 mm was used. It should be noted that the in-plane position of the linear scale could easily be determined by locating the edges of the linear scale to coincide with those of the CCD image from the visual feedback system. Meanwhile, highly precise in-plane positioning of the linear scale was not required since the lateral resolution of the CCD camera was larger than that of the positioning stages employed in the experiments.

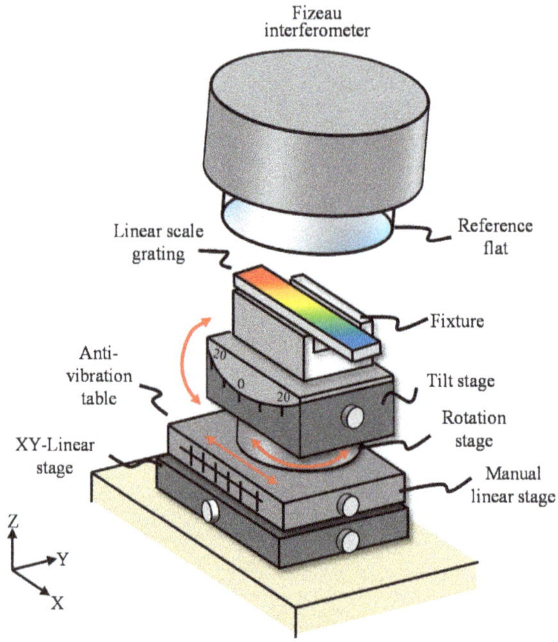

Figure 9. Experimental setup with a commercial Fizeau interferometer.

In the experiment, the linear scale having a nominal pitch of 1.6 μm over an area of 5 mm × 105 mm was employed as the measurement specimen. The linear scale was measured through three shots for each diffraction order and the diffracted beams in different sub-apertures were measured by reciprocating the linear stage. First, the positive first-order diffracted beam of the right part of the linear scale was measured by tilting the linear scale clockwise. Then, the scale grating was translated forward with a distance of 12 mm to measure the positive first-order diffracted beam from the middle part of the linear scale. At last, the positive first-order diffracted beam from the left part of the linear scale was measured by moving the scale slightly forward with a distance of 13 mm. Once the measurements of the positive first-order diffracted beams in each sub-aperture were finished, the linear scale was moved back to the initial position. The linear scale was then tilted counter-clockwise to measure the negative first-order diffracted beams. The measurement procedure of the negative first-order diffracted beams in each sub-aperture was the same as the measurement of the positive first-order diffracted beams. In each measurement, the number of the observed interference fringes in the visual feedback system was reduced to a minimum through adjusting the manual tilt stage to reduce the influence of the angular misalignment from the Littrow angle. In addition, to reduce the influence of environmental noise, three repetitive measurements were conducted in each position. Excluding the warm-up and pre-adjustment time, it took approximately 20 min to conduct all the measurements including the translation of the linear stage and the tilt operation of the linear scale grating. The measured phase outputs were then processed offline for the analysis with the self-calibration stitching method.

4.2. Experiment Results and Discussions

Figure 10 shows the observed positive first-order diffracted beams from each sub-aperture, while Figure 11 shows the measured negative first-order diffracted beams from sub-aperture 1 to sub-aperture 3, respectively.

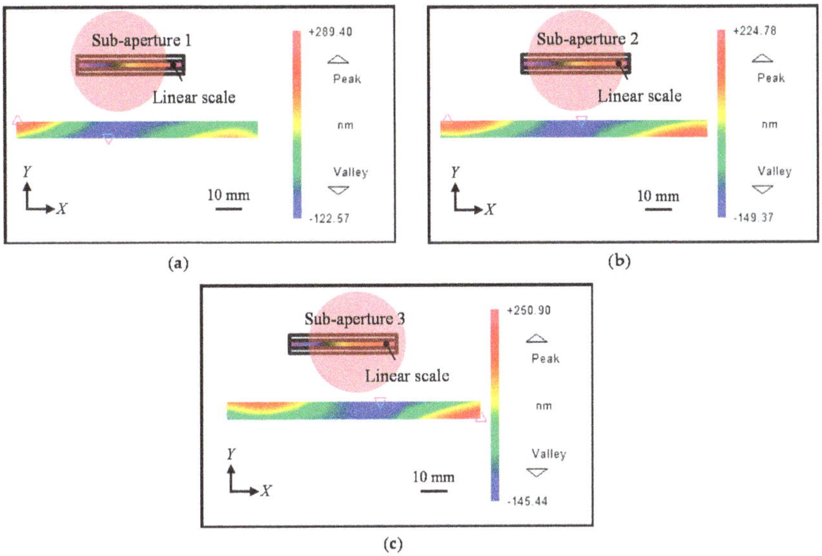

Figure 10. Measured positive first-order diffracted wavefront from the linear scale grating. (**a**) X-directional positive first-order diffracted beam from Sub-aperture 1; (**b**) X-directional positive first-order diffracted beam from Sub-aperture 2; (**c**) X-directional positive first-order diffracted beam from Sub-aperture 3.

Sub-aperture 2 was arranged as the reference aperture since it has the maximum overlap area with the other two sub-apertures. The error coefficients (a, b, c) of sub-aperture two were then calculated using Equations (8) and (9) for different stitching strategies. The calculated error coefficients of each sub-aperture are summarized in Table 1.

Table 1. Error coefficients for each sub-aperture.

	a	b	c
$I_{X+1,1}$	−2.06	−8.24	−0.59
$I_{X+1,3}$	1.95	8.37	−0.39
$I_{X-1,1}$	−2.29	−10.44	−0.66
$I_{X-1,3}$	2.25	9.98	−0.33
$e_{X,1}$	28.80	280.04	8.47
$e_{X,3}$	−38.62	−204.10	−7.02

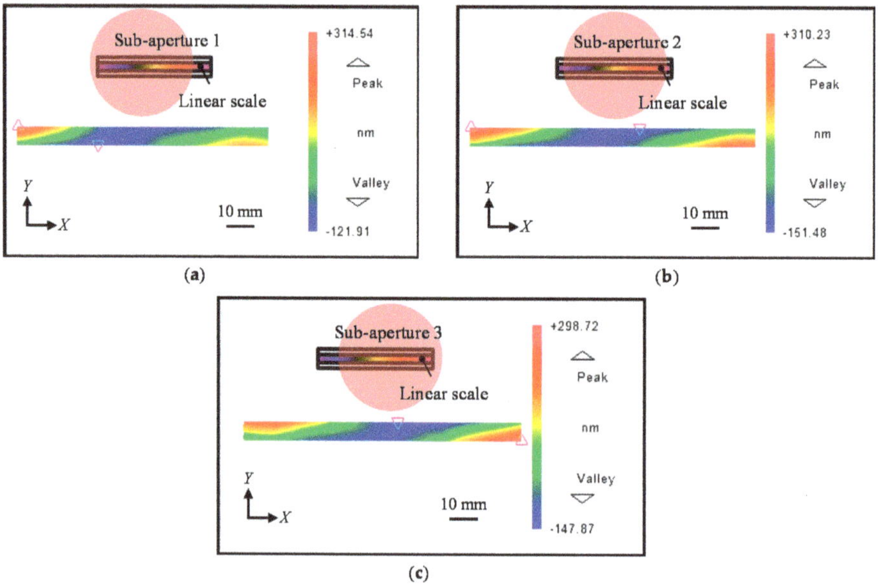

Figure 11. Measured negative first-order diffracted wavefront from the linear scale grating. (**a**) X-directional negative first-order diffracted beam from Sub-aperture 1; (**b**) X-directional negative first-order diffracted beam from Sub-aperture 2; (**c**) X-directional negative first-order diffracted beam from Sub-aperture 3.

Figure 12 shows the pitch deviation results obtained in the previously described two different strategies. The pitch deviation calculated from the stitched phase maps is shown in Figure 12a, which indicates that the pitch deviation over the whole area of the linear scale had a peak-to-valley (PV) value of 343 nm. Meanwhile, Figure 12b shows the pitch deviation of the linear scale calculated by stitching the pitch deviation in each sub-aperture, which also had a PV value of 343 nm. Figure 12c shows the difference between the 2D pitch deviation maps obtained by the two different methods. From the small differential result, it is noted that almost the same results were obtained by using the two different self-calibration stitching methods.

To verify the stitching calibration results, a one-shot experiment was also conducted. By zooming out the observation area of the CCD camera, most of the scale area (about 5 mm × 101 mm) was covered by the illumination area of the one-shot measurement. The positive and negative first-order diffracted beams of the linear scale were evaluated by tilting the linear scale clockwise and counter-clockwise, as shown in Figure 13a,b, respectively. Figure 14 presents the calculated pitch deviation of the linear scale by using the measured wavefronts from the one-shot measurement and the results obtained by using the self-calibrated stitching method. Figure 14a shows the measured pitch deviation with the one-shot measurement had a PV value of 314 nm, which is slightly smaller than that of the self-calibrated stitching result. Nevertheless, the high correspondence of the topography features in the two 2D pitch deviation maps verified the feasibility of the proposed method for pitch deviation stitching calibration.

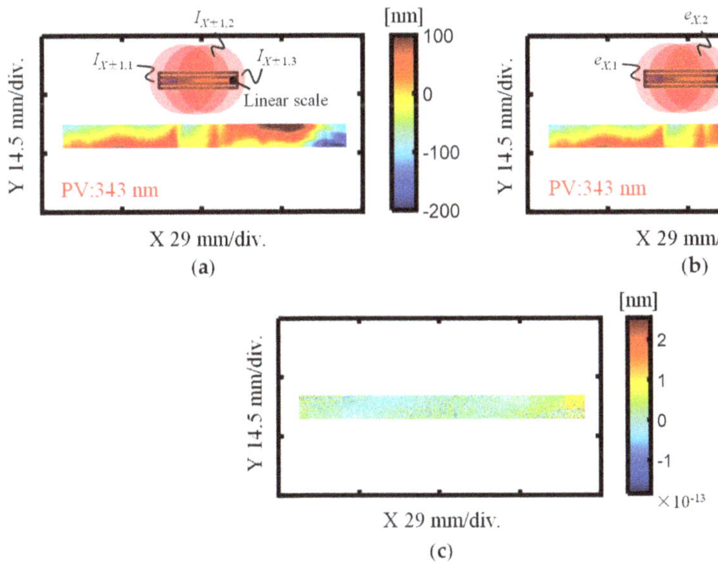

Figure 12. Evaluated *X*-directional pitch deviation of the linear scale through different stitching strategies. (**a**) Pitched deviation evaluated from stitched first-order diffracted beams; (**b**) Pitched deviation evaluated by stitching the pitch deviation from each sub-aperture; (**c**) Difference obtained from the two results in (**a**,**b**).

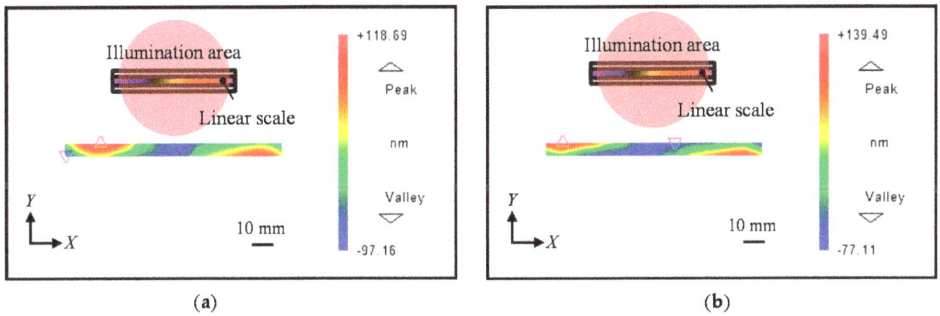

Figure 13. Measured first-order diffracted beams from the linear scale with one-shot measurement. (**a**) *X*-directional positive first-order diffracted beam; (**b**) *X*-directional negative first-order diffracted beam.

To further verify the detail, the averaged *X*-direction cross-section of the two results was calculated and compared with each other. Figure 15a shows the comparison of the averaged *X*-directional cross-sections of the two calibration results in Figure 14. The pitch deviation was then reconstructed by considering the cosine value of the Littrow angle with a 20-order polynomial function using the least-square technique [17]. The difference was then calculated with the two reconstruction results, as shown in Figure 15b. From the figures, the two averaged cross-sections show good correspondence with each other, and the difference was within the range of ± 50 nm over the whole calibration area, verifying the capability of the proposed self-calibration stitching method for long-range linear scale pitch deviation evaluation.

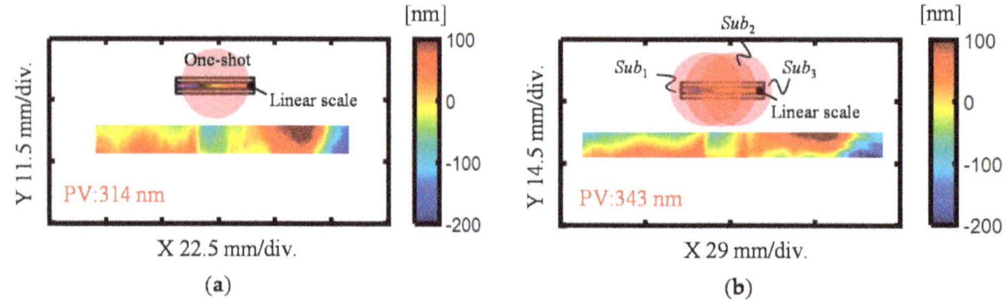

Figure 14. Evaluated pitch deviation of the linear scale through one-shot and three-shot measurements. (**a**) Pitch deviation of the linear scale evaluated with a one-shot measurement; (**b**) Pitch deviation of the linear scale evaluated with a three-shot measurement and stitching method.

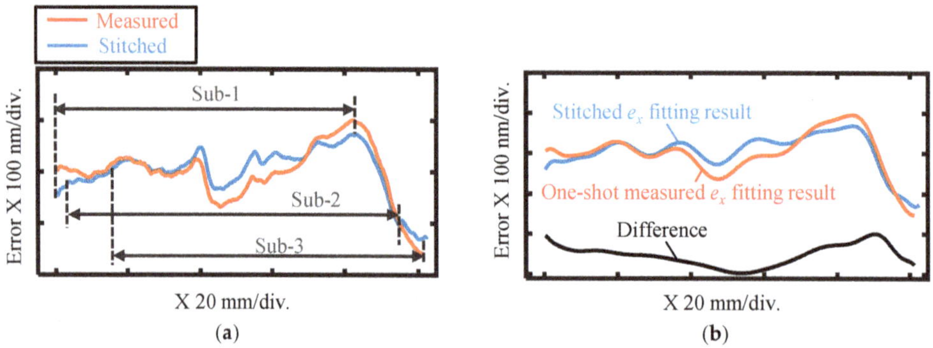

Figure 15. Comparison of the *X*-directional averaged cross-sections of the measured and stitched pitch deviation through one-shot and three-shot measurements. (**a**) *X*-directional averaged cross-section of the measured and stitched pitch deviation results; (**b**) Reconstruction results of the two averaged pitch deviation cross-sections and their difference.

The pitch deviation of a long-range linear scale was successfully calibrated using the proposed stitching calibration method. The same results are obtained by using different stitching strategies with the self-calibration method. Considering the lengths of the industrial used commercial linear scale are mostly within the range of 300 mm [9], these could be measured with a commercial interferometer with three to four shots. The measurement results obtained using the proposed self-calibrated stitching method are thus representative, indicating the feasibility of the proposed method to measure a 200 mm or 300 mm long linear scale. On the other hand, it was found that there was a difference between the stitched pitch deviation with three-shot measurement and the evaluated pitch deviation with one-shot measurement. The difference could be mainly caused by the *X*-directional straightness error of the linear stage, which should be calibrated before the stitching measurement. The positioning error related to the straightness error and the linear stage could also influence the stitching accuracy, although it is usually ignored since the lateral resolution of the interferometer is worse than that of the precision positioning stage [35,36]. It should be pointed out that systematic errors such as the calibration error of the phase shifter as well as the possible speckle effect could also influence the measurement results and reduce the stitching accuracy. Well-designed PSI algorithms and speckle reduction methods could be applied to address these issues [37,38]. Other uncertainties such as the environmental noise and the uncertainty related to the interferometer are estimated to be several nanometers and are not considered in the analysis [19].

It should be noted that the main focus of this paper is to verify the feasibility of applying the proposed self-calibration stitching method for the long-range linear scale pitch deviation calibration. The proposed method could be extended to calibrate the *Z*-directional out-of-flatness error of the scale grating as well as the form errors of the two-dimensional planar scale grating with an *XY* motorized precision stage [39]. The measurement time could increase in this case and the automation of the *XY* stage, as well as the self-calibration stitching program, is expected to facilitate the calibration process. Future research will include the comparison of the calibrated pitch deviation of the linear scale and the nonlinear component error of the linear optical encoder.

5. Conclusions

A self-calibration stitching method based on the Fizeau interferometer, in which the pitch deviation of the linear scale can be evaluated while removing the influence from the reference flat error, has been proposed to evaluate the pitch deviation of a long-range linear scale. The stitching interferometry method has never been applied to evaluate the *X*-directional pitch deviation of the scale grating while eliminating the accumulative second-order effect of the systematic error of the interferometer. Therefore, in this paper, theoretical analysis and simulation have been carried out to develop and test the feasibility of the self-calibration stitching method. Following the theoretical analysis and simulation verification, experiments are conducted by using a long-range linear scale with a length of 105 mm. The pitch deviation of the linear scale can be obtained by directly stitching the pitch deviation from the sub-apertures or evaluated from the firstly stitched order phase outputs, which result to be the same. The PV value of the *X*-directional pitch deviation of the linear scale is evaluated to be 343 nm over its whole area. The pitch deviation evaluation result with three-shot measurement is further compared with the result obtained within one-shot measurement. The high correspondence of the topography in the two assessed pitch deviation maps indicates the capability of the proposed method for long-range linear scale pitch deviation calibration. Meanwhile, the small difference in the averaged *X*-directional cross-sections from the two results also show they have good correspondence with each other. As the first step of the self-calibration stitching of the pitch deviation of the long-range linear scale grating, theoretical analysis has been conducted and primary experiment results have verified the feasibility of the proposed method for the long-range linear scale pitch deviation evaluation. Comprehensive uncertainty analysis of the measurement results, and an extension of the proposed method for the *Z*-directional out-of-flatness measurement of the scale grating as well as the comparison of the evaluated pitch deviation with the optical encoder error will be conducted as future work.

Author Contributions: Conceptualization, W.G., Y.S. and X.X.; Methodology, X.X.; Software, X.X.; Validation, X.X.; Formal Analysis, X.X.; Investigation, X.X.; Resources, X.X.; Data Curation, X.X., H.M. and Y.S.; Writing-Original Draft Preparation, X.X. and Y.S.; Writing-Review and Editing, W.G. and Y.S.; Visualization, W.G. and Y.S.; Supervision, W.G.; Project Administration, W.G.; Funding Acquisition, W.G. and Y.S. All authors have read and agreed to the published version of the manuscript.

Funding: This work is supported by the Japan Society for the Promotion of Science (JSPS) Grant Number 20H00211.

Institutional Review Board Statement: Not applicable.

Informed Consent Statement: Not applicable.

Conflicts of Interest: The authors declare no conflict of interest.

References

1. Gao, W.; Kim, S.-W.; Bosse, H.; Haitjema, H.; Chen, Y.; Lu, X.; Knapp, W.; Weckenmann, A.; Estler, W.; Kunzmann, H. Measurement technologies for precision positioning. *CIRP Ann.* **2015**, *64*, 773–796. [CrossRef]
2. Gao, W.; Haitjema, H.; Fang, F.; Leach, R.; Cheung, C.; Savio, E.; Linares, J. On-machine and in-process surface metrology for precision manufacturing. *CIRP Ann.* **2019**, *68*, 843–866. [CrossRef]
3. Schwenke, H.; Schmitt, R.; Jatzkowski, P.; Warmann, C. On-the-fly calibration of linear and rotary axes of machine tools and CMMs using a tracking interferometer. *CIRP Ann.* **2009**, *58*, 477–480. [CrossRef]
4. Fan, K.-C.; Li, R.-J.; Xu, P. Design and Verification of Micro/Nano-Probes for Coordinate Measuring Machines. *Nanomanufacturing Metrol.* **2018**, *2*, 1–15. [CrossRef]

5. Linear Encoder Products—Sealed and Enclosed—HEIDENHAIN. Available online: https://www.heidenhain.us/product/linear-encoders-for-length-measurement/ (accessed on 10 July 2021).
6. Gao, W.; Araki, T.; Kiyono, S.; Okazaki, Y.; Yamanaka, M. Precision nano-fabrication and evaluation of a large area sinusoidal grid surface for a surface encoder. *Precis. Eng.* **2003**, *27*, 289–298. [CrossRef]
7. Kunzmann, H.; Pfeifer, T.; Flügge, J. Scales vs. Laser Interferometers Performance and Comparison of Two Measuring Systems. *CIRP Ann.* **1993**, *42*, 753–767. [CrossRef]
8. Teimel, A. Technology and applications of grating interferometers in high-precision measurement. *Precis. Eng.* **1992**, *14*, 147–154. [CrossRef]
9. Fleming, A. A review of nanometer resolution position sensors: Operation and performance. *Sens. Actuators A Phys.* **2013**, *190*, 106–126. [CrossRef]
10. Bosse, H.; Häßler-Grohne, W.; Flügge, J.; Köning, R. Final report on CCL-S3 supplementary line scale comparison Nano3. *Metrologia* **2003**, *40*, 04002. [CrossRef]
11. Sawabe, M.; Maeda, F.; Yamaryo, Y.; Simomura, T.; Saruki, Y.; Kubo, T.; Sakai, H.; Aoyagi, S. A new vacuum interferometric comparator for calibrating the fine linear encoders and scales. *Precis. Eng.* **2004**, *28*, 320–328. [CrossRef]
12. Coveney, T. A review of state-of-the-art 1D length scale calibration instruments. *Meas. Sci. Technol.* **2019**, *31*, 042002. [CrossRef]
13. Chen, X.; Koenders, L. A novel pitch evaluation of one-dimensional gratings based on a cross-correlation filter. *Meas. Sci. Technol.* **2014**, *25*, 044007. [CrossRef]
14. Dai, G.; Koenders, L.; Fluegge, J.; Hemmleb, M. Fast and accurate: High-speed metrological large-range AFM for surface and nanometrology. *Meas. Sci. Technol.* **2018**, *29*, 054012. [CrossRef]
15. Klapetek, P.; Valtr, M.; Picco, L.; Payton, O.; Martinek, J.; Yacoot, A.; Miles, M. Large area high-speed metrology SPM system. *Nanotechnology* **2015**, *26*, 065501. [CrossRef] [PubMed]
16. Gao, W.; Kimura, A. A fast evaluation method for pitch deviation and out-of-flatness of a planar scale grating. *CIRP Ann.* **2010**, *59*, 505–508. [CrossRef]
17. Chen, X.; Shimizu, Y.; Xiong, X.; Chen, Y.-L.; Gao, W. Self-calibration of Fizeau interferometer and planar scale gratings in Littrow setup. *Opt. Express* **2017**, *25*, 21567–21582. [CrossRef] [PubMed]
18. Xiong, X.; Matsukuma, H.; Shimizu, Y.; Gao, W. Evaluation of the pitch deviation of a linear scale based on a self-calibration method with a Fizeau interferometer. *Meas. Sci. Technol.* **2020**, *31*, 094002. [CrossRef]
19. Xiong, X.; Shimizu, Y.; Chen, X.; Matsukuma, H.; Gao, W. Uncertainty Evaluation for Measurements of Pitch Deviation and Out-of-Flatness of Planar Scale Gratings by a Fizeau Interferometer in Littrow Configuration. *Appl. Sci.* **2018**, *8*, 2539. [CrossRef]
20. Yang, S.; Zhang, G. A review of interferometry for geometric measurement. *Meas. Sci. Technol.* **2018**, *29*, 102001. [CrossRef]
21. Vannoni, M.; Martín, I.F. Large aperture Fizeau interferometer commissioning and preliminary measurements of a long x-ray mirror at European X-ray Free Electron Laser. *Rev. Sci. Instrum.* **2016**, *87*, 51901. [CrossRef]
22. Bray, M. Stitching interferometry and absolute surface shape metrology: Similarities. *Int. Symp. Opt. Sci. Technol.* **2001**, *4451*, 375–383. [CrossRef]
23. Huang, L.; Idir, M.; Zuo, C.; Wang, T.; Tayabaly, K.; Lippmann, E. Two-dimensional stitching interferometry based on tilt measurement. *Opt. Express* **2018**, *26*, 23278–23286. [CrossRef] [PubMed]
24. Peng, J.; Xu, H.; Yu, Y.; Chen, M. Stitching interferometry for cylindrical optics with large angular aperture. *Meas. Sci. Technol.* **2015**, *26*, 025204. [CrossRef]
25. Huang, L.; Wang, T.; Tayabaly, K.; Kuhne, D.; Xu, W.; Xu, W.; Vescovi, M.; Idir, M. Stitching interferometry for synchrotron mirror metrology at National Synchrotron Light Source II (NSLS-II). *Opt. Lasers Eng.* **2020**, *124*, 105795. [CrossRef]
26. Yumoto, H.; Koyama, T.; Matsuyama, S.; Yamauchi, K.; Ohashi, H. Stitching interferometry for ellipsoidal x-ray mirrors. *Rev. Sci. Instrum.* **2016**, *87*, 051905. [CrossRef]
27. Chen, S.; Dai, Y.; Li, S.; Peng, X.; Wang, J. Error reductions for stitching test of large optical flats. *Opt. Laser Technol.* **2012**, *44*, 1543–1550. [CrossRef]
28. Bray, M. Stitching interferometry: The practical side of things. *Opt. Eng. Appl.* **2009**, *7426*, 74260Q. [CrossRef]
29. Bray, M. Stitching interferometry: How and why it works. *Opt. Syst. Des. Prod.* **1999**, *3739*, 259–273. [CrossRef]
30. Murphy, P.; Forbes, G.; Fleig, J.; Dumas, P.; Tricard, M. Stitching Interferometry: A Flexible Solution for Surface Metrology. *Opt. Photon-News* **2003**, *14*, 38–43. [CrossRef]
31. Chen, S.; Wu, C.; Tie, G.; Zhai, D. Stitching test of large flats by using two orthogonally arranged wavefront interferometers. *Appl. Opt.* **2017**, *56*, 9193. [CrossRef]
32. Zhao, C.; Burge, J.H. Stitching of off-axis sub-aperture null measurements of an aspheric surface. *Opt. Eng. Appl.* **2008**, *7063*, 706316. [CrossRef]
33. Peng, J.; Wang, Q.; Peng, X.; Yu, Y. Stitching interferometry of high numerical aperture cylindrical optics without using a fringe-nulling routine. *J. Opt. Soc. Am. A* **2015**, *32*, 1964–1972. [CrossRef] [PubMed]
34. Okada, K.; Otsubo, M.; Tsujiuchi, J. Measurement of large plane surface shapes by connecting small-aperture interferograms. *Opt. Eng.* **1994**, *33*, 608–613. [CrossRef]
35. Wiegmann, A.; Stavridis, M.; Walzel, M.; Siewert, F.; Zeschke, T.; Schulz, M.; Elster, C. Accuracy evaluation for sub-aperture interferometry measurements of a synchrotron mirror using virtual experiments. *Precis. Eng.* **2011**, *35*, 183–190. [CrossRef]

36. Xue, J.; Huang, L.; Gao, B.; Kaznatcheev, K.; Idir, M. One-dimensional stitching interferometry assisted by a triple-beam interferometer. *Opt. Express* **2017**, *25*, 9393–9405. [CrossRef]
37. De Groot, P.J. Correlated errors in phase-shifting laser Fizeau interferometry. *Appl. Opt.* **2014**, *53*, 4334. [CrossRef] [PubMed]
38. Crimmins, T.R. Geometric filter for speckle reduction. *Appl. Opt.* **1985**, *24*, 1438–1443. [CrossRef] [PubMed]
39. Xiong, X.; Quan, L.; Shimizu, Y.; Matsukuma, H.; Gao, W. Self-calibration of a variable-line-spacing grating for an absolute optical encoder with a Fizeau interferometer. *Meas. Sci. Technol.* **2021**, *32*, 064005. [CrossRef]

sensors MDPI

Article

A Simple Yet Effective Preanalytical Strategy Enabling the Application of Aptamer-Conjugated Gold Nanoparticles for the Colorimetric Detection of Antibiotic Residues in Raw Milk

Víctor Díaz-García *, Braulio Contreras-Trigo, Camila Rodríguez, Pablo Coelho and Patricio Oyarzún *

Facultad de Ingeniería y Tecnología, Universidad San Sebastián, Sede Concepción, Concepción 4080871, Chile; bcontrerast@docente.uss.cl (B.C.-T.); rodriguezcamila@gmail.com (C.R.); pablo.coelho@uss.cl (P.C.)
* Correspondence: victor.diazg@uss.cl (V.D.-G.); patricio.oyarzun@uss.cl (P.O.)

Abstract: The misuse of antibiotics in the cattle sector can lead to milk contamination, with concomitant effects on the dairy industry and human health. Biosensors can be applied in this field; however, the influence of the milk matrix on their activity has been poorly studied in light of the preanalytical process. Herein, aptamer-conjugated gold nanoparticles (nanoaptasensors) were investigated for the colorimetric detection in raw milk of four antibiotics used in cattle. The effect of milk components on the colorimetric response of the nanoaptasensors was analyzed by following the selective aggregation of the nanoparticles, using the absorption ratio A_{520}/A_{720}. A preanalytical strategy was developed to apply the nanoaptasensors to antibiotic-contaminated raw milk samples, which involves a clarification step with Carrez reagents followed by the removal of cations through dilution, chelation (EDTA) or precipitation ($NaHCO_3$). The colorimetric signals were detected in spiked samples at concentrations of antibiotics as low as 0.25-fold the maximum residue limits (MRLs) for kanamycin (37.5 μg/L), oxytetracycline (25 μg/L), sulfadimethoxine (6.25 μg/L) and ampicillin (1 μg/L), according to European and Chilean legislation. Overall, we conclude that this methodology holds potential for the semiquantitative analysis of antibiotic residues in raw milk obtained directly from dairy farms.

Keywords: nanoaptasensor; preanalytical processing; clarification; antibiotics; gold nanoparticles; aptamer

Citation: Díaz-García, V.; Contreras-Trigo, B.; Rodríguez, C.; Coelho, P.; Oyarzún, P. A Simple Yet Effective Preanalytical Strategy Enabling the Application of Aptamer-Conjugated Gold Nanoparticles for the Colorimetric Detection of Antibiotic Residues in Raw Milk. *Sensors* **2022**, *22*, 1281. https://doi.org/10.3390/s22031281

Academic Editors: Maria Lepore and Ines Delfino

Received: 5 December 2021
Accepted: 29 December 2021
Published: 8 February 2022

1. Introduction

Bovine milk is one of the most important nutrient-rich food sources in the human diet [1], the global production of which reached 540,925 million tons in July 2021 [2]. The quality and safety of milk and dairy products are of paramount importance for the food industry and public health [3]. There is a rising concern surrounding the presence of antibiotic residues in raw milk, which are extensively used in dairy farms for the treatment and prevention of bacterial diseases affecting cattle as well as to improve animal performance [4]. Exposure to low levels of antibiotics in milk and milk derivatives is known to cause serious harmful effects in human health [5]. In addition, the usage of antibiotics in food-producing animals is a well-recognized factor in triggering the emergence of antimicrobial resistance, which contributes to select bacterial strains and antimicrobial resistance genes that can be further transferred to the human microbiome [6]. Bacterial multidrug resistance to antibiotics is today among the biggest threats to global health, accounting for over 700,000 deaths each year due to infections caused by antibiotic-resistant bacteria (ARB) [7]. ARB are expected to cause 10 million deaths per year by 2050 at an economic cost of USD 100 trillion [8]. Finally, antibiotic-contaminated raw milk may interfere with fermentation processes of dairy products by inhibiting the growth of lactic acid bacteria (starter cultures) [9].

The Codex Alimentarius Commission develops international food standards, including reference maximum residue limits (MRLs) for veterinary drugs. In the European

Union (EU) the MRLs of antibiotic residues in milk are fixed by Commission Regulation N° 37/2010 [10], which are comparable to those set in Chilean legislation [11]. The Ministry of Health in Chile implements a national plan to control the presence of antibiotic residues in dairy products, while the Agriculture and Livestock Service (SAG, for its acronym in Spanish) is responsible for ensuring the sanitary inspection and control of veterinary pharmaceuticals in dairy cattle farms. Accordingly, raw milk contaminated with antibiotic residues at levels above the MRLs must be discarded to prevent them from entering the human food chain and their impacts on food safety and public health. Therefore, rapid screening kits (e.g., SNAPR beta-lactam tests) to monitor the presence of antibiotic residues in bovine raw milk have become a major requirement for farmers and the dairy industry. However, these tests are qualitative and unsuitable for the simultaneous detection of multiple groups of antibiotics. There is thus growing scientific and industrial interest in supporting the development of fast, sensitive and quantitative biosensing technologies to determine antibiotic residues in raw milk [12].

Nanotechnology is currently driving innovation in a pleyade of fields, including the sensing of environmental and food contaminants, industrial advanced materials and nanocatalysis applications in the chemical industry, as well as for water purification and the detection of explosives, among others [13–17]. Nanobiosensors are nanoscale sensors that include a biological recognition molecule (bioreceptor), which allow for improved analytical performances along with rapid and sensitive detection of analytes in the ppb (μg/L) concentration range [18,19]. A promising class of bioreceptors is DNA aptamers, which consist of short single-stranded oligonucleotides that provide high affinity and specificity for non-nucleotide molecules, including those with low molecular weight, toxic or nonimmunogenic [20]. The combination of AuNPs with aptamers in nanoaptasensors (NAS) is widely used to investigate the detection of a variety of analytes [21–24], with a growing number of reports addressing the detection of antibiotics in the field of food safety [25,26]. Surface plasmon resonance (SPR) is an optical property of AuNPs that allows the development of label-free NAS capable of analyzing multiple analytes in real time [27]. SPR causes a sharp and intense absorption band in the visible range, enabling colorimetric detection by following a red-to-purple–blue shift of the absorption spectrum during the aggregation of the nanoparticles. Optical NAS are a preferred sensing technique due to their non-invasive nature, high sensitivity, direct readout and easy coupling with other technologies, with recent developments in optical technology lowering the cost of the instrumentation [28,29]. However, most works address colorimetric detection exclusively from the point of view of the analytical behavior of the biosensor. Consequently, significant gaps remain in the literature regarding the preanalytical processes that are required to implement this technology in real-world applications with complex matrices.

Milk is a heterogeneous fluid composed of multi-dispersed phases of emulsion (fat–water), colloidal suspension (protein–water) and solution (salts–water), whose chemical complexity interferes with the analysis of antibiotic residues [30]. Previous reports shed light on the capacity of AuNP-based colorimetric aptasensors to detect a few of the antibiotic residues in commercial milk, including kanamycin [31], tetracycline [32], oxytetracycline [33] and streptomycin [34,35]. Preanalytical techniques described in these studies are often laborious and involve the use of chemical agents such as ethyl acetate, trichloroacetic acid or trifluoracetic acid. In addition, solvent extraction is especially challenging for nanobiosensor technology since pH and ionic environment severely influence their analytical behavior [36].

Herein, we investigated the effects of the main raw milk components on the activity of aptamer-conjugated AuNPs in the colorimetric detection of antibiotics belonging to four different groups used in cattle: kanamycin (aminoglycosides), oxytetracycline (tetracyclines), sulfadimethoxine (sulfonamides) and ampicillin (beta-lactams). A methodology was proposed to address preanalytical and analytical variables affecting the colorimetric detection of the four antibiotics in this matrix. The utility of this strategy was demonstrated according to the MRLs of veterinary drugs in food for human consumption accepted in EU and Chilean legislation.

2. Materials and Methods

2.1. Chemicals and Reagents

Kanamycin, oxytetracycline, sulfadimethoxine, ampicillin, tetrachloroauric acid solution ($HAuCl_4{\cdot}3H_2O$), Carrez clarification reagent kit, Total Protein Kit, Micro Lowry reagent kit and Sephadex G-25 resin were purchased from Merk (Darmstadt, Germany). Ethyl acetate, lactose, ethylenediaminetetraacetic acid (EDTA), sodium citrate and all salt solutions were purchased from Winkler (Santiago, Chile). Aptamers were purchased from Integrated DNA Technologies, Inc. (Coralville, IA, USA).

2.2. Synthesis of Gold Nanoparticles (AuNPs)

The synthesis of AuNPs was carried out according to the standard citrate reduction method [24]. Briefly, 100 mL of 1 mM tetrachloroauric acid solution ($HAuCl_4{\cdot}3H_2O$) was prepared with nanopure water (18 M $\Omega{\cdot}$cm). The solution was isovolumetrically heated to boiling point under stirring and refluxed with a three-neck round flask connected to the condenser. Then, 10 mL of 38.8 mM trisodium citrate solution at pH 11 was preheated to 60 °C and quickly added to the boiling solution of $HAuCl_4$ under vigorous stirring [35]. After the color of the solution turned deep red the mixture was refluxed for an additional 30 min and cooled down to room temperature without stirring for 2 h. The resulting nanoparticle suspension was filtered with Millipore nylon filters (0.45 µm) and preserved in the dark at 4°C. The concentrations of AuNPs and NAS were calculated according to the Beer–Lambert law by measuring the absorbance at 520 nm (extinction coefficient of 2.01×10^8 M^{-1} cm^{-1}) [37,38].

2.3. AuNPs Characterization

The spectroscopic characterization of the synthesized AuNPs was carried out with an Epoch™ microplate spectrophotometer (Biotek Instruments, Winooski, VT, USA). The size distribution and surface charge characterization (pZ) of the AuNPs were determined with a Zetasizer Nano-ZS90 dynamic light scattering (DLS) analyzer (Malvern Instruments, Westboroug, MA, USA). The characterization of the size and morphology of the nanoparticles was carried out by transmission electron microscopy (TEM) with a 4 Å resolution (TEM; JEOL-JEM 1200EX-II, Tokyo, Japan), using a Gatan CCD camera for image acquisition (model 782; Gatan, Inc., Pleasanton, CA, USA).

2.4. Synthesis of NAS and the Determination of Detection Parameters

AuNPs were functionalized with aptamers specific for the antibiotics listed in Table 1. Thiol-modified aptamers (C3-S-S-Aptamer) were reduced by incubation with dithiothreitol 0.1 M in a phosphate buffer of pH 8 for 3 h at 37 °C. Then, 3′-SH-aptamers were purified by gel filtration with Sephadex G-25 and incubated with the AuNPs in a phosphate buffer (10 mM, pH 7.4) for 2 days at room temperature and darkness [39]. AuNP–aptamer molar ratios of 1:20, 1:40 and 1:60 were investigated to determine the best parameters and conditions for the maximization of the colorimetric signal produced by the NAS in the detection of antibiotics at concentrations equal to their MRLs. The resulting NAS were activated by being heated at 80 °C for 10 min and then cooled down at room temperature for 10 min to induce linear conformation on the nanoparticle surface. This step is paramount to stabilize the nanoparticles and to allow the aptamers to interact with the antibiotics.

Table 1. Antibiotics used in the experiments, including their maximum acceptable levels in raw milk (MRLs) according to EU and Chilean legislation, and aptamers used in the NAS to recognize each antibiotic.

Antibiotic	Maximum Residual Limit (MRL)	Aptamer Sequence (ssDNA)	Aptamer Ref.
Kanamycin	150 μg/L *	5′-TGGGGGTTGAGGCTAAGCC GA-3′ (21b)	[40]
Oxytetracycline	100 μg/L *	5′-CGTACGGAATTCGCTAGCG GGCGGGGGTGCTGGGGGAAT GGAGTGCTGCGTGCTGCGGGG ATCCGAGCTCCACGTG-3′ (76b)	[41]
Sulfadimethoxine	25 μg/L † 100 μg/L *	5°′-GAGGGCAACGAGTGTTTA TAGA-3′ (22b)	[42]
Ampicillin	4 μg/L *,†	5′-GCGGGCGGTTGTATAGC GG-3′ (19b)	[43]

* EU legislation [10]. † Chilean legislation [11].

A typical assay in microplate wells consisted of 200 μL of antibiotic solution incubated with 100 μL of the activated NAS (4 nM) at 60 °C for 10 min, subsequently cooled down at room temperature. Then, 60 μL of NaCl 1 M was added into the solutions and incubated for 30 min to monitor the aggregation process. AuNP aggregation data were analyzed spectrophotometrically by measuring the shift of the plasmon resonance peak from 520 nm to 620 nm (A_{520}/A_{620}). The absorbance values were calibrated by subtracting the value in nanopure water (blank) and expressed as the difference in colorimetric signal between the control (without antibiotic) and the antibiotic solution, using concentration ranges around their MRLs (Equation (1)):

$$\frac{A520}{A620}\text{Control} - \text{Treatment} = \text{Signal}\frac{A520}{A620}\text{NAS without antibiotics} - \text{Signal}\frac{A520}{A620}\text{NAS with antibiotics(MRL)} \quad (1)$$

The aggregation of the NAS followed, during 60 min in a mixture of 100 μL of the activated NAS and 200 μL of nanopure water (without antibiotics). The absorbance was measured at 520 nm and 720 nm during 60 min (the absorption ratio A_{520}/A_{720}) upon the addition of the saline solution. Further experiments monitoring the SPR shift to 720 nm were also expressed according to Equation (1).

2.5. Preanalytical Processing of Raw Milk

Bulk tank raw milk samples (50 mL) were spiked with the antibiotics in final concentrations of 0.5×, 1×, 2× and 4× the corresponding MRLs and then homogenized for 30 min. The samples were clarified by using both Carrez reagents and ethyl acetate as follows:

- Carrez clarification: Five hundred milliliters of Carrez I reagent was added into 10 mL of raw milk without antibiotics (control) or with kanamycin, oxytetracycline, sulfadimethoxine or ampicillin, and vortexed for 1 min. Then, 500 mL of Carrez II reagent was added and vortexed for 1 min until the mixture was homogeneous, which was subsequently centrifuged at 1000× *g* for 5 min. The supernatant (milk whey) was recovered and immediately used for the detection of antibiotics with the NAS.
- Ethyl acetate clarification: Four milliliters of raw milk was centrifuged for 20 min at 1000 and 10 °C to separate the fat. Then, 2 mL of the supernatant was diluted with 2 mL of nanopure water and stirred for 10 min in a vortex. Seven milliliters of ethyl acetate was added, vortexed for 15 min and centrifuged for 15 min at 1500× *g* and 4 °C, which gave rise to a three-phase mixture. The bottom layer (milk whey) was recovered and centrifugated again before being stored at 4 °C.

2.6. Milk and Milk Whey Characterization

The proximate analysis of raw milk and whey (by Carrez method) was determined in accordance with the standard methods of the Association of Official Analytical Chemists (AOAC methods) [44]: fat (AOAC 945.16), ash (AOAC 920.181), crude fiber (AOAC 962.09), total protein (total nitrogen × 6.25) (AOAC 978.02), total carbohydrates (AOAC 929.09) and lactose (AOAC 982.14). The protein concentration was determined by the Lowry protein assay (Peterson's modification) with protein precipitation, using the Total Protein Kit, Micro Lowry, in accordance with the manufacturer's protocol. Optical density was measured at 650 nm in 96-well plates using and Epoch^{TM} microplate spectrophotometer. The protein removal efficiency was determined according to Equation (2):

$$\text{Protein removal efficiency} = \frac{C_{RM} - C_{MW}}{C_{RM}} \times 100 \quad (2)$$

where C_{MW} is the protein concentration of the milk whey and C_{RM} is the concentration of the raw milk.

2.7. Colorimetric Detection of Antibiotics in Clarificatecd Raw Milk

Ten milliliters of raw milk was spiked (contaminated) with kanamycin, oxytetracycline, sulfadimethoxine and ampicillin at final concentrations of 0.25×, 0.5×, 1×, 2× and 4× the MRL of each antibiotic and then incubated for 30 min at room temperature. The colorimetric detection of the antibiotics was thus investigated in solutions containing the main soluble components of bovine raw milk (lactose and ionic species) to determine their specific effect on the aggregation process of AuNPs (Table 2). Raw milk (with and without antibiotics) was clarified using the Carrez method, after which samples of the resulting whey were assessed with the NAS by following the aggregation according to the methodology described in Section 2.4.

Table 2. Solutions employed to study the effect of soluble milk constituents on the aggregation of the NAS.

Element	Concentration	Synthetic Milk Whey	Lactose Solution	Ionic Solution
Lactose	5% $^{w}/_{v}$	X	X	
Ca^{2+}	30.120 mM (60.232 mEq/L)	X		X
Mg^{2+}	4.750 mM (9.506 mEq/L)	X		X
Fe^{3+}	0.011 mM (0.032 mEq/L)	X		X
PO_4^{3-}	30.582 mM (91.745 mEq/L)	X		X
Na^{+}	23.587 mM (23.587 mEq/L)	X		X
K^{+}	38,930 mM (38.930 mEq/L)	X		X
Zn^{2+}	0.066 mM (0.131 mEq/L)	X		X
$Citrate^{3-}$	9.292 mM (18.584 mEq/L)	X		X
Cu^{2+}	0.002 mM (0.005 mEq/L)	X		X

The symbol (X) indicates the presence of each particular milk component in the solutions.

2.8. Colorimetric Detection of Antibiotics in Cation-Removed Milk Whey

Raw milk samples were incubated with the antibiotics at concentrations of 0.25×, 0.5×, 1×, 2× and 4× their MRLs and subsequently subjected to Carrez clarification (see Section 2.5). Prior to performing the experiments of antibiotics detection the whey samples (1.5 mL) were divided into three isovolumetric fractions to remove cations by different methods: (i) by adding 80 μL of $NaHCO_3$ 1 M into the solutions to a final concentration of 30 mM, which were subsequently incubated (10 min at 60 °C), cooled down at room temperature and centrifugated at 20,000× *g* for 10 min to recover the supernatant; (ii) by adding 80 μL of EDTA to a final concentration of 5 mM; and (iii) by diluting 1:1 with nanopure water to attenuate the effects of cations and lactose in the salt-induced aggregation of AuNPs. The absorbance values were normalized with respect to the absorbance ratio measured for the whey without antibiotics (control), according to the following equation:

$$\text{Fold A520/A720} = \frac{\text{signal}\frac{A520}{A720}\text{of NAS with antibiotic}}{\text{signal}\frac{A520}{A720}\text{of NAS without antibiotic (control)}} \quad (3)$$

2.9. Statistical Analysis

Data shown are the average ± standard error of at least three independent experiments. Statistical significance was determined at a 95% confidence interval, using a nonparametric Mann–Whitney U test for the comparison of two groups.

3. Results and Discussions

3.1. NAS and Detection Principle

The NAS developed herein are based on previously reported aptamers specific to four antibiotics used in cattle (kanamycin, oxytetracycline, sulfadimethoxine and ampicillin). Numerous works in the nanobiosensor field are based on aptamers bound to the nanoparticles through electrostatic interactions, which is a suitable approach for the detection of antibiotics in a saline buffer solution or highly clarified matrices. The recognition of antibiotics by adsorbed aptamers causes them to detach from the AuNPs with the concomitant loss of stability and subsequent aggregation [41]. However, this strategy becomes less applicable in complex matrices containing chemical species that can interfere with the analytical method. Accordingly, the experimental approach followed in this work involved the covalent conjugation of the ssDNA aptamers on the AuNPs' surface (through thiol–gold interactions) to prevent their release from the nanoparticle surface. It is worth mentioning that this reaction has been well-characterized in numerous works, using different instrumental techniques such as X-ray photoelectron spectroscopy (XPS), X-ray diffraction (XRD) and atomic force microscopy (AFM). The resulting interaction between the thiol group and the gold surface is strong and stable, providing a robust mechanism to link aptamers onto AuNPs [45–48]. Importantly, the aptamers adopt a flexible random coil linear structure that allows their bases to interact with the nanoparticle surface through van der Waals forces [49]. Figure 1 provides a schematic description of the detection reaction, showing the aptamers coating the AuNPs' surface and inhibiting salt-induced aggregation due to electrostatic repulsions among the nanoparticles. However, in the presence of the antibiotics the aptamers adopt a folded structure that leads to a decrease in surface protection and the subsequent aggregation of the AuNPs upon the addition of NaCl. The aggregation process is proportional to the antibiotic concentration and can be followed through the decrease in the absorption ratio (A_{520}/A_{620}).

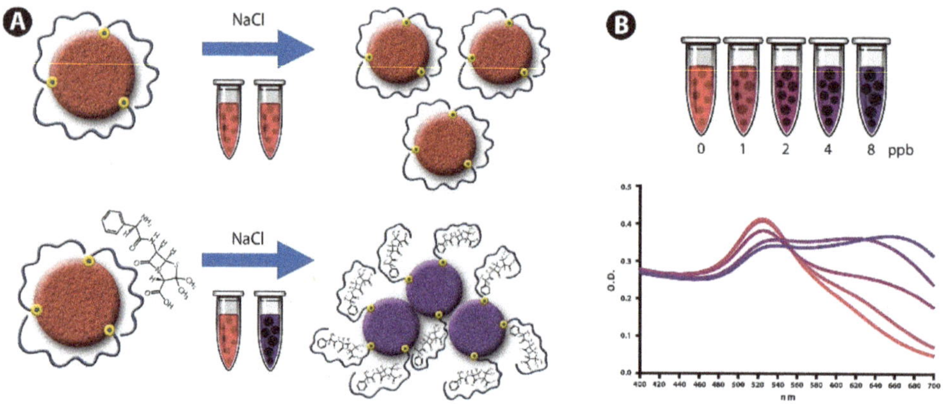

Figure 1. Schematic illustration showing the sensing principle of aptamer-conjugated AuNPs for the colorimetric detection of antibiotic residues present in raw milk. (**A**) Covalently conjugated aptamers in a random coiled linear structure inhibit salt-induced aggregation, while the conformational change induced by antibiotic interaction decreases surface protection and allows the aggregation of AuNPs upon the addition of NaCl. (**B**) Absorption spectra of AuNPs showing antibiotic-induced aggregation of the nanoparticles and the resulting shift in the plasmon resonance peak (from red to blue–purple).

3.2. Determination of the Detection Parameters of NAS

The detection parameters of (i) AuNP:aptamer molar ratio; (ii) absorption ratio; and (iii) incubation time were evaluated in water to select conditions that maximized the colorimetric signal generated by the NAS. AuNP:aptamer molar ratios of 1:20, 1:40 and 1:60 were compared for each NAS in the presence of each antibiotic at concentrations equal to the corresponding MRL in the EU (Table 1). In the case of sulfadimethoxine the MRL considered was 25 μg/L (instead of 100 μg/L), which is the highest concentration of sulfonamides legally permitted in Chile for milk [11]. Differences in susceptibility to salt-induced aggregation at different molar ratios (AuNP functionalization) could be a consequence of the variable degrees of surface coverage in each NAS. Figure 2 shows the variation in the colorimetric signal produced by the NAS in detection of the four antibiotics upon the addition of NaCl. Dispersed and aggregated NAS resulting from the detection process are presented in Figure 2, with TEM images (inset) showing typical spherical particles with diameters of ~15 nm [24,27]. The highest colorimetric response was determined in the case of kanamycin at a molar ratio of 1:60, while for oxytetracycline, sulfadimethoxine and ampicillin the best molar ratio was 1:20. These parameters were selected for subsequent antibiotic detection assays.

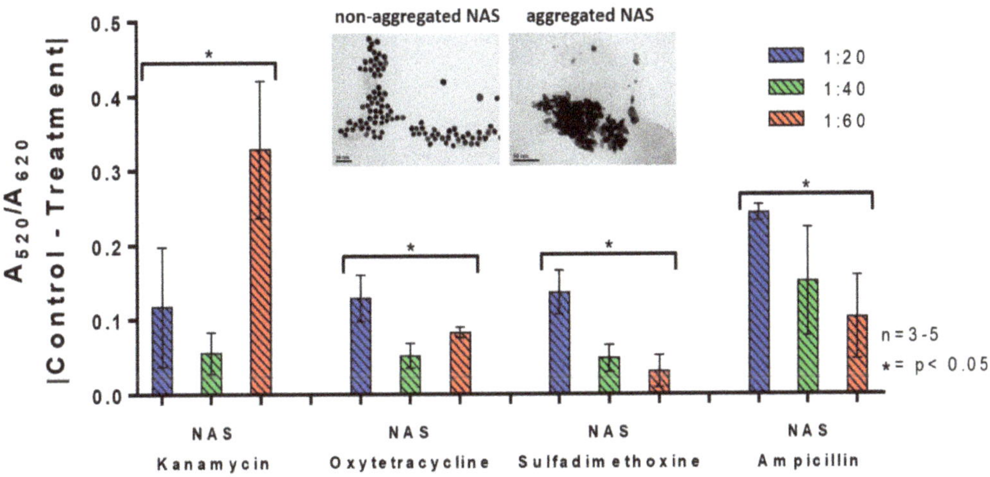

Figure 2. Colorimetric response of the NAS at molar ratios (AuNP:aptamer) of 1:20 (blue), 1:40 (green) and 1:60 (red) for the detection of antibiotics in water at their maximum concentration of residues permitted in milk (MRLs) for kanamycin, oxytetracycline, sulfadimethoxine and ampicillin. Treatments correspond to antibiotic-containing samples at the MRL concentrations and controls include nanopure water instead of antibiotics. The inset shows TEM images of dispersed NAS (before the detection of antibiotics) and aggregated NAS (after the detections of antibiotics). The absorbance readings were transformed according to Equation (1). Results were averaged from 3 to 5 independent experiments. Data were analyzed using a nonparametric Mann–Whitney U test. Asterisks denote statistically significant differences between the treatments and controls. * = $p < 0.05$.

We hypothesized that differences in the colorimetric response of NAS at different molar ratios are associated with the effect of this parameter on the degree of surface coverage of the AuNPs by the aptamers (i.e., reactive surface). Thus, nanoparticles partially coated on their surface would have fewer reactive sites available to interact with the antibiotics in comparison with mostly coated nanoparticles. By calculating the surface coverages for each mol of NAS and considering the aptamer length for NAS-Kan (21b), NAS-Oxy (76b), NAS-Sul (22b) and NAS-Amp (19b), the total number of moles of DNA bases that would be interacting with each NAS is 1260b (NAS-Kan), 1520b (NAS-Oxy), 440b (NAS-Sul) and

380b (NAS-Amp) (Table S1). Interestingly, both NAS-Amp and NAS-Sul share a low total number of bases available on their surfaces and a lower concentration of antibiotics in the assays (MRLs of 4 µg/L and 25 µg/L, respectively) in comparison with NAS-Oxy and NAS-Kan. By contrast, the latter groups of NAS have a higher number of bases coating the nanoparticle surface and a higher concentration of antibiotics (100 µg/L and 150 µg/L, respectively) in common. This analysis lends support to the relationship between reactive surface and the analyte concentration, given that NAS with a low proportion of aptamer coverage are more likely to become saturated at low concentrations and vice versa. Kim et al., for example, explored different molar ratios for the colorimetric detection of oxytetracycline in water, finding an optimal molar ratio of 1:50. However, this work was conducted with aptamers bound to AuNPs via electrostatic interactions, and the authors discussed the need to determine in each NAS the best molar ratio for optimal detection [35].

The shift of the SPR peak (from 520 nm) associated with the aggregation of AuNPs spans a wide spectral region (until 800 nm), despite most of the works in the field of NAS following the shift between 520 nm and 620 or 650 nm [50]. Accordingly, our group previously developed an electro-opto-mechanic device for high-resolution AuNP spectral data in a wavelength range from 400 to 800 nm, which was proven to improve the analytical performance of NAS with the aid of machine learning tools [51]. With this in mind, AuNP aggregation data were analyzed spectrophotometrically between 400 and 750 nm, using two absorption ratios (A_{520}/A_{620} and A_{520}/A_{720}) to select reading parameters that maximize the colorimetric signal associated with NAS detection and aggregation. As shown in Figure 3, the ratio A_{520}/A_{720} outperformed A_{520}/A_{620} in generating a color intensity three–four times greater in terms of the shift of the SPR peak. We proposed that the reason for this result is the greater difference between the absorbance values at each wavelength in the spectra before and after NaCl-induced AuNP aggregation. This phenomenon can be clearly seen in the Figure S1 of Supplementary Materials, where the differences in absorbance at 720 nm between the red (before aggregation) and blue (after aggregation) curves are greater than the difference at 620 nm. The relevant parameter is thus the difference between absorbances (not the absolute value); therefore, a greater difference at 720 nm is expected to decrease the detection limit by providing a stronger colorimetric signal. To the best of our knowledge this is the first study proposing the absorption ratio A_{520}/A_{720} nm as a way to improve the sensitivity of the NAS.

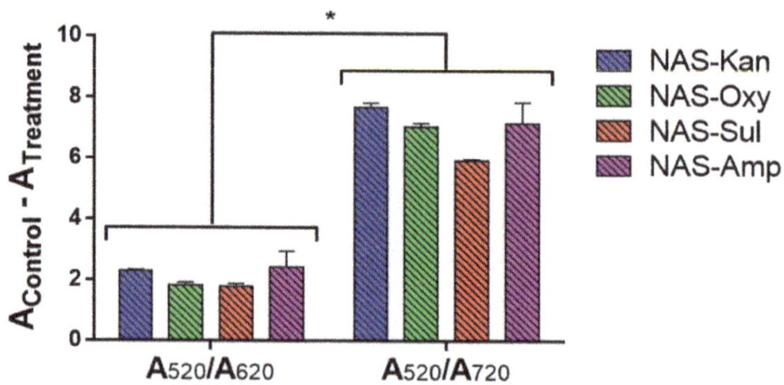

Figure 3. Comparison of the absorption ratios A_{520}/A_{620} and A_{520}/A_{720} from spectral variation associated with the aggregation of AuNPs as a colorimetric signal generated by the NAS in the detection of kanamycin (blue), oxytetracycline (green), sulfadimethoxine (red) and ampicillin (purple). Treatments correspond to antibiotic-containing samples at the MRL concentrations of each antibiotic, while controls include nanopure water instead of antibiotics. The absorbance readings were transformed according to Equation (1). Results were averaged from 3 to 5 independent experiments. Data were analyzed using a nonparametric Mann–Whitney U test. Asterisks denote statistically significant differences between the treatments and controls. * = $p < 0.05$.

Figure 4 presents the spectral variation across time resulting from the salt-induced aggregation of AuNPs for the four NAS. The curves exhibit an exponential decay, with a sharp decrease in the absorption ratio over the first 5 min followed by a slower exponential decline between 30 and 60 min. Based on these results the incubation time with NaCl was established as 30 min prior to proceeding with spectroscopic characterization.

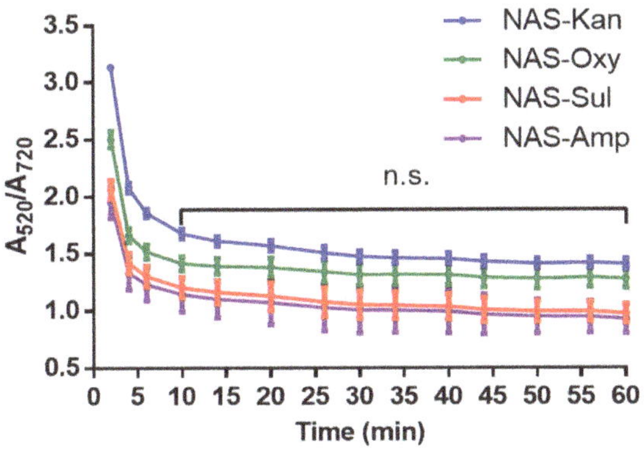

Figure 4. Kinetics of the aggregation of AuNPs in response to NAS detecting kanamycin (red), oxytetracycline (green), sulfadimethoxine (blue) and ampicillin (purple) upon the addition of NaCl. The results were averaged from 3 independent experiments for each NAS ($n = 3$). Non-significant differences are shown as n.s. Data were analyzed using a nonparametric Mann–Whitney U test.

The previously determined parameters were employed in antibiotic detection experiments. As shown in Figure 5, the NAS allowed for the discrimination of trace-level concentrations of the antibiotics kanamycin (0, 37.5, 75 and 150 µg/L), oxytetracycline (0, 50, 100 and 200 µg/L), sulfadimethoxine (above 12.5 µg/L) and ampicillin (above 1 µg/L).

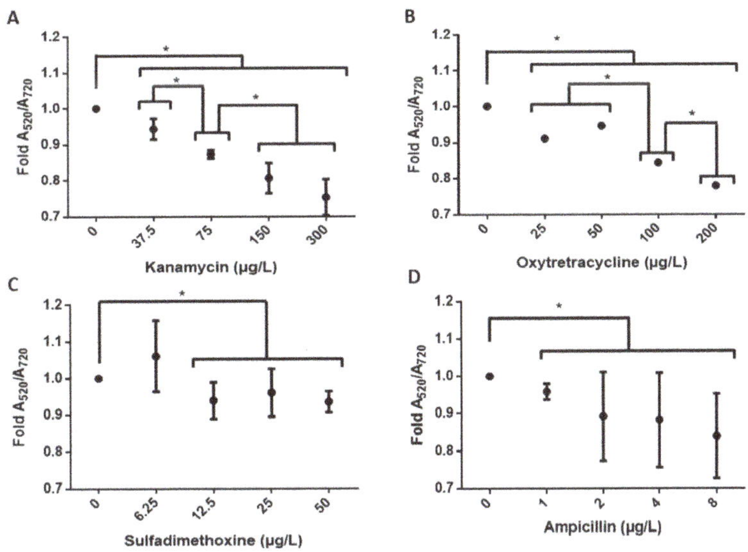

Figure 5. Colorimetric detection of the antibiotics in water for (**A**) kanamycin, (**B**) oxytetracycline, (**C**) sulfadimethoxine and (**D**) ampicillin, based on the A_{520}/A_{720} absorption ratio. Results were averaged from three independent experiments ($n = 3$). Data were analyzed using a nonparametric Mann–Whitney U test. Asterisks denote statistically significant differences between the treatments and controls. * = $p < 0.05$.

These results suggest the need to include spectral information above 700 nm to better analyze the aggregation of AuNPs, in contrast with the most common approach of following the shift of the SPR band between 520 nm (red) and 620 or 650 nm (purple–blue) [37,41,42].

3.3. *Raw Milk Clarification*

The aggregation behavior of non-functionalized AuNPs in raw milk was initially determined in order to explore the activity of the NAS in this matrix. Figure S2 accounts for the absorption spectra of dispersed (~15 nm) and aggregated nanoparticles in the raw milk, which showed no spectral change in the plasmon resonance peak (at 520 nm) when incubating AuNP-containing raw milk with NaCl 0.1 M. The aggregation of AuNPs was followed through the A_{520}/A_{620} ratio by increasing the NaCl concentration from 0.1 to 0.5 M. No significant differences between the control (without NaCl) and treated samples (with NaCl) were observed. The lack of aggregation of AuNPs in the milk suggests that the nanoparticles are strongly stabilized by electrostatic interactions with lactose or ion species, which interfere with the activity of NAS.

Two preanalytical treatments were employed for protein and fat removal from raw milk. The procedure based on the Carrez reagent produces two layers consisting of the supernatant (whey) and a precipitate of proteins and fat (Figure 6A). Instead, ethyl acetate forms three distinct layers corresponding to ethyl acetate (upper layer), a white layer of insoluble constituents (middle layer) and milk whey (bottom layer) (Figure 6B). The Carrez method outperformed in effectivity and simplicity the treatment with ethyl acetate to clarify raw milk, allowing around 98% of the protein content to be removed with the full elimination of fats (Table 3).

The Carrez method is often employed in the preanalytical processing of complex matrices. However, the absorption spectra of the four NAS in the resulting milk whey still showed no variations, accounting for a strong stabilization of the nanoparticles caused by ionic species in this solution (Figure S3). The NAS are prevented from generating a colorimetric response under these conditions, requiring further clarification steps.

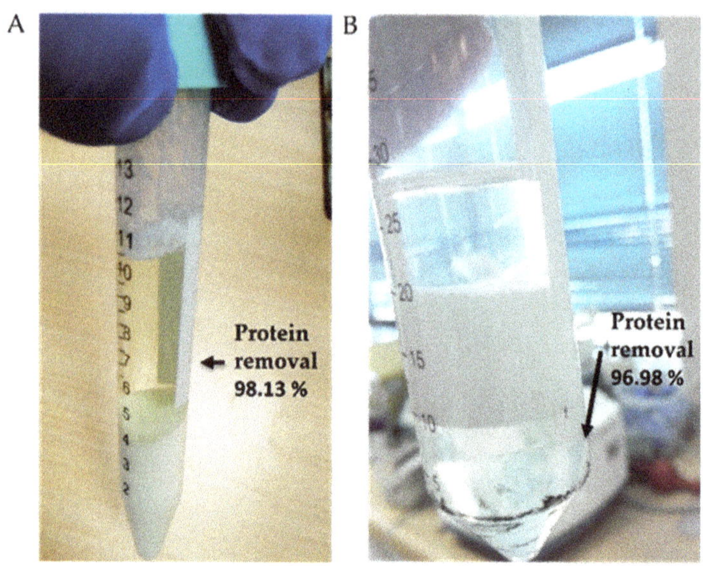

Figure 6. Phases and protein removal resulting from raw milk clarification using (**A**) the Carrez reagent and (**B**) ethyl acetate.

Table 3. Proximate analysis of raw milk and whey obtained with the Carrez method.

Components	Raw Milk	Milk Whey
Lipids	5.3%	0.0%
Proteins	3.2%	0.2%
Raw fiber	0.4%	0.6%
Total carbohydrates	4.7%	4.5%
Lactose	5.0%	5.3%
Ashes	0.7%	1.2%
pH	5.94	6.05

3.4. *Effect of Ion and Lactose on the Aggregation of AuNPs*

In the milk serum sodium and potassium ions (Na^+ and K^+) form weak ion pairs with chloride, citrate and phosphate, mainly remaining as free ions while divalent cations (Ca^{2+} and Mg^{2+}) are mostly complexed with citrate [52]. These cations are thus capable of interacting with negative charges of the citrate layer adsorbed onto AuNPs (citrate-capped AuNPs) and with phosphate groups from the DNA aptamers, thereby competing with electrostatic interactions that stabilize the nanoparticles and inducing their aggregation. This phenomenon occurs due to the displacement of the capping citrate layer and the consequent loss of the repulsive charges that counteract the van der Waal attractive forces between gold particles. Lactose, on the other hand, is the mayor carbohydrate component of milk and its interaction with AuNPs is also expected to influence aggregation. Given these complex interactions, synthetic milk whey (SMW) mimicking the ions and lactose composition of bovine raw milk was prepared with the aim of understanding the specific effects of these components on the activity of NAS.

Figure 7 shows that the SMW (red line) and the ionic solution (blue line) induce the aggregation of AuNPs even before the addition of NaCl, according to the spectral shift of the plasmon resonance peak. The lactose solution (green line) prevented the salt-induced aggregation of AuNPs from taking place upon incubation both without NaCl (Figure 7A) and with NaCl (Figure 7B). This result suggests that lactose–AuNP interactions contribute to the stabilization of nanoparticles and would therefore inhibit salt-induced aggregation, a finding that is in agreement with previous works that show that this molecule is capable of interacting with metals cations such as iron [53] and calcium [54]. A recent study also reports the interaction of gold with the acetalized OH group of lactose [55]. Figure 7A,C point to the strong effect of milk ions on the aggregation of AuNPs as long as the aggregation of AuNPs caused by the ionic solution is similar to that induced by NaCl incubation, even decreasing the concentration of the ions four-fold. The activity of NAS becomes suppressed due to this effect (Figure S3).

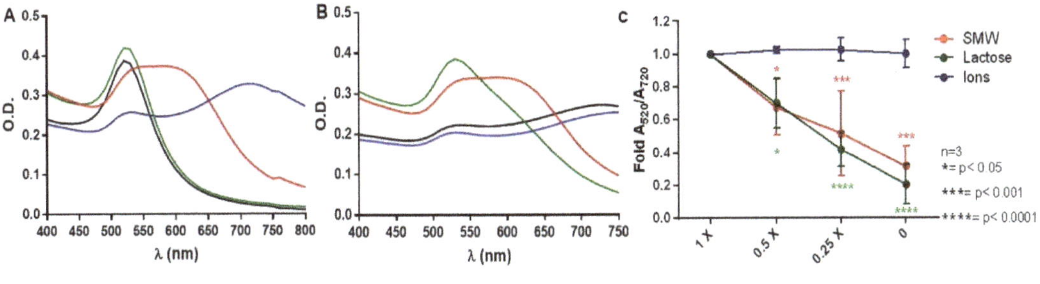

Figure 7. Spectroscopic characterization of the aggregation of AuNPs in synthetic whey (red line), lactose (green line), solution of ion species (blue line) and water (black line) during the incubation of AuNPs at (**A**) 10 min without the addition of NaCl addition and (**B**) 30 min after the addition of NaCl addition. (**C**) Effect of the solution dilution on the aggregation of AuNPs. The curves are based on the average of the results from three independent experiments ($n = 3$). Statistically significant differences compared with the controls and different treatments are indicated as * $p < 0.05$, *** $p < 0.001$ and **** $p < 0.0001$.

The concentration of ions in the milk whey includes 132.423 mEq/L of cations (mono, di and trivalent) and 110.329 mEq/L of anions, where citrate and PO_4^{3-} bind and chelate metallic cations in the solution (Table 2). Accordingly, ≈22.1 mEq/L of free cations are in theory involved in the aggregation of NAS. On the other hand, the aggregation of NAS induced by NaCl (monovalent cation) occurs at concentrations of 100 mEq/L of cations, while the aggregation of AuNPs starts at above 50 mEq/L (Figure S4). This analysis suggests that cations with a higher valence generate a stronger effect on the aggregation of AuNPs under equal conditions of normal concentration (Eq/L). Therefore, both cations and lactose need to be removed from the milk whey to make the aggregation of AuNPs possible and allow the NAS to be applied in this matrix.

3.5. Cation Removal and Antibiotic Detection

Figure 8 shows that both the dilution and treatment of milk whey with chelating (EDTA) and precipitating ($NaHCO_3$) agents allow the NAS to colorimetrically detect the corresponding antibiotics in concentrations as low as 0.25× their corresponding MRL. EDTA acts by sequestering divalent cations present in whey (Ca^{2+} and Mg^{2+}), while $NaHCO_3$ precipitates Ca^{2+} (as $CaCO_3$) and Mg^{2+} (as $Mg(OH)_2$). The colorimetrical signal is inversely correlated with the concentration of the different antibiotics (0–600 µg/L), where a decrease in the signal accounts for a larger presence of antibiotics in raw milk. Therefore, preanalytical steps must be considered as a part of the protocol to eliminate interference effects of cationic species, though the specific treatment with EDTA or $NaHCO_3$ would depend on the antibiotic.

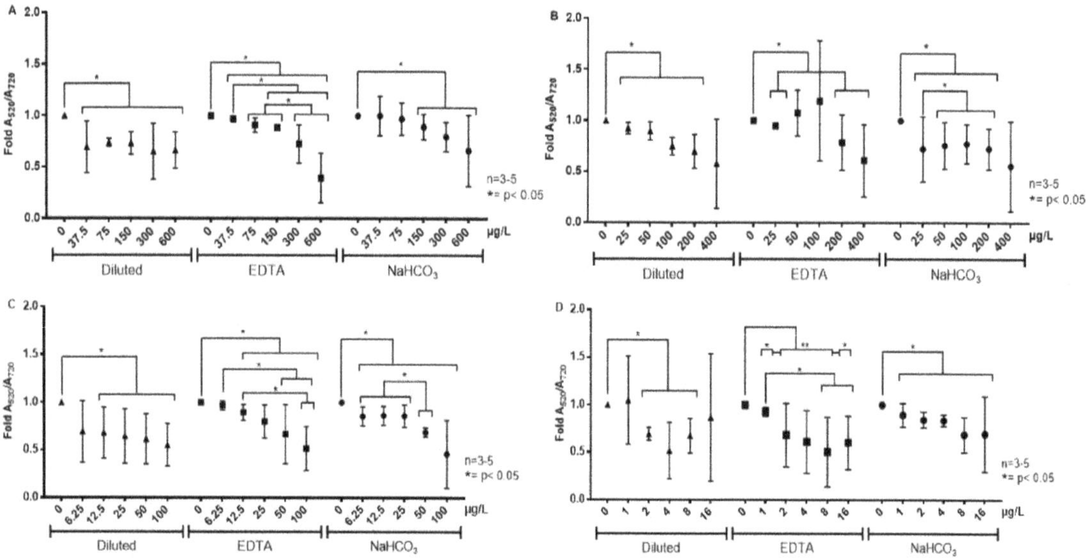

Figure 8. Colorimetric detection of antibiotics in cation-removed milk whey with NAS specific for (**A**) kanamycin, (**B**) oxytetracycline, (**C**) sulfadimethoxine and (**D**) ampicillin. The interferents were removed by dilution (▲), EDTA treatment (■) and $NaHCO_3$ treatment (•). The results were averaged from three independent experiments (n = 3). Statistically significant differences compared with the controls and different treatments are indicated. * $p < 0.05$ and ** $p < 0.01$.

As summarized in Table 4, the methods employed to remove cations from the milk whey (dilution, chelation or precipitation) enabled the NAS to perform the sensitive detection of kanamycin (from 37.5 µg/L), oxytetracycline (from 25 µg/L), sulfadimethoxine (from 6.25 µg/L) and ampicillin (from 1 µg/L) in the clarified matrix. However, some of the treatments are associated with a higher degree of data dispersion (colorimetric signal) and these differences were further analyzed to select the best preanalytical methodology. In the case of kanamycin, the EDTA treatment allowed a highly sensitive detection to be achieved and qualitative discrimination between concentrations lower and higher than its MRL (150 µg/L). By contrast, the dilution and $NaHCO_3$ treatments only allowed for the detection of the presence of antibiotics at the MRL concentration. For oxytetracycline both the dilution with water and $NaHCO_3$ treatments enabled sensitive detection (from 25 to 200 µg/L), even though dilution was associated with less data variation and thus with a better detection performance. The $NaHCO_3$ treatment was also suitable to perform the sensitive detection of sulfadimethoxine (from 6.25 µg/L) and to discriminate concentrations under the MRL and up to two-fold the MRL (50 µg/L). Instead, the dilution and EDTA treatments allowed for detection (from 12.5 µg/L) but without discrimination capability. Finally, the EDTA and $NaHCO_3$ treatments were associated with the sensitive detection of ampicillin (from 1 µg/L). However, EDTA outperformed $NaHCO_3$ in discriminating among different concentrations, despite it producing a rather high data dispersion. Therefore, the methodologies developed herein provide suitable conditions to discriminate at least 0.25 times the MRLs of each antibiotic, making it possible to implement the semiquantitative detection of them in samples of contaminated raw milk.

Table 4. Selected methods for the removal of cationic interferents, which provide the minimum detection limit of antibiotic for each NAS.

Nanoaptasensor	Best Method for Interference Elimination	Detection Limit
Kanamycin	EDTA	37.50 µg/L (0.25 MRL)
Oxytetracycline	Dilution	25.00 µg/L (0.25 MRL)
Sulfadimethoxine	$NaHCO_3$	6.25 µg/L (0.25 MRL)
Ampicillin	EDTA	1.00 µg/L (0.25 MRL)

In summary, the experimental strategy developed in this work involves three simple steps: (i) removal of fats and proteins with the Carrez reagents; (ii) elimination of ionic interferents by either dilution, chelation (EDTA) or precipitation ($NaHCO_3$), depending on the target antibiotic; and (iii) incubation with the NAS to generate the colorimetric response (Figure 9).

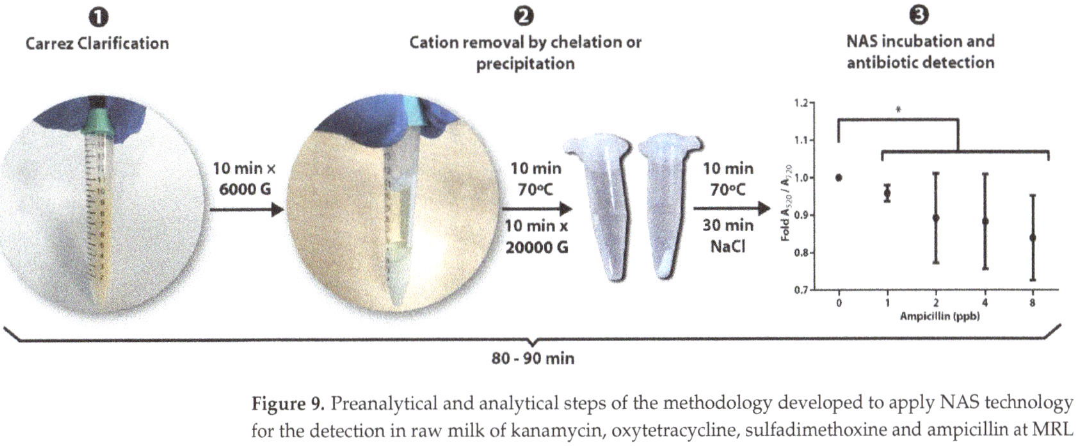

Figure 9. Preanalytical and analytical steps of the methodology developed to apply NAS technology for the detection in raw milk of kanamycin, oxytetracycline, sulfadimethoxine and ampicillin at MRL concentration levels.

Table 5 shows that the methodology developed reached low detection limits (nM range) and assay times (70 min) that are comparable with similar methods, but avoided the use of organic solvents and the requirement of complex pretreatment procedures. The proposed approach allowed for the gain of knowledge on the influence of the milk matrix on NAS activity, focusing on the preanalytical strategy required to implement the detection of antibiotic residues in raw milk directly obtained and assayed in a dairy farm. In addition, this is the first study addressing the effect of raw milk constituents on the colorimetric response of NAS in the process of detecting antibiotics belonging to four different groups (aminoglycoside, tetracycline, sulfonamide and beta-lactam).

Table 5. Colorimetric AuNP-based aptasensors for the determination of antibiotics in milk.

Milk Sample	Antibiotic Residue	Preanalytical Time	Assay Time	LOD	Range	Pretreatment/Antibiotic Addition	Ref.
Supermarket milk	Ampicillin (AMP)	1.5 h	80 min	10 ng/mL (28.6 nM)	1–100 ng/mL (2.86–286 nM)	Ethyl acetate addition/centrifugation at 5000 rpm to obtain supernatant/nitrogen blow-down at 40 °C/pellet resuspension in water (total dilution: 2-fold)/**AMP addition (after pretreatment)**.	[43]
N.S.	Tetracycline (TET)	N.S.	25 min	45.8 nM	10–400 nM	Milk dilution with water (1:5)/acetic acid addition/centrifugation (total dilution: 5-fold)/**TET addition** (protocol details unspecified).	[56]
Milk powder	Tetracycline (TET)	N.S.	60 min	122 nM	10–500 nM	Acetic acid addition/centrifugation/**TET addition** (protocol details unspecified).	[32]
N.S.	Kanamycin (KAN)	35 min	100 min	1 nM	1–8 nM 100–500 nM	**KAN addition (before pretreatment)**/acetic acid addition/incubation at 45 °C/centrifugation at 10,000 rpm/filtration 0.22 μm/pH adjustment (total dilution unspecified).	[31]
N.S.	Streptomycin (STR)	N.A.	70 min	73.1 nM	30–1030 nM	Milk dilution with water (total dilution: 50-fold)/**STR addition (after pretreatment)**.	[34]
Supermarket milk	Streptomycin (STR)	30 min	60 min	86 nM	100–500 nM	Milk dilution with water (total dilution: 5-fold)/**STR addition (after dilution)**/EDTA and trifluoracetic acid addition/centrifugation at 6000 rpm/supernatant collection.	[35]
Raw milk from a dairy farm	Ampicillin Oxytetracycline Sulfadimethoxine Kanamycin	40 min	30 min	1 μg/L (2.9 nM) 25 μg/L (54.3 nM) 6.25 μg/L (20.9 nM) 37.5 μg/L (77.4 nM)	1–16 μg/L (2.9–46.4 nM) 25–200 μg/L (54.3–434.4 nM) 6.25–100 μg/L (20.9–334.4 nM) 37.5–600 μg/L (77.4–1238.4 nM)	**Antibiotic addition (before pretreatment)**/Carrez reagents/centrifugation at 1000 rpm/dilution, $NaHCO_3$ or EDTA treatments/centrifugation at 1000 rpm/supernatant collection (total dilution: 1.16-fold).	This study

N.S. = Details not specified by the authors. N.A. = Not applicable.

In contrast to our approach, previous works report the determination of antibiotics on processed (commercial) milk, which is typically subjected to treatments involving the use of chemical agents such as ethyl acetate [43], trichloroacetic acid [31,33] or trifluoracetic acid [35], followed by centrifugation/resuspension. In some cases, final steps involve the adjustment of pH or the removal of solvents by nitrogen blow-down [57]. Milk dilution has also been reported elsewhere as a simple clarification procedure to deal with matrix complexity by using dilution factors ranging from 5-fold [35] to 50-fold [34]. However, in these works milk samples are typically spiked with antibiotics after dilution.

A realistic condition to apply NAS technology must consider the analyte already present in the sample of raw milk, because dilution could lower its concentration below the detection limit of the method. However, with the exception of Zhou et al. [31] the approach followed in the rest of the works consists of contaminating the milk after and not before the clarification step (Table 5). This is the reason why our work focuses on antibiotic-spiked raw milk samples, which were employed as the starting point from which to investigate a preanalytical strategy and analytical techniques necessary to advance NAS technology toward its application for real samples obtained in dairy farms. Table 5 highlights that this is the first study addressing this problem in raw milk, while other works have focused on milk powder [32], supermarket milk [35,43] or non-specified information [31,34,56]. Finally, a number of approaches have been proposed to generate and scale-up, at low costs, the production of AuNPs, as well as for the recovery of gold from laboratory wastes [58,59], which together contribute to the economic viability of this technology at an industrial level.

4. Conclusions

The contamination of raw milk with antibiotic residues is an issue of worldwide concern due to its impacts on the dairy industry and human health. The application of NAS for the detection of antibiotics in bovine raw milk involves preanalytical and analytical challenges related with the chemical complexity of this matrix. In this context, the bioconjugation of AuNPs with thiolated ssDNA aptamers provided suitable experimental conditions with which to apply NAS for the colorimetric detection of antibiotics in raw milk. For the first time a methodology was proposed to enable the highly sensitive colorimetric detection of antibiotics belonging to four different groups, namely kanamycin (aminoglycosides), oxytetracycline (tetracyclines), sulfadimethoxine (sulfonamides) and ampicillin (beta-lactams), using a straightforward preanalytical process. The method developed herein consists of a clarification treatment with Carrez reagents, followed by the removal of cations from milk whey through dilution, chelation (EDTA) or precipitation ($NaHCO_3$). In all cases the methodology allowed for the semiquantitative detection of the colorimetric signals generated by the NAS at concentrations as low as 0.25-fold the MRL of the antibiotics. However, the specific treatment to address ionic interference (dilution/chelation/precipitation) depends on the particular antibiotic to be determined in the raw milk. Furthermore, the results proved that analyzing the spectral shift at 720 nm (A_{520}/A_{720}) improved the analytical performance of the NAS in comparison with the typical absorption ratio (A_{520}/A_{620}). Overall, this methodology combines simplicity and sensitivity for the four antibiotics, holding the potential to be applied for semiquantitative analyses of antibiotic residues in raw milk obtained directly from dairy farms.

Supplementary Materials: The following are available online at https://www.mdpi.com/article/10.3390/s22031281/s1, Figure S1: Spectral variation associated with the aggregation of AuNPs of the NAS, Figure S2: Aggregation of AuNPs in raw milk, Figure S3: Detection of antibiotics in milk whey using the NAS, Figure S4: Aggregation of AuNPs versus NAS in water, Table S1: Molar ratio between AuNPs-aptamers and AuNPs-aptamers bases.

Author Contributions: Conceptualization, P.O., V.D.-G. and P.C.; methodology, V.D.-G., B.C.-T. and C.R.; formal analysis, P.O., V.D.-G., P.C. and B.C.-T.; investigation, all authors; data curation, V.D.-G. and B.C.-T.; writing—original draft preparation, V.D.-G. and P.O.; writing—review and editing, P.O., V.D.-G., B.C.-T. and P.C.; supervision, P.O. and V.D.-G.; project administration, P.O.; funding acquisition, all authors. All authors have read and agreed to the published version of the manuscript.

Funding: This research was funded by the National Agency for Research and Development (ANID) of Chile, through its programs *Convocatoria Nacional Subvención a la Instalación en la Academia* (grant number PAI77180078) and *Fondecyt Posdoctorado* (grant number 3190734).

Institutional Review Board Statement: Not applicable.

Informed Consent Statement: Not applicable.

Data Availability Statement: The data presented in this study are available on request from the corresponding author. The data are not publicly available due to privacy restrictions.

Acknowledgments: We thank the Agriculture and Livestock Service of Chile (SAG) for valuable discussions and orientations.

Conflicts of Interest: The authors declare no conflict of interest.

References

1. Haug, A.; Høstmark, A.T.; Harstad, O.M. Bovine Milk in Human Nutrition—A Review. *Lipids Health Dis.* **2007**, *6*, 25. [CrossRef] [PubMed]
2. USDA Foreign Agricultural Service. Dairy: World Markets and Trade. Washington, DC, USA. Available online: https://Apps.Fas.Usda.Gov/Psdonline/Circulars/Dairy.Pdf (accessed on 13 September 2021).
3. Fusco, V.; Chieffi, D.; Fanelli, F.; Logrieco, A.F.; Cho, G.-S.; Kabisch, J.; Böhnlein, C.; Franz, C.M.A.P. Microbial Quality and Safety of Milk and Milk Products in the 21st Century. *Compr. Rev. Food Sci. Food Saf.* **2020**, *19*, 2013–2049. [CrossRef] [PubMed]
4. Boeckel, T.P.V.; Brower, C.; Gilbert, M.; Grenfell, B.T.; Levin, S.A.; Robinson, T.P.; Teillant, A.; Laxminarayan, R. Global Trends in Antimicrobial Use in Food Animals. *PNAS* **2015**, *112*, 5649–5654. [CrossRef] [PubMed]
5. Hao, H.; Cheng, G.; Iqbal, Z.; Ai, X.; Hussain, H.I.; Huang, L.; Dai, M.; Wang, Y.; Liu, Z.; Yuan, Z. Benefits and Risks of Antimicrobial Use in Food-Producing Animals. *Front. Microbiol.* **2014**, *5*, 288. [CrossRef]
6. Sharma, C.; Rokana, N.; Chandra, M.; Singh, B.P.; Gulhane, R.D.; Gill, J.P.S.; Ray, P.; Puniya, A.K.; Panwar, H. Antimicrobial Resistance: Its Surveillance, Impact, and Alternative Management Strategies in Dairy Animals. *Front. Vet. Sci.* **2018**, *4*, 237. [CrossRef]
7. Limmathurotsakul, D.; Dunachie, S.; Fukuda, K.; Feasey, N.A.; Okeke, I.N.; Holmes, A.H.; Moore, C.E.; Dolecek, C.; van Doorn, H.R.; Shetty, N.; et al. Improving the Estimation of the Global Burden of Antimicrobial Resistant Infections. *Lancet Infect. Dis.* **2019**, *19*, e392–e398. [CrossRef]
8. O'Neill, J. Tackling Drug-Resistant Infections Globally: Final Report and Recommendations. HM Government and Welcome Trust: UK. 2016. Available online: https://Amr-Review.Org/Sites/Default/Files/160518_Final%20paper_with%20cover.Pdf. (accessed on 18 October 2021).
9. Chiesa, L.M.; DeCastelli, L.; Nobile, M.; Martucci, F.; Mosconi, G.; Fontana, M.; Castrica, M.; Arioli, F.; Panseri, S. Analysis of Antibiotic Residues in Raw Bovine Milk and Their Impact toward Food Safety and on Milk Starter Cultures in Cheese-Making Process. *LWT* **2020**, *131*, 109783. [CrossRef]
10. Commission Regulation (EU) No 37/2010 of 22 December 2009 on Pharmacologically Active Substances and Their Classification Regarding Maximum Residue Limits in Foodstuffs of Animal Origin. Official Journal of the European Union. 2010 L15: 1–72. Available online: https://Ec.Europa.Eu/Health/Sites/Default/Files/Files/Eudralex/Vol-5/Reg_2010_37/Reg_2010_37_en.Pdf (accessed on 18 October 2021).
11. MINSAL Subsecretaría de Salud Pública. Resolución 1560 Exenta Fija Límites Máximos de Residuos de Medicamentos Veterinarios En Alimentos Destinados al Consumo Humano. Santiago, Chile. Available online: https://www.Bcn.Cl/Leychile/Navegar?IdNorma=1135977 (accessed on 18 October 2021).
12. Gaudin, V. The Growing Interest in Development of Innovative Optical Aptasensors for the Detection of Antimicrobial Residues in Food Products. *Biosensors* **2020**, *10*, 21. [CrossRef]
13. Somwanshi, S.; Somvanshi, S.; Kharat, P. Nanocatalyst: A Brief Review on Synthesis to Applications. *J. Phys. Conf. Ser.* **2020**, *1644*, 012046. [CrossRef]
14. Jain, K.; Patel, A.S.; Pardhi, V.P.; Flora, S.J.S. Nanotechnology in Wastewater Management: A New Paradigm Towards Wastewater Treatment. *Molecules* **2021**, *26*, 1797. [CrossRef]
15. McClements, D. Nanotechnology Approaches for Improving the Healthiness and Sustainability of the Modern Food Supply. *ACS Omega* **2020**, *5*, 29623–29630. [CrossRef] [PubMed]
16. Zuliani, A.; Ranjan, P.; Luque, R.; Van der Eycken, E.V. Heterogeneously Catalyzed Synthesis of Imidazolones via Cycloisomerizations of Propargylic Ureas Using Ag and Au/Al SBA-15 Systems. *ACS Sustain. Chem. Eng.* **2019**, *7*, 5568–5575. [CrossRef]
17. Moram, S.S.B.; Shaik, A.K.; Byram, C.; Hamad, S.; Soma, V.R. Instantaneous Trace Detection of Nitro-Explosives and Mixtures with Nanotextured Silicon Decorated with Ag–Au Alloy Nanoparticles Using the SERS Technique. *Anal. Chim. Acta* **2020**, *1101*, 157–168. [CrossRef] [PubMed]
18. Srivastava, A.K.; Dev, A.; Karmakar, S. Nanosensors and Nanobiosensors in Food and Agriculture. *Environ. Chem. Lett.* **2018**, *16*, 161–182. [CrossRef]
19. Prajapati, S.; Padhan, B.; Amulyasai, B.; Sarkar, A. Chapter 11—Nanotechnology-Based Sensors. In *Biopolymer-Based Formulations*; Pal, K., Banerjee, I., Sarkar, P., Kim, D., Deng, W.-P., Dubey, N.K., Majumder, K., Eds.; Elsevier: Amsterdam, The Netherlands, 2020; pp. 237–262. ISBN 978-0-12-816897-4. [CrossRef]
20. Díaz-García, V.; Retamal-Morales, G. Aptamer and Riboswitches: A Novel Tool for the Need of New Antimicrobial Active Compounds. In *Antibiotic Materials in Healthcare*; Kokkarachedu, V., Kanikireddy, V., Sadiku, R., Eds.; Academic Press: Cambridge, MA, USA, 2020; pp. 231–247. [CrossRef]

21. Hu, X.; Chang, K.; Wang, S.; Sun, X.; Hu, J.; Jiang, M. Aptamer-Functionalized AuNPs for the High-Sensitivity Colorimetric Detection of Melamine in Milk Samples. *PLoS ONE* **2018**, *13*, e0201626. [CrossRef]
22. Bai, W.; Zhu, C.; Liu, J.; Yan, M.; Yang, S.; Chen, A. Gold Nanoparticle-Based Colorimetric Aptasensor for Rapid Detection of Six Organophosphorous Pesticides. *Environ. Toxicol. Chem.* **2015**, *34*, 2244–2249. [CrossRef]
23. Yin, X.; Wang, S.; Liu, X.; He, C.; Tang, Y.; Li, Q.; Liu, J.; Su, H.; Tan, T.; Dong, Y. Aptamer-Based Colorimetric Biosensing of Ochratoxin A in Fortified White Grape Wine Sample Using Unmodified Gold Nanoparticles. *Anal. Sci.* **2017**, *33*, 659–664. [CrossRef]
24. Priyadarshini, E.; Pradhan, N. Gold Nanoparticles as Efficient Sensors in Colorimetric Detection of Toxic Metal Ions: A Review. *Sens. Actuators B Chem.* **2017**, *238*, 888–902. [CrossRef]
25. Liu, G.; Lu, M.; Huang, X.; Li, T.; Xu, D. Application of Gold-Nanoparticle Colorimetric Sensing to Rapid Food Safety Screening. *Sensors* **2018**, *18*, 4166. [CrossRef]
26. Zahra, Q.; Luo, Z.; Ali, R.; Khan, M.I.; Li, F.; Qiu, B. Advances in Gold Nanoparticles-Based Colorimetric Aptasensors for the Detection of Antibiotics: An Overview of the Past Decade. *Nanomaterials* **2021**, *11*, 840. [CrossRef]
27. Zhang, F.; Liu, J. Label-Free Colorimetric Biosensors Based on Aptamers and Gold Nanoparticles: A Critical Review. *Anal. Sens.* **2021**, *1*, 30–43. [CrossRef]
28. Amendola, V.; Pilot, R.; Frasconi, M.; Maragò, O.M.; Iatì, M.A. Surface Plasmon Resonance in Gold Nanoparticles: A Review. *J. Phys. Condens. Matter* **2017**, *29*, 203002. [CrossRef]
29. Yan, S.; Foroughi, M.; Safaei, M.; Jahani, S.; Ebrahimpour, N.; Borhani, F.; Zade, N.; Aramesh-Boroujeni, Z.; Foong, L. A review: Recent advances in ultrasensitive and highly specific recognition aptasensors with various detection strategies. *Int. J. Biol. Macrom.* **2020**, *155*, 184–207. [CrossRef]
30. Andrew, S.M. Effect of Fat and Protein Content of Milk from Individual Cows on the Specificity Rates of Antibiotic Residue Screening Tests. *J. Dairy Sci.* **2000**, *83*, 2992–2997. [CrossRef]
31. Zhou, N.; Zhang, J.; Tian, Y. Aptamer-Based Spectrophotometric Detection of Kanamycin in Milk. *Anal. Methods* **2014**, *6*, 1569–1574. [CrossRef]
32. He, L.; Luo, Y.; Zhi, W.; Wu, Y.; Zhou, P.; He, L.; Luo, Y.; Zhi, W.; Wu, Y.; Zhou, P. A Colorimetric Aptamer Biosensor Based on Gold Nanoparticles for the Ultrasensitive and Specific Detection of Tetracycline in Milk. *Aust. J. Chem.* **2013**, *66*, 485–490. [CrossRef]
33. Kim, C.-H.; Lee, L.-P.; Min, J.-R.; Lim, M.-W.; Jeong, S.-H. An Indirect Competitive Assay-Based Aptasensor for Detection of Oxytetracycline in Milk. *Biosens. Bioelectron.* **2014**, *51*, 426–430. [CrossRef] [PubMed]
34. Emrani, A.S.; Danesh, N.M.; Lavaee, P.; Ramezani, M.; Abnous, K.; Taghdisi, S.M. Colorimetric and Fluorescence Quenching Aptasensors for Detection of Streptomycin in Blood Serum and Milk Based on Double-Stranded DNA and Gold Nanoparticles. *Food Chem.* **2016**, *190*, 115–121. [CrossRef]
35. Zhao, J.; Wu, Y.; Tao, H.; Chen, H.; Yang, W.; Qiu, S. Colorimetric Detection of Streptomycin in Milk Based on Peroxidase-Mimicking Catalytic Activity of Gold Nanoparticles. *RSC Adv.* **2017**, *7*, 38471–38478. [CrossRef]
36. Adami, A.; Mortari, A.; Morganti, E.; Lorenzelli, L. Microfluidic Sample Preparation Methods for the Analysis of Milk Contaminants. *J. Sens.* **2015**, *2016*, e2385267. [CrossRef]
37. Contreras-Trigo, B.; Díaz-García, V.; Guzmán-Gutierrez, E.; Sanhueza, I.; Coelho, P.; Godoy, S.E.; Torres, S.; Oyarzún, P. Slight pH Fluctuations in the Gold Nanoparticle Synthesis Process Influence the Performance of the Citrate Reduction Method. *Sensors* **2018**, *18*, 2246. [CrossRef]
38. Maye, M.M.; Han, L.; Kariuki, N.N.; Ly, N.K.; Chan, W.-B.; Luo, J.; Zhong, C.-J. Gold and Alloy Nanoparticles in Solution and Thin Film Assembly: Spectrophotometric Determination of Molar Absorptivity. *Anal. Chim. Acta* **2003**, *496*, 17–27. [CrossRef]
39. Hill, H.D.; Mirkin, C.A. The Bio-Barcode Assay for the Detection of Protein and Nucleic Acid Targets Using DTT-Induced Ligand Exchange. *Nat. Protoc.* **2006**, *1*, 324–336. [CrossRef] [PubMed]
40. Song, K.-M.; Cho, M.; Jo, H.; Min, K.; Jeon, S.H.; Kim, T.; Han, M.S.; Ku, J.K.; Ban, C. Gold Nanoparticle-Based Colorimetric Detection of Kanamycin Using a DNA Aptamer. *Anal. Biochem.* **2011**, *415*, 175–181. [CrossRef] [PubMed]
41. Kim, Y.S.; Kim, J.H.; Kim, I.A.; Lee, S.J.; Jurng, J.; Gu, M.B. A Novel Colorimetric Aptasensor Using Gold Nanoparticle for a Highly Sensitive and Specific Detection of Oxytetracycline. *Biosens. Bioelectron.* **2010**, *26*, 1644–1649. [CrossRef]
42. Chen, A.; Jiang, X.; Zhang, W.; Chen, G.; Zhao, Y.; Tunio, T.M.; Liu, J.; Lv, Z.; Li, C.; Yang, S. High Sensitive Rapid Visual Detection of Sulfadimethoxine by Label-Free Aptasensor. *Biosens. Bioelectron.* **2013**, *42*, 419–425. [CrossRef]
43. Song, K.-M.; Jeong, E.; Jeon, W.; Cho, M.; Ban, C. Aptasensor for Ampicillin Using Gold Nanoparticle Based Dual Fluorescence-Colorimetric Methods. *Anal. Bioanal. Chem.* **2012**, *402*, 2153–2161. [CrossRef]
44. Horwitz, W.; Latimer, G.W. *Official Methods of Analysis of AOAC International*; AOAC International: Gaithersburg, MA, USA, 2005.
45. Xue, Y.; Li, X.; Li, H.; Zhang, W. Quantifying Thiol–Gold Interactions towards the Efficient Strength Control. *Nat. Commun.* **2014**, *5*, 4348. [CrossRef]
46. Adeel, M.; Rahman, M.; Lee, J.-J. Label-Free Aptasensor for the Detection of Cardiac Biomarker Myoglobin Based on Gold Nanoparticles Decorated Boron Nitride Nanosheets. *Biosens. Bioelectron.* **2019**, *126*, 143–150. [CrossRef]
47. Peng, Y.; Li, L.; Mu, X.; Guo, L. Aptamer-Gold Nanoparticle-Based Colorimetric Assay for the Sensitive Detection of Thrombin. *Sens. Actuators B Chem.* **2013**, *177*, 818–825. [CrossRef]

48. Azzam EM, S.; El-Farargy AF, M.; Hegazy, M.A.; Abd El-Aal, A.A. Detection of Heavy Metal Ions Using Synthesized Amino Thiol Surfactants Assembled on Gold Nanoparticles. *J. Disp. Sci. Technol.* **2014**, *35*, 175–184. [CrossRef]
49. Li, H.; Rothberg, L. Colorimetric Detection of DNA Sequences Based on Electrostatic Interactions with Unmodified Gold Nanoparticles. *PNAS* **2004**, *101*, 14036–14039. [CrossRef] [PubMed]
50. Ma, Q.; Wang, Y.; Jia, J.; Xiang, Y. Colorimetric aptasensors for determination of tobramycin in milk and chicken eggs based on DNA and gold nanoparticles. *Food Chem.* **2018**, *30*, 98–103. [CrossRef] [PubMed]
51. Gutiérrez, P.; Godoy, S.E.; Torres, S.; Oyarzún, P.; Sanhueza, I.; Díaz-García, V.; Contreras-Trigo, B.; Coelho, P. Improved Antibiotic Detection in Raw Milk Using Machine Learning Tools over the Absorption Spectra of a Problem-Specific Nanobiosensor. *Sensors* **2020**, *20*, 4552. [CrossRef]
52. Holt, C. The Milk Salts and Their Interaction with Casein. In *Advanced Dairy Chemistry: Lactose, Water, Salts and Vitamins*; Fox, P.F., Ed.; Springer: Boston, MA, USA, 1997; Volume 3, pp. 233–256. [CrossRef]
53. Bachran, K.; Bernhard, R.A. Interaction of Iron(II) with Lactose. *J. Agric. Food Chem.* **1980**, *28*, 536–540. [CrossRef]
54. Reynolds, A.J.; Haines, A.H.; Russell, D.A. Gold Glyconanoparticles for Mimics and Measurement of Metal Ion-Mediated Carbohydrate–Carbohydrate Interactions. *Langmuir* **2006**, *22*, 1156–1163. [CrossRef]
55. Ho, T.T.-T.; Dang, C.-H.; Huynh, T.K.-C.; Hoang, T.K.-D.; Nguyen, T.-D. In Situ Synthesis of Gold Nanoparticles on Novel Nanocomposite Lactose/Alginate: Recyclable Catalysis and Colorimetric Detection of Fe(III). *Carbohydr. Polym.* **2021**, *251*, 116998. [CrossRef]
56. He, L.; Luo, Y.; Zhi, W.; Zhou, P. Colorimetric Sensing of Tetracyclines in Milk Based on the Assembly of Cationic Conjugated Polymer-Aggregated Gold Nanoparticles. *Food Anal. Methods* **2013**, *6*, 1704–1711. [CrossRef]
57. Chen, M.; Gan, N.; Zhou, Y.; Li, T.; Xu, Q.; Cao, Y.; Chen, Y. A novel aptamer- metal ions- nanoscale MOF based electrochemical biocodes for multiple antibiotics detection and signal amplification. *Sens. Actuators B Chem.* **2017**, *242*, 1201–1209. [CrossRef]
58. Giannone, G.; Santi, M.; Ermini, M.L.; Cassano, D.; Voliani, V. A Cost-Effective Approach for Non-Persistent Gold Nano-Architectures Production. *Nanomaterials* **2020**, *10*, 1600. [CrossRef]
59. Gan, P.P.; Ng, S.H.; Huang, Y.; Li, S.F. Green synthesis of gold nanoparticles using palm oil mill effluent (POME): A low-cost and eco-friendly viable approach. *Bioresour. Technol.* **2012**, *113*, 132–135. [CrossRef] [PubMed]

sensors

MDPI

Article

Plasmonic Strain Sensors Based on Au-TiO_2 Thin Films on Flexible Substrates

Marco S. Rodrigues *, Joel Borges and Filipe Vaz

CF-UM-UP, Centro de Física das Universidades do Minho e do Porto, Universidade do Minho, 4710-057 Braga, Portugal; joelborges@fisica.uminho.pt (J.B.); fvaz@fisica.uminho.pt (F.V.)
* Correspondence: marcopsr@gmail.com or mprodrigues@fisica.uminho.pt

Abstract: This study aimed at introducing thin films exhibiting the localized surface plasmon resonance (LSPR) phenomenon with a reversible optical response to repeated uniaxial strain. The sensing platform was prepared by growing gold (Au) nanoparticles throughout a titanium dioxide dielectric matrix. The thin films were deposited on transparent polymeric substrates, using reactive magnetron sputtering, followed by a low temperature thermal treatment to grow the nanoparticles. The microstructural characterization of the thin films' surface revealed Au nanoparticle with an average size of 15.9 nm, an aspect ratio of 1.29 and an average nearest neighbor nanoparticle at 16.3 nm distance. The plasmonic response of the flexible nanoplasmonic transducers was characterized with custom-made mechanical testing equipment using simultaneous optical transmittance measurements. The higher sensitivity that was obtained at a maximum strain of 6.7%, reached the values of 420 nm/ε and 110 pp/ε when measured at the wavelength or transmittance coordinates of the transmittance-LSPR band minimum, respectively. The higher transmittance gauge factor of 4.5 was obtained for a strain of 10.1%. Optical modelling, using discrete dipole approximation, seems to correlate the optical response of the strained thin film sensor to a reduction in the refractive index of the matrix surrounding the gold nanoparticles when uniaxial strain is applied.

Keywords: gold nanoparticles; localized surface plasmon resonance; flexible optical sensors; plasmonic thin films

Citation: Rodrigues, M.S.; Borges, J.; Vaz, F. Plasmonic Strain Sensors Based on Au-TiO_2 Thin Films on Flexible Substrates. *Sensors* **2022**, *22*, 1375. https://doi.org/10.3390/s22041375

Academic Editor: Nikolay Kazanskiy

Received: 11 January 2022
Accepted: 7 February 2022
Published: 11 February 2022

1. Introduction

The world market for strain sensors has been thriving in the last decade and is expected to continue growing in the next years, supported by the development of smart cities, buildings automation, and wearable devices [1]. Strain sensors can measure or detect strain, pressure, vibration, impact and deflection on an object after a change in their optical or electrical response. Although most of the commercialized and investigated strain sensors rely on a change in the electrical properties of a material as the transduction mechanism [2–5], the use of optical phenomena in sensing applications as increased in basic research, health applications and industry, mostly due to its high sensitivity; aging, and high temperature, stability; safety to be used in flammable and explosive atmospheres; and blindness to surrounding electric noise [6–8]. Furthermore, the advancements in low-energy optoelectronic components have enabled the construction of miniaturized and portable optical devices [9].

Among the several optical transduction mechanisms, the use of the localized surface plasmon resonance (LSPR) phenomenon, related to the interaction of light with the free electrons of noble metal nanoparticles [10–18], has already revealed its capabilities in (bio)chemical sensing [19–22] and gas sensing [23–26]. Moreover, regarding the detection of physical and mechanical stimuli, such as temperature and force, some pioneering works have been published, most of them related to theoretical approaches in ultra-thin gold films [27–38] and arrays of plasmonic nanoparticles [39–48]. These and other works led to the development of a new field in plasmonics and nano-optomechanics, also related to mechanochromism [49], named thereafter as plasmomechanics by Maurer et al. [36].

In very simple terms, plasmomechanics studies the reciprocal interactions of light, plasmonic structures, forces applied (from the macro- to the nano-scale) and localized thermal effects [50]. In plasmomechanical systems, the electromagnetic field confinement in metallic nanostructures with subwavelength dimensions, enables to generate large resonance frequency shifts for extremely small displacements. Furthermore, the resonant excitation of these plasmonic nanostructures leads to a strong absorption of the incident light, generating photo-induced thermal effect that can be exploited for actuation of mechanical systems at the nanoscale. Therefore, these systems offer optical and mechanical degrees of freedom for different kinds of applications including active control of nano-mechanical motion transduction and amplification of weak external stimuli [51].

When using plasmomechanics for strain sensing with materials containing noble metal nanoparticles, the transduction mechanism can rely on LSPR band changes induced by external physical stimuli. For example, an applied force may change the nanoparticles' shape (form) (Figure 1a) and distribution (Figure 1b), or the refractive index (Figure 1c) of the surrounding medium where they might be embedded [36].

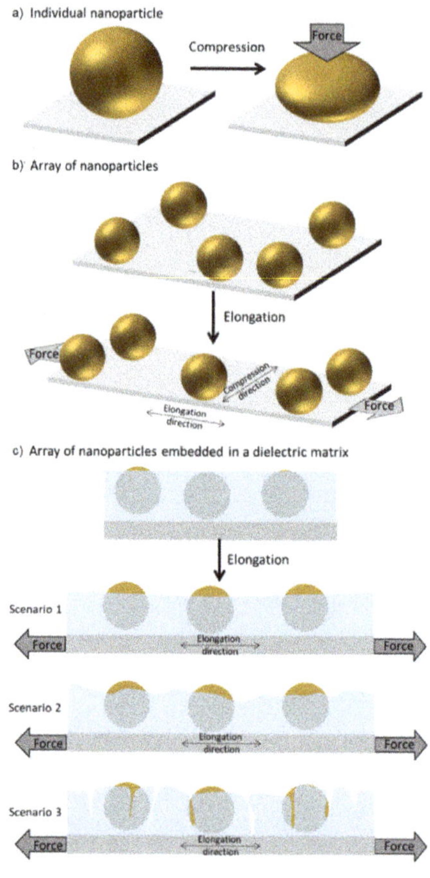

Figure 1. Simplified schematics of deformation mechanisms induced to (**a**) individual nanoparticles being compressed, (**b**) arrays of nanoparticles in a substrate that is being stretched and (**c**) arrays of nanoparticles embedded in a dielectric matrix that is being stretched, where the thickness (scenario 1), the roughness (scenario 2) and the cohesion (scenario 3) of the matrix is changed.

In this work, nanoplasmonic thin films with gold (Au) embedded in titanium dioxide (TiO_2) were deposited on polydimethylsiloxane (PDMS) and characterized. The correlation between the tensile deformation of the flexible nanomaterial and its optical (plasmonic) response was studied. Different mechanisms for strain sensing with nanoparticles exhibiting the LSPR phenomenon were simulated using the discrete dipole approximation (DDA). The obtained results were then used to explain the optical behavior of nanocomposite thin films, when uniaxial strain was applied. The plasmonic sensing platforms' repeatability was evaluated under several load/unload strain cycles.

2. Materials and Methods

The sensing platform consists of an optically transparent polydimethylsiloxane (PDMS) substrate that was functionalized with a nanocomposite gold and titanium dioxide (Au-TiO_2) thin film (Figure 2a). The thin film was deposited by reactive DC magnetron sputtering [52,53]. In order to obtain an optical LSPR response a thermal annealing protocol was applied (150 °C for 10 min) as explained in detail elsewhere [54].

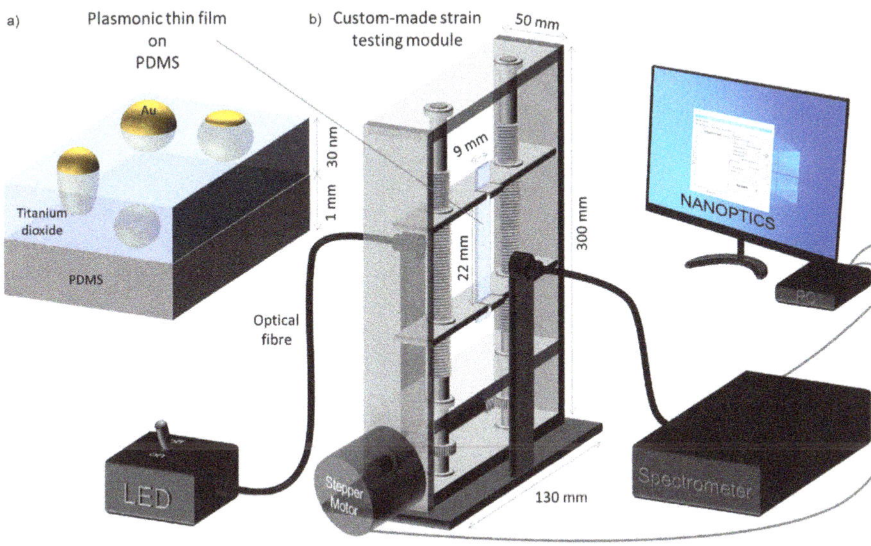

Figure 2. Simplified schematics of (**a**) the sensing platform, a nanoplasmonic thin film (composed by gold nanoparticles dispersed in TiO_2) deposited on transparent PDMS, and (**b**) the strain testing module coupled to an optical transmittance measurement system.

The in-depth chemical composition profile was evaluated by Rutherford Backscattering Spectrometry (RBS). The measurements were conducted in a chamber, where three detectors were installed, a standard at 140° and two pin-diode detectors located symmetrically to each other, both at 165°. Spectra were collected for 2 MeV $4He^+$ beams at normal incidence (0°). The obtained RBS data were analyzed with IBA DataFurnace NDF v10.0a [55,56], and the double scattering and pileup were calculated with the algorithms given in N. P. Barradas et al. [57] and in N. P. Barradas et al. [58], respectively.

To study the microstructure of the thin films, scanning electron microscopy (SEM) was used with a FEI Quanta 650 FEG (Field Emission Gun) environmental SEM microscope operating at 20 kV from INL (International Iberian Nanotechnology Laboratory). Then, a custom MATLAB algorithm was employed to study the features of the Au nanoparticles (Feret diameter, nearest neighbor and aspect ratio) [59].

The plasmonic response of the Au-TiO_2 thin films was investigated in a custom-made strain testing module coupled with an optical system, as depicted in Figure 2b.

While the strain module stretches the sensor, its optical transmittance spectrum is measured using a modular spectrometer (SPEC RES + UV/Vis, SARSPEC—with a diffraction grating), and a LED light source (LS-LED, SARSPEC) to excite the localized surface plasmons. Precise strain was controlled by a stepper motor (200 steps per turn) connected to a 20:1 worm reduction gear system, with a total 4000 steps per turn, thus giving the strain movement a resolution of 0.15 μm per step. Two protocols were used to apply stress strains to the sensors, by (i) varying the strain speed keeping the maximum strain constant; and (ii) varying the maximum strain keeping the strain speed constant.

The optical transmittance measurement was synchronized with the applied strain and a full spectrum (from 450 nm to 750 nm) was recorded every 0.5 s with an integration time of 3 ms and an average of 20 spectra. These transmittance spectra data were then analyzed using the software NANOPTICS [60] to follow the LSPR band minimum wavelength and transmittance coordinates.

Optical modelling and simulation of gold nanoparticles was conducted with the discrete dipole approximation (DDA) method [7,61,62] using the software nanoDDSCAT+ [63] freely available at "nanohub.org" (Figure S1).

For all the simulations, the extinction spectra were calculated with the following principal parameters: two dipoles per nanometer for each shape; unpolarized incident light; Au dielectric function from Johnson and Christy [64]; surrounding external refractive index set either to 1.5, 2.0, or 2.5, as a "semi-infinite" layer; and 5 nm steps in all spectra windows. For individual nanoparticles with different sizes and aspect ratios, the predefined configurable ellipsoid shapes in DDSCAT tool (Figure S1-III) were used. For the simulations using arrays of nanoparticles, a 16 nm spherical nanoparticle was designed with Blender tool (Figure S1-I), then converted to dipoles with DDA convert tool (Figure S1-II), and exported to DDSCAT tool (Figure S1-III), where, along with the previously described parameters, a 2-D object periodicity was selected, with the y and z spacings of interest.

3. Results and Discussion

3.1. Nanoplasmonic Thin Film Characterization

The RBS analysis of the as-deposited thin film revealed a uniform in-depth Au concentration of 22 at.%, across a total estimated thickness of 30 nm.

To study the distribution of Au nanoparticles dispersed in the TiO_2 matrix, the nanoplasmonic thin film deposited on the flexible substrate and annealed at 150 °C was analyzed using SEM (Figure 3). The micrographs revealed the Au nanoparticles partially embedded in the TiO_2 matrix, Figure 3a, and equally distributed throughout the observed areas. After image processing in a custom MATLAB algorithm, the SEM micrograph was thresholded to black (TiO_2 matrix) and white (nanoparticles), as depicted in Figure 3b, thus allowing to analyze the Au nanoparticles distributions, presented in Figure 3c. The nanoparticles' analysis revealed a 10.2% of area covered by Au, with a nanoparticle density of 451 nanoparticles/μm^2 and an average size of 15.9 nm. The fact that 55% of the nanoparticles possessed a Feret diameter between 10 and 20 nm is also noteworthy. Regarding the nearest neighbor distance, 51% of the nanoparticles possess a nearest Au nanoparticle neighbor between 10 to 20 nm, giving an average nearest neighbor of 16.3 nm. When analyzing the aspect ratio (AR), an average of 1.29 was obtained, with 58% of the nanoparticles having an AR between 1 and 1.2, and 83% with an AR below 1.4. Following these results, 16 nm Au nanoparticles were used for the starting conditions in the simulations presented in Section 3.2, as well as a starting nearest neighbor distance of 16 nm in Sections 3.2.3 and 3.2.4.

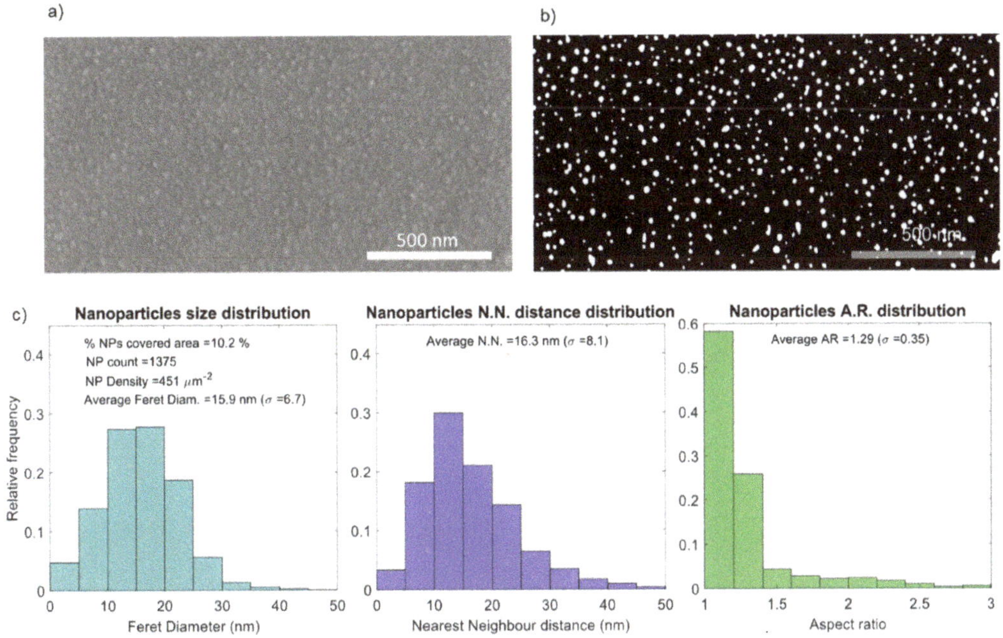

Figure 3. Au nanoparticles analysis in the Au-TiO_2 thin film deposited on PDMS and annealed at 150 °C, showing the (**a**) SEM top-view micrograph of the Au-TiO_2 thin film's surface, (**b**) the thresholded micrograph, revealing the Au nanoparticles in white, as well as (**c**) a statistical analysis of the exposed nanoparticles, with the distribution histograms of nanoparticles diameter (average of 15.9 nm), nearest neighbor distance (average of 16.3 nm) and aspect ratio (average of 1.29).

3.2. *Gold Nanoparticles' Optical Modelling under Different Mechanical Strain Conditions*

The LSPR phenomenon arises from the interaction of light with metallic nanoparticles of a size of about one order of magnitude lower than the wavelength of the incident light. This interaction depends on the size, shape and distribution of the nanoparticles [65–67] but also on the material surrounding them [59,68], and it can be characterized by studying their extinction spectra, both experimentally and theoretically. In this section, the optical extinction spectra of isolated and arrays of Au nanoparticles was simulated considering the parameters from the previous section, during the application of several mechanical stimulus. The results were then used to be compared to the experimental results in Section 3.3.

Among the numerous methods available for optical modelling, DDA was used because it allows the simulation of nanoparticles with different shapes [7,61] and organized in 1D or 2D networks [62].

3.2.1. Compressing Single Nanoparticles

Nanoparticles with different shapes have distinct optical responses as it can be confirmed in the supplementary material for a Au sphere, an oblate and prolate spheroid, and an ellipsoid with similar effective volumes (Figure S2). While the sphere presents an extinction maximum at a wavelength of 535 nm (when surrounded by a semi-infinite layer with a refractive index of 1.5), for the oblate and prolate spheroid, and for the ellipsoid, the extinction maximum occurs due to the longitudinal vibration modes of the localized surface plasmons [7] that gave rise to a second peak at higher wavelengths, 575 nm, 610 nm and 635 nm, respectively. On the other hand, since a single Au nanoparticle can be deformed elastically [69], if it is compressed and transformed into an oblate spheroid with an AR of 1.4 (as simulated in Figure S3), it gives rise to a second peak in the extinction spectrum at higher wavelengths (Figure 4). As the AR of this oblate spheroid increases to 1.8 (due to the compression of the nanoparticle), the second peak shifts to higher wavelengths. If the maximum compression is limited to the elastic regime of the nanoparticle's deformation, this process will be reversible; this effect can be used, for e.g., in optical strain sensing.

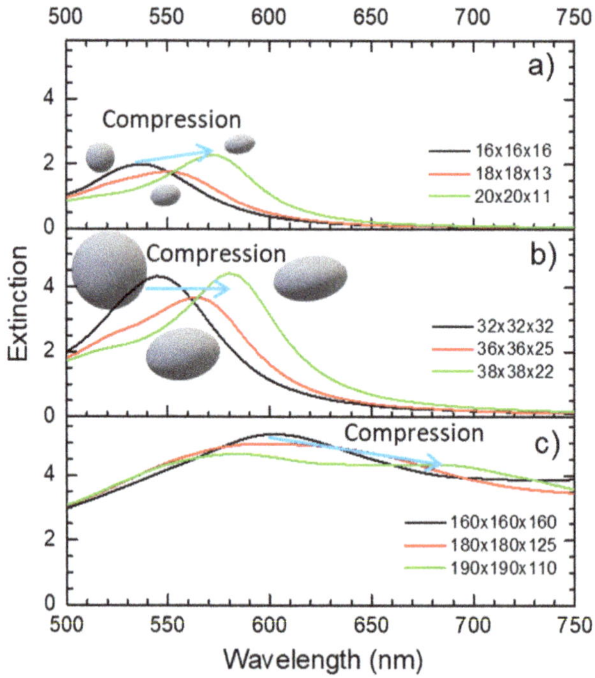

Figure 4. Optical extinction spectra of nanoparticles with different diameters (**a**) 16 nm, (**b**) 32 nm and (**c**) 160 nm, and after being compressed, changing from a sphere (initial state) to an oblate spheroid with an AR of 1.4, and then to an oblate spheroid with an AR of 1.8, calculated with nanoDDSCAT+. The blue arrow shows the shift direction of the extinction spectra (red-shift) when the nanoparticle is compressed.

3.2.2. Biaxial Elongation of Gold Nanoparticles Network

For pairs or arrays of nanoparticles, the LSPR coupling effect between nanoparticles that are close together leads to a strong interaction that can be used for strain and deflection sensing. When considering a biaxial elongation of an array of Au nanoparticles (with 16 nm diameter, for example) where the nearest neighbor starts at a distance of 1 nm, as depicted in Figure S4, the strong coupling effect quickly vanishes, leading to a blue shift of the LSPR band, as can be observed in Figure 5. When the nearest neighbor is at a distance of 24 nm (strain of 135%-ε = 1.35), the coupling between the 16 nm nanoparticles is very weak, leading to a total LSPR band shift of 84 nm. Further increasing the nearest neighbor distance from 24 to 44 nm (strain of 253%-ε = 2.53) the LSPR band blue-shifts only 3 nm, and the coupling effect practically disappears. In fact, as depicted in Figure 5, the extinction spectrum of the array of nanoparticles for distances higher than 44 nm is similar to the single nanoparticle spectrum. The inverse process (approximation of nanoparticles) would give rise to (an opposite) red-shift.

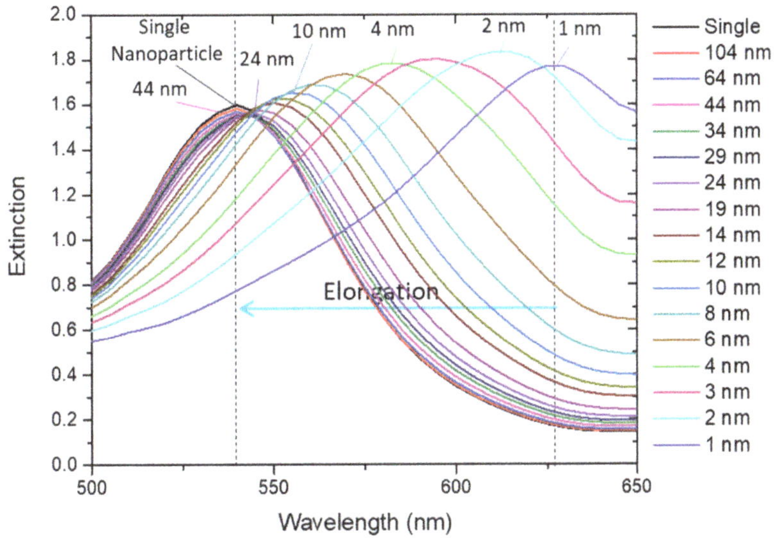

Figure 5. Optical extinction spectra of a network of gold nanoparticles being biaxially elongated, calculated with nanoDDSCAT+. The blue arrow shows the shift direction of the extinction spectra (blue-shift) when the array of nanoparticles is biaxially elongated.

3.2.3. Uniaxial Elongation of a Gold Nanoparticles Network in Different Surrounding Refractive Indices

For this case, an array of Au nanoparticles with 16 nm diameter, a starting nearest neighbor distance of 16 nm (surface to surface) and surrounded by different refractive indices (1.5, 2.0 and 2.5) was simulated during uniaxial elongation, as depicted in Figure S5. These parameters were chosen considering the nanoparticles' distribution analyzed in Section 3.1 and previous works from the authors [54,70].

From the simulation results, shown in Figure 6, it seems obvious that the refractive index of the material surrounding the nanoparticles holds a strong influence in the position of the LSPR band. In the simple case, i.e., for a single nanoparticle, when the refractive index decreased from 2.5 to 1.5 (Figure 6a,c) it led to a blue-shift of 125 nm. For the arrays of nanoparticles suffering uniaxial elongation (separation in the direction of the elongation), a maximum strain of 156% (ε = 1.56) led to blue-shifts of 15 nm, 8 nm and 5 nm for surrounding refractive indices of 2.5, 2.0 and 1.5 (Figure 6a, Figure 6b, Figure 6c), respectively.

Therefore, if one could separate the nanoparticles while lowering the refractive index of the surrounding matrix, a strain sensor with enhanced sensitivity would be obtained.

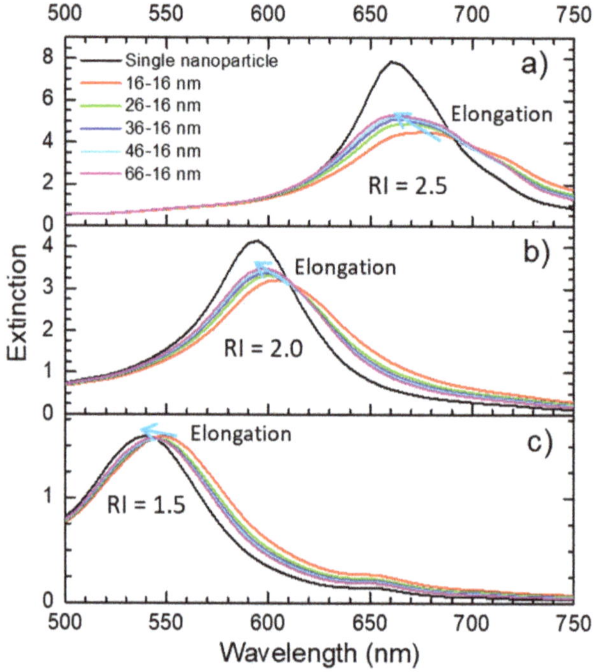

Figure 6. Optical extinction spectra of a network of nanoparticles being uniaxially elongated in a refractive index of (**a**) 2.5, (**b**) 2.0 and (**c**) 1.5, calculated with nanoDDSCAT+. The blue arrow shows the shift direction of the extinction spectra (blue-shift) when the array of nanoparticles is elongated.

3.2.4. Elongation of a Gold Nanoparticles Array, Considering Transverse Deformation

Since there is a transverse deformation when a flexible material such as PDMS is elongated in a tensile test, a Poisson's ratio of 0.499 (from PDMS [71]) was used in the simulations to calculate the elongation and compression of the nanoparticles' array. Following the previous sections, a similar array of nanoparticles was used (i.e., same initial relative positions) surrounded by a refractive index of 1.5 and a maximum strain of 56% ($\varepsilon = 0.56$), as depicted in Figure S6.

The extinction spectra presented in Figure 7 indicate that under these conditions, the red-shift of the LSPR band, due to the enhanced coupling between the nanoparticles in the transverse direction to the tensile test direction, overcame the expected blue-shift due to the lower coupling between the nanoparticles in the elongation direction. In this example, applying a maximum strain of 56% ($\varepsilon = 0.56$), led to a total LSPR band red-shift of 10 nm.

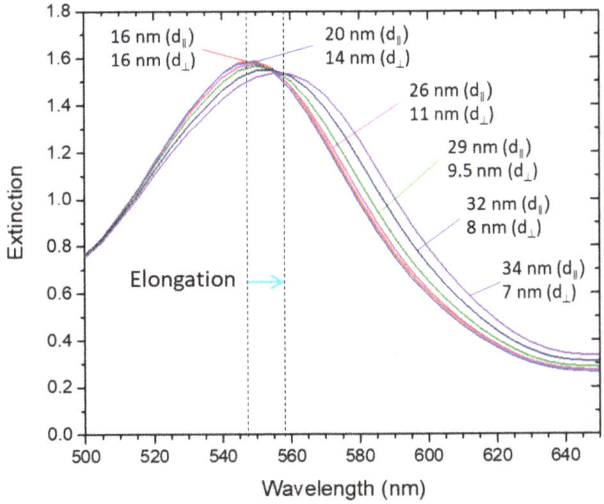

Figure 7. Optical extinction spectra of a network of nanoparticles being uniaxially elongated and transversely compressed, calculated with nanoDDSCAT+. The blue arrow shows the shift direction of the extinction spectra (red-shift) when the array of nanoparticles is elongated while transverse compression occurs.

3.3. Experimental Uniaxial Elongation of Nanoplasmonic Au-TiO$_2$ Thin Films Deposited on PDMS

For the experimental tensile tests, a Au–TiO$_2$ thin film deposited on flexible and transparent PDMS was used. After a low temperature annealing process at 150 °C for 10 min, the Au nanoparticles were formed (see Section 3.1) and the Au–TiO$_2$ thin film revealed its plasmonic nature by exhibiting the LSPR band, as depicted in Figure 8.

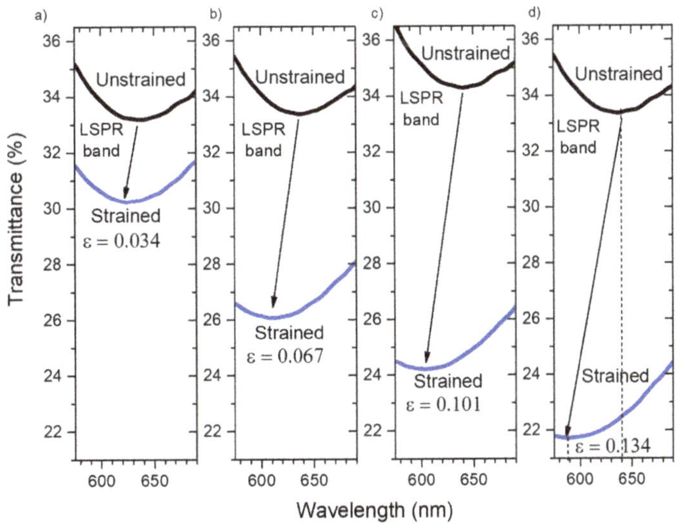

Figure 8. Optical transmittance of the unstrained (black line) and strained (blue line) thin film, with a maximum strain of (**a**) 3.4%, (**b**) 6.7%, (**c**) 10.1% and (**d**) 13.4%. The black arrow shows the shift direction of the extinction spectra (blue-shift and lower transmittance shift) when the nanoplasmonic sensors are strained.

After applying different strains to the flexible sensor, the optical transmittance of the plasmonic sensor suffered an overall reduction, and the LSPR band minimum was blue-shifted, Figure 8. The transmittance coordinate of the LSPR band minimum was reduced by 3 pp, 7 pp, 10 pp and 12 pp, while the wavelength coordinate was blue-shifted by 12 nm, 25 nm, 37 nm and 50 nm, at maximum strains of 3.4% ($\varepsilon = 0.034$, Figure 8a), 6.7% ($\varepsilon = 0.067$, Figure 8b), 10.1% ($\varepsilon = 0.101$, Figure 8c) and 13.4% ($\varepsilon = 0.134$, Figure 8d), respectively.

When comparing the simulation results with this optical LSPR band response to the maximum applied strain of 0.134 (13.4%), it seems that the most important effect is not related to changes in nanoparticle's coupling. Indeed, and since the results simulated in Section 3.2.4 are the ones with the most comparable conditions, one would expect that the LSPR band would be red-shifting due to a higher coupling effect in the transverse direction [72]. Instead, Figure 8 shows that there was a considerable blue-shift, with the minimum wavelength coordinate varying from 640 nm to 590 nm.

These results seem to indicate that the refractive index surrounding the nanoparticles might be the main effect, as it can be confirmed by the simulation of the optical behavior evolution in Section 3.2.3.

Following these results, several strain cycles were applied to the same sensor, by changing the strain speed, from 15 μm/s up to 310 μm/s, Figure 9, and the maximum strain, from 3.4% ($\varepsilon = 0.034$) up to 13.4% ($\varepsilon = 0.134$), Figure 10. The response of the plasmonic thin film sensor was replicated for all the tested conditions with no drifts nor hysteresis in the monitored cycles.

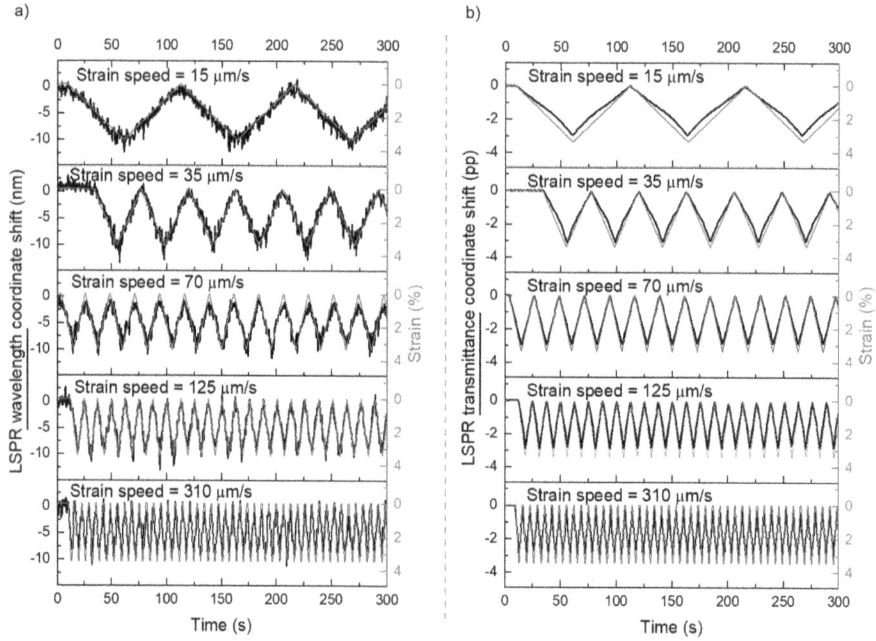

Figure 9. Response of the nanoplasmonic sensor, with the LSPR band minimum (**a**) wavelength and (**b**) transmittance coordinates monitoring during strain cycles at different speeds (from 15 μm/s up to 310 μm/s) and with a maximum strain of 3.4%. In green (right axis) the corresponding strain monitoring for each test.

Figure 9 shows the real time response of the plasmonic platform, after consecutive tensile tests with a maximum strain of 3.4% ($\varepsilon = 0.034$) by following the variations of the LSPR band wavelength and transmittance coordinates. These results show that the response of the sensor is consistent if the deformation speed is changed, which means that the response time of the sensor is lower, or in the order of the spectra acquisition time (0.5 s). Regarding the quality of the signals, both the wavelength and the transmittance coordinates showed an acceptable signal-to-noise ratio, although the transmittance signal is fairly better.

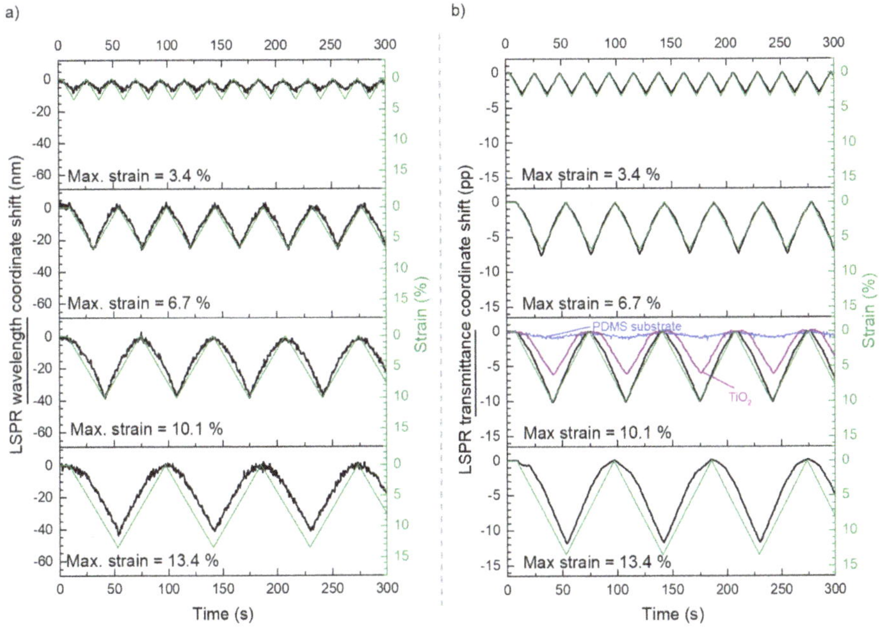

Figure 10. Cyclic response of the nanoplasmonic sensor, with the LSPR band minimum (**a**) wavelength and (**b**) transmittance coordinates monitoring during strain cycles with different maximum strains (from 3.4% to 13.3%) and at a speed of 70 μm/s. In green (right axis) the corresponding strain monitoring for each test. (**b**) also presents in blue and purple, the transmittance at 650 nm of the PDMS substrate and the TiO_2 thin film deposited on PDMS respectively, for a maximum strain of 10.1% and a strain speed of 70 μm/s.

To test the strain limits of these sensors, several cycles were applied with different maximum strains, as depicted in Figure 10, using a constant strain speed. The signals obtained from the wavelength (Figure 10a) and the transmittance (Figure 10b) coordinates of the LSPR band minimum also offered a consistent optical response that seemed to saturate for a maximum strain of 13.4% ($\varepsilon = 0.134$). Nonetheless, the sensor seems to be recovering for all the used strains. Similarly to the tensile tests with speed variations in Figure 9, the transmittance coordinate of the LSPR band minimum has a superior signal-to-noise ratio than the wavelength coordinate signal.

Furthermore, to compare the influence of the PDMS substrate and the TiO_2 matrix deposited on the PDMS substrate, a maximum strain of 10.1% ($\varepsilon = 0.101$) and a strain speed of 70 μm/s was used to test them. Since these materials have no plasmonic extinction band, a transmittance near the LSPR band minimum (650 nm) was analyzed. The results can be observed in Figure 10b. The transmittance of the PDMS substrate (Figure 10b in blue) only varied a small fraction of the plasmonic sensor (less than 10%), most probably due to some scattering as a result of the elongation of the polymer along the tensile direction, and compression on the perpendicular direction. Regarding the TiO_2 matrix deposited on PDMS, the transmittance variation (Figure 10b in purple) was considerably higher than for the PDMS substrate, but still lower than for the Au-TiO_2 thin film, reaching about 60% of the change in transmittance of the plasmonic sensor.

To quantify and compare the response of the plasmonic sensor to different strains, similarly to the gauge factor that is used in piezoresistive sensors [73] and surface-enhanced Raman scattering (SERS) piezoplamonics [74], the transmittance gauge factor (GF_T) was defined as follows:

$$GF_T = \frac{\frac{T_{initial} - T_{strained}}{T_{strained}}}{\varepsilon}, \quad (1)$$

where $T_{initial}$ and $T_{strained}$ are the transmittance coordinate at the LSPR band minimum before and after the application of a maximum strain ε.

Furthermore, the sensitivity of both the transmittance (S_T) and the wavelength (S_λ) signals of the LSPR band minimum were calculated using the following equations [43]:

$$S_T = \frac{\Delta T}{\varepsilon}, \quad (2)$$

$$S_\lambda = \frac{\Delta \lambda}{\varepsilon}, \quad (3)$$

where ΔT and $\Delta\lambda$ are the variations of the transmittance and the wavelength coordinate of the LSPR band minimum, respectively, after an applied strain ε.

The estimated sensitivities and gauge factors are represented in Figure 11, as a function of the maximum strain applied. It is obvious that, for both coordinates, the maximum sensitivity was reached for strains between of 6.7% (ε = 0.067) and 10.1% (ε = 0.101). At a strain of 6.7% the sensitivity for the wavelength coordinate reached 420 nm/ε (Figure 11a), while for the transmittance a sensitivity 110 pp/ε was achieved (Figure 11b). These results are close to those found in the literature for gold nanoparticles' arrays [43], but, for this work, a change of the refractive index surrounding the nanoparticles seems to be the reversible event that tunes the LSPR band of the sensor for different applied strains.

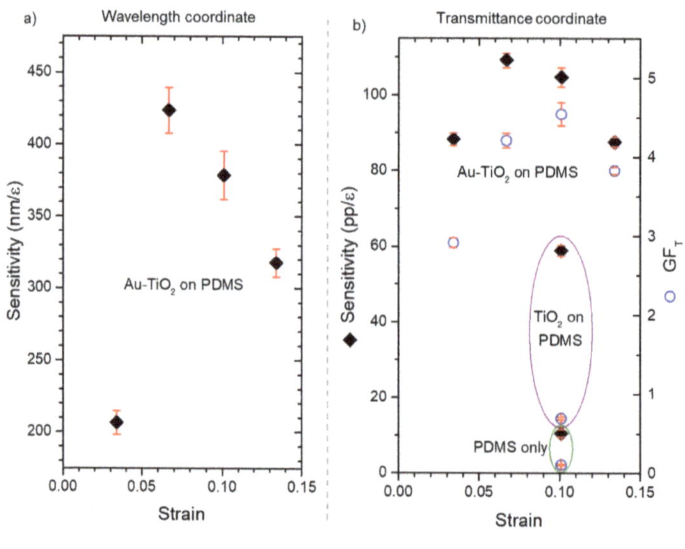

Figure 11. Sensitivity of (**a**) the wavelength and (**b**) the transmittance coordinates of the LSPR band minimum of the nanoplasmonic sensors, plotted against each maximum strain applied (3.4%, 6.7%, 10.1% and 13.4%). (**b**) Also shows the transmittance sensitivity (at 650 nm) of the PDMS substrate and a TiO_2 thin film deposited on PMDS. Furthermore, the transmittance gauge factor (GF_T) considering the transmittance coordinates of the nanoplasmonic sensor, the TiO_2 on PDMS and the PDMS substrate are presented in (**b**).

When analyzing the transmittance gauge factor (Figure 11b) it is clear that the transmittance signal was enhanced in the plasmonic sensor, until a maximum strain of 10.1% (ε = 0.101) was attained, leading to a gauge factor of 4.5. The gauge factor of the PDMS substrate and the TiO_2 thin film on PDMS, were considerably lower, 0.1 and 0.7 respectively. Therefore, when a strain of 10.1% was applied, the plasmonic thin film's gauge factor was still 45 and 6.5 times higher than the gauge factor of PDMS and TiO_2, respectively.

4. Conclusions

A novel flexible plasmonic strain sensor was prepared by functionalizing a PDMS substrate with a TiO_2 thin film containing Au nanoparticles. Preliminary tensile tests revealed that the LSPR band of these sensors could be modified by applying different strains. After correlating these findings with the simulations results in different strain scenarios, it became plausible that, in these sensors, the LSPR band is adjusted mainly due to changes in the refractive index surrounding the Au nanoparticles, probably caused by some induced porosity. Several strain protocols were applied to the sensors with consistent repeatability, and the maximum sensitivity was achieved for strains between 6.7% and 10.1%.

Supplementary Materials: The following supporting information can be downloaded at: https://www.mdpi.com/article/10.3390/s22041375/s1, Figure S1: nanoDDSCAT+ software interface tools, Figure S2: Optical extinction efficiency coefficient estimation of several shapes with nanoDDSCAT+, Figure S3: Nanoparticles being compressed change their aspect ratios (AR = 1, AR = 1.4 and AR = 1.8) but keep the same effective volume., Figure S4: Array of 16 nm diameter nanoparticles being biaxially elongated, Figure S5: Array of 16 nm diameter nanoparticles being uniaxially elongated, Figure S6: Array of 16 nm diameter nanoparticles being uniaxially elongated with transverse compression.

Author Contributions: Conceptualization, F.V. and J.B.; methodology, M.S.R. and J.B.; software, M.S.R.; validation, F.V. and J.B.; formal analysis, M.S.R.; investigation, M.S.R.; resources, F.V. and J.B.; data curation, M.S.R. and J.B.; writing—original draft preparation, M.S.R. and J.B.; writing—review and editing, J.B. and F.V.; visualization, M.S.R. and J.B.; supervision, F.V. and J.B.; project administration, F.V.; funding acquisition, J.B. and F.V. All authors have read and agreed to the published version of the manuscript.

Funding: This research was funded by the Portuguese Foundation for Science and Technology (FCT) in the framework of the Strategic Funding UIDB/04650/2020 and by the projects NANO4BIO: POCI-01-0145-FEDER-032299, with FCT reference PTDC/FIS-MAC/32299/2017, and CO2Plasmon with FCT reference EXPL/CTM-REF/0750/2021.

Institutional Review Board Statement: Not applicable.

Informed Consent Statement: Not applicable.

Data Availability Statement: Raw data are available upon request to the corresponding author.

Conflicts of Interest: The authors declare no conflict of interest.

References

1. Transparency Market Research. *Ultra-Thin Highly Sensitive Strain Sensors Market—Global Industry Analysis, Size, Share, Growth, Trends, and Forecast, 2021–2031*; Transparency Market Research: New York, NY, USA, 2021.
2. Cheng, H.-W.; Yan, S.; Shang, G.; Wang, S.; Zhong, C.-J. Strain sensors fabricated by surface assembly of nanoparticles. *Biosens. Bioelectron.* **2021**, *186*, 113268. [CrossRef] [PubMed]
3. Shintake, J.; Piskarev, Y.; Jeong, S.H.; Floreano, D. Ultrastretchable Strain Sensors Using Carbon Black-Filled Elastomer Composites and Comparison of Capacitive Versus Resistive Sensors. *Adv. Mater. Technol.* **2018**, *3*, 1700284. [CrossRef]
4. Aslanidis, E.; Skotadis, E.; Moutoulas, E.; Tsoukalas, D. Thin Film Protected Flexible Nanoparticle Strain Sensors: Experiments and Modeling. *Sensors* **2020**, *20*, 2584. [CrossRef] [PubMed]
5. Tolvanen, J.; Hannu, J.; Jantunen, H. Stretchable and Washable Strain Sensor Based on Cracking Structure for Human Motion Monitoring. *Sci. Rep.* **2018**, *8*, 13241. [CrossRef]
6. Cunha, L.T.; Pedrosa, P.; Tavares, C.J.; Alves, E.; Vaz, F.; Fonseca, C. The role of composition, morphology and crystalline structure in the electrochemical behaviour of TiNx thin films for dry electrode sensor materials. *Electrochim. Acta* **2009**, *55*, 59–67. [CrossRef]

7. Rodrigues, M.S.; Borges, J.; Lopes, C.; Pereira, R.M.S.; Vasilevskiy, M.I.; Vaz, F. Gas Sensors Based on Localized Surface Plasmon Resonances: Synthesis of Oxide Films with Embedded Metal Nanoparticles, Theory and Simulation, and Sensitivity Enhancement Strategies. *Appl. Sci.* **2021**, *11*, 5388. [CrossRef]
8. Delfino, I.; Diano, N.; Lepore, M. Advanced Optical Sensing of Phenolic Compounds for Environmental Applications. *Sensors* **2021**, *21*, 7563. [CrossRef]
9. Cappi, G.; Spiga, F.M.; Moncada, Y.; Ferretti, A.; Beyeler, M.; Bianchessi, M.; Decosterd, L.; Buclin, T.; Guiducci, C. Label-Free Detection of Tobramycin in Serum by Transmission-Localized Surface Plasmon Resonance. *Anal. Chem.* **2015**, *87*, 5278–5285. [CrossRef]
10. Maier, S.A. *Plasmonics: Fundamentals and Applications*; Springer: New York, NY, USA, 2007; ISBN 978-0-387-37825-1.
11. Mayer, K.M.; Hafner, J.H. Localized surface plasmon resonance sensors. *Chem. Rev.* **2011**, *111*, 3828–3857. [CrossRef]
12. Stockman, M.I. Nanoplasmonics: Past, present, and glimpse into future. *Opt. Express* **2011**, *19*, 22029–22106. [CrossRef]
13. Stockman, M.I.; Kneipp, K.; Bozhevolnyi, S.I.; Saha, S.; Dutta, A.; Ndukaife, J.; Kinsey, N.; Reddy, H.; Guler, U.; Shalaev, V.M.; et al. Roadmap on plasmonics. *J. Opt.* **2018**, *20*, 043001. [CrossRef]
14. Borges, J.; Pereira, R.M.S.; Rodrigues, M.S.; Kubart, T.; Kumar, S.; Leifer, K.; Cavaleiro, A.; Polcar, T.; Vasilevskiy, M.I.; Vaz, F. Broadband Optical Absorption Caused by the Plasmonic Response of Coalesced Au Nanoparticles Embedded in a TiO_2 Matrix. *J. Phys. Chem. C* **2016**, *120*, 16931–16945. [CrossRef]
15. Song, M.; Wang, D.; Peana, S.; Choudhury, S.; Nyga, P.; Kudyshev, Z.A.; Yu, H.; Boltasseva, A.; Shalaev, V.M.; Kildishev, A. V Colors with plasmonic nanostructures: A full-spectrum review. *Appl. Phys. Rev.* **2019**, *6*, 41308. [CrossRef]
16. Deng, Y.; Cao, G.; Wu, Y.; Zhou, X.; Liao, W. Theoretical Description of Dynamic Transmission Characteristics in MDM Waveguide Aperture-Side-Coupled with Ring Cavity. *Plasmonics* **2015**, *10*, 1537–1543. [CrossRef]
17. Deng, Y.; Cao, G.; Yang, H.; Zhou, X.; Wu, Y. Dynamic Control of Double Plasmon-Induced Transparencies in Aperture-Coupled Waveguide-Cavity System. *Plasmonics* **2018**, *13*, 345–352. [CrossRef]
18. Zhou, F.; Qin, F.; Yi, Z.; Yao, W.; Liu, Z.; Wu, X.; Wu, P. Ultra-wideband and wide-angle perfect solar energy absorber based on Ti nanorings surface plasmon resonance. *Phys. Chem. Chem. Phys.* **2021**, *23*, 17041–17048. [CrossRef]
19. Aćimović, S.S.; Šípová, H.; Emilsson, G.; Dahlin, A.B.; Antosiewicz, T.J.; Käll, M. Superior LSPR substrates based on electromagnetic decoupling for on-a-chip high-throughput label-free biosensing. *Light Sci. Appl.* **2017**, *6*, e17042. [CrossRef]
20. Chang, C.-Y.; Lin, H.-T.; Lai, M.-S.; Shieh, T.-Y.; Peng, C.-C.; Shih, M.-H.; Tung, Y.-C. Flexible Localized Surface Plasmon Resonance Sensor with Metal–Insulator–Metal Nanodisks on PDMS Substrate. *Sci. Rep.* **2018**, *8*, 11812. [CrossRef]
21. Barbosa, A.I.; Borges, J.; Meira, D.I.; Costa, D.; Rodrigues, M.S.; Rebelo, R.; Correlo, V.M.; Vaz, F.; Reis, R.L. Development of label-free plasmonic Au-TiO_2 thin film immunosensor devices. *Mater. Sci. Eng. C* **2019**, *100*, 424–432. [CrossRef]
22. Wu, X.; Zheng, Y.; Luo, Y.; Zhang, J.; Yi, Z.; Wu, X.; Cheng, S.; Yang, W.; Yu, Y.; Wu, P. A four-band and polarization-independent BDS-based tunable absorber with high refractive index sensitivity. *Phys. Chem. Chem. Phys.* **2021**, *23*, 26864–26873. [CrossRef]
23. Rodrigues, M.S.; Borges, J.; Proença, M.; Pedrosa, P.; Martin, N.; Romanyuk, K.; Kholkin, A.L.; Vaz, F. Nanoplasmonic response of porous Au-TiO_2 thin films prepared by oblique angle deposition. *Nanotechnology* **2019**, *30*, 225701. [CrossRef] [PubMed]
24. Proença, M.; Rodrigues, M.S.; Borges, J.; Vaz, F. Optimization of Au:CuO Nanocomposite Thin Films for Gas Sensing with High-Resolution Localized Surface Plasmon Resonance Spectroscopy. *Anal. Chem.* **2020**, *92*, 4349–4356. [CrossRef] [PubMed]
25. Proença, M.; Rodrigues, M.S.; Vaz, F.; Borges, J. Carbon Monoxide (CO) Sensor Based on Au Nanoparticles Embedded in a CuO Matrix by HR-LSPR Spectroscopy at Room Temperature. *IEEE Sens. Lett.* **2021**, *5*, 1–3. [CrossRef]
26. Rodrigues, M.S.; Borges, J.; Vaz, F. Enhancing the Sensitivity of Nanoplasmonic Thin Films for Ethanol Vapor Detection. *Materials* **2020**, *13*, 870. [CrossRef]
27. Thijssen, R.; Verhagen, E.; Kippenberg, T.J.; Polman, A. Plasmon Nanomechanical Coupling for Nanoscale Transduction. *Nano Lett.* **2013**, *13*, 3293–3297. [CrossRef]
28. Algorri, J.F.; García-Cámara, B.; García-García, A.; Urruchi, V.; Sánchez-Pena, J.M. Fiber Optic Temperature Sensor Based on Amplitude Modulation of Metallic and Semiconductor Nanoparticles in a Liquid Crystal Mixture. *J. Light. Technol.* **2015**, *33*, 2451–2455. [CrossRef]
29. Lio, G.E.; Palermo, G.; Caputo, R.; De Luca, A. Opto-mechanical control of flexible plasmonic materials. *J. Appl. Phys.* **2019**, *125*, 82533. [CrossRef]
30. Topcu, G.; Guner, T.; Inci, E.; Demir, M.M. Colorimetric and plasmonic pressure sensors based on polyacrylamide/Au nanoparticles. *Sens. Actuators A Phys.* **2019**, *295*, 503–511. [CrossRef]
31. Lee, S.; Song, H.; Ahn, H.; Kim, S.; Choi, J.; Kim, K. Fiber-Optic Localized Surface Plasmon Resonance Sensors Based on Nanomaterials. *Sensors* **2021**, *21*, 819. [CrossRef]
32. Mansouri, M.; Mir, A.; Farmani, A.; Izadi, M. Numerical Modeling of an Integrable and Tunable Plasmonic Pressure Sensor with Nanostructure Grating. *Plasmonics* **2021**, *16*, 27–36. [CrossRef]
33. Mahi, N.; Lévêque, G.; Saison, O.; Marae-Djouda, J.; Caputo, R.; Gontier, A.; Maurer, T.; Adam, P.-M.; Bouhafs, B.; Akjouj, A. In Depth Investigation of Lattice Plasmon Modes in Substrate-Supported Gratings of Metal Monomers and Dimers. *J. Phys. Chem. C* **2017**, *121*, 2388–2401. [CrossRef]
34. Han, X.; Liu, Y.; Yin, Y. Colorimetric Stress Memory Sensor Based on Disassembly of Gold Nanoparticle Chains. *Nano Lett.* **2014**, *14*, 2466–2470. [CrossRef] [PubMed]

35. Correa-Duarte, M.A.; Salgueiriño-Maceira, V.; Rinaldi, A.; Sieradzki, K.; Giersig, M.; Liz-Marzán, L.M. Optical strain detectors based on gold/elastomer nanoparticulated films. *Gold Bull.* **2007**, *40*, 6–14. [CrossRef]
36. Maurer, T.; Marae-Djouda, J.; Cataldi, U.; Gontier, A.; Montay, G.; Madi, Y.; Panicaud, B.; Macias, D.; Adam, P.M.; Lévêque, G.; et al. The beginnings of plasmomechanics: Towards plasmonic strain sensors. *Front. Mater. Sci.* **2015**, *9*, 170–177. [CrossRef]
37. Cataldi, U.; Caputo, R.; Kurylyak, Y.; Klein, G.; Chekini, M.; Umeton, C.; Bürgi, T. Growing gold nanoparticles on a flexible substrate to enable simple mechanical control of their plasmonic coupling. *J. Mater. Chem. C* **2014**, *2*, 7927–7933. [CrossRef]
38. Palermo, G.; Cataldi, U.; Condello, A.; Caputo, R.; Bürgi, T.; Umeton, C.; De Luca, A. Flexible thermo-plasmonics: An opto-mechanical control of the heat generated at the nanoscale. *Nanoscale* **2018**, *10*, 16556–16561. [CrossRef]
39. Yoo, D.; Johnson, T.W.; Cherukulappurath, S.; Norris, D.J.; Oh, S.-H. Template-Stripped Tunable Plasmonic Devices on Stretchable and Rollable Substrates. *ACS Nano* **2015**, *9*, 10647–10654. [CrossRef]
40. Laible, F.; Gollmer, D.A.; Dickreuter, S.; Kern, D.P.; Fleischer, M. Continuous reversible tuning of the gap size and plasmonic coupling of bow tie nanoantennas on flexible substrates. *Nanoscale* **2018**, *10*, 14915–14922. [CrossRef]
41. Jia, P.; Kong, D.; Ebendorff-Heidepriem, H. Flexible Plasmonic Tapes with Nanohole and Nanoparticle Arrays for Refractometric and Strain Sensing. *ACS Appl. Nano Mater.* **2020**, *3*, 8242–8246. [CrossRef]
42. Marae-Djouda, J.; Gontier, A.; Caputo, R.; Lévêque, G.; Bercu, B.; Madi, Y.; Montay, G.; Adam, P.-M.; Molinari, M.; Stagon, S.; et al. Dense Brushes of Tilted Metallic Nanorods Grown onto Stretchable Substrates for Optical Strain Sensing. *ACS Appl. Nano Mater.* **2018**, *1*, 2347–2355. [CrossRef]
43. Chang, K.-H.; Cheng, J.-S.; Lu, T.-W.; Lee, P.-T. Engineering surface lattice resonance of elliptical gold nanodisk array for enhanced strain sensing. *Opt. Express* **2018**, *26*, 33215–33225. [CrossRef] [PubMed]
44. Feng, D.; Zhang, H.; Xu, S.; Tian, L.; Song, N. Fabrication of Plasmonic Nanoparticles on a Wave Shape PDMS Substrate. *Plasmonics* **2017**, *12*, 1627–1631. [CrossRef]
45. Minati, L.; Chiappini, A.; Armellini, C.; Carpentiero, A.; Maniglio, D.; Vaccari, A.; Zur, L.; Lukowiak, A.; Ferrari, M.; Speranza, G. Gold nanoparticles 1D array as mechanochromic strain sensor. *Mater. Chem. Phys.* **2017**, *192*, 94–99. [CrossRef]
46. Fu, R.; Gómez, D.E.; Shi, Q.; Yap, L.W.; Lyu, Q.; Wang, K.; Yong, Z.; Cheng, W. Orientation-Dependent Soft Plasmonics of Gold Nanobipyramid Plasmene Nanosheets. *Nano Lett.* **2021**, *21*, 389–396. [CrossRef] [PubMed]
47. Mizuno, A.; Ono, A. Static and dynamic tuning of surface plasmon resonance by controlling interparticle distance in arrays of Au nanoparticles. *Appl. Surf. Sci.* **2019**, *480*, 846–850. [CrossRef]
48. Kwon, Y.H.; Fernandes, J.; Kim, J.-J.; Chen, J.; Jiang, H. Micro-Actuated Tunable Hierarchical Silver Nanostructures to Measure Tensile Force for Biomedical Wearable Sensing Applications. *Micromachines* **2021**, *12*, 476. [CrossRef]
49. Clough, J.M.; Weder, C.; Schrettl, S. Mechanochromism in Structurally Colored Polymeric Materials. *Macromol. Rapid Commun.* **2021**, *42*, 2000528. [CrossRef]
50. Roxworthy, B.J.; Aksyuk, V.A. Electrically tunable plasmomechanical oscillators for localized modulation, transduction, and amplification. *Optica* **2018**, *5*, 71–79. [CrossRef]
51. Koya, A.N.; Cunha, J.; Guerrero-Becerra, K.A.; Garoli, D.; Wang, T.; Juodkazis, S.; Proietti Zaccaria, R. Plasmomechanical Systems: Principles and Applications. *Adv. Funct. Mater.* **2021**, *31*, 2103706. [CrossRef]
52. Torrell, M.; Cunha, L.; Cavaleiro, A.; Alves, E.; Barradas, N.P.; Vaz, F. Functional and optical properties of Au:TiO_2 nanocomposite films: The influence of thermal annealing. *Appl. Surf. Sci.* **2010**, *256*, 6536–6542. [CrossRef]
53. Rodrigues, M.S.; Borges, J.; Gabor, C.; Munteanu, D.; Apreutesei, M.; Steyer, P.; Lopes, C.; Pedrosa, P.; Alves, E.; Barradas, N.P.; et al. Functional behaviour of TiO_2 films doped with noble metals. *Surf. Eng.* **2015**, *32*, 554–561. [CrossRef]
54. Rodrigues, M.S.; Meira, D.I.; Lopes, C.; Borges, J.; Vaz, F. Preparation of Plasmonic Au-TiO_2 Thin Films on a Transparent Polymer Substrate. *Coatings* **2020**, *10*, 227. [CrossRef]
55. Barradas, N.P.; Jeynes, C.; Jackson, S.M. RBS/Simulated annealing analysis of buried SiCOx layers formed by implantation of O into cubic silicon carbide. *Nucl. Instrum. Methods Phys. Res. Sect. B Beam Interact. Mater. At.* **1998**, *136–138*, 1168–1171. [CrossRef]
56. Barradas, N.P.; Jeynes, C. Advanced physics and algorithms in the IBA DataFurnace. *Nucl. Instrum. Methods Phys. Res. Sect. B Beam Interact. Mater. At.* **2008**, *266*, 1875–1879. [CrossRef]
57. Barradas, N.P.; Pascual-Izarra, C. Double scattering in RBS analysis of PtSi thin films on Si. *Nucl. Instrum. Methods Phys. Res. Sect. B Beam Interact. Mater. At.* **2005**, *228*, 378–382. [CrossRef]
58. Barradas, N.P.; Reis, M.A. Accurate calculation of pileup effects in PIXE spectra from first principles. *X-ray Spectrom.* **2006**, *35*, 232–237. [CrossRef]
59. Pereira-Silva, P.; Borges, J.; Rodrigues, M.S.; Oliveira, J.C.; Alves, E.; Barradas, N.P.; Dias, J.P.; Cavaleiro, A.; Vaz, F. Nanocomposite Au-ZnO thin films: Influence of gold concentration and thermal annealing on the microstructure and plasmonic response. *Surf. Coat. Technol.* **2020**, *385*, 125379. [CrossRef]
60. Rodrigues, M.S.; Pereira, R.M.S.; Vasilevskiy, M.I.; Borges, J.; Vaz, F. NANOPTICS: In-depth analysis of NANomaterials for OPTICal localized surface plasmon resonance Sensing. *SoftwareX* **2020**, *12*, 100522. [CrossRef]
61. Draine, B.T.; Flatau, P.J. Discrete-Dipole Approximation for Scattering Calculations. *J. Opt. Soc. Am. A* **1994**, *11*, 1491–1499. [CrossRef]
62. Draine, B.T.; Flatau, P.J. Discrete-dipole approximation for periodic targets: Theory and tests. *J. Opt. Soc. Am. A* **2008**, *25*, 2693–2703. [CrossRef]

63. Sobh, A.N.; White, S.; Smith, J.; Sobh, N.; Jain, P.K. nanoDDSCAT+ 2019. Available online: https://nanohub.org/resources/ddaplus (accessed on 25 October 2021).
64. Johnson, P.B.; Christy, R.W. Optical Constants of the Noble Metals. *Phys. Rev. B* **1972**, *6*, 4370–4379. [CrossRef]
65. Borges, J.; Rodrigues, M.S.; Lopes, C.; Costa, D.; Ferreira, A.; Pereira, R.M.S.; Costa, M.F.; Vasilevskiy, M.I.; Vaz, F. Ag fractals formed on top of a porous TiO_2 thin film. *Phys. Status Solidi Rapid Res. Lett.* **2016**, *10*, 530–534. [CrossRef]
66. Domingues, R.P.; Rodrigues, M.S.; Lopes, C.; Pedrosa, P.; Alves, E.; Barradas, N.P.; Borges, J.; Vaz, F. Thin films composed of metal nanoparticles (Au, Ag, Cu) dispersed in AlN: The influence of composition and thermal annealing on the structure and plasmonic response. *Thin Solid Films* **2019**, *676*, 12–25. [CrossRef]
67. Proença, M.; Borges, J.; Rodrigues, M.S.; Meira, D.I.; Sampaio, P.; Dias, J.P.; Pedrosa, P.; Martin, N.; Bundaleski, N.; Teodoro, O.M.N.D.; et al. Nanocomposite thin films based on Au-Ag nanoparticles embedded in a CuO matrix for localized surface plasmon resonance sensing. *Appl. Surf. Sci.* **2019**, *484*, 152–168. [CrossRef]
68. Borges, J.; Rodrigues, M.S.; Kubart, T.; Kumar, S.; Leifer, K.; Evaristo, M.; Cavaleiro, A.; Apreutesei, M.; Pereira, R.M.S.; Vasilevskiy, M.I.; et al. Thin films composed of gold nanoparticles dispersed in a dielectric matrix: The influence of the host matrix on the optical and mechanical responses. *Thin Solid Films* **2015**, *596*, 8–17. [CrossRef]
69. Mordehai, D.; Lee, S.-W.; Backes, B.; Srolovitz, D.J.; Nix, W.D.; Rabkin, E. Size effect in compression of single-crystal gold microparticles. *Acta Mater.* **2011**, *59*, 5202–5215. [CrossRef]
70. Koneti, S.; Borges, J.; Roiban, L.; Rodrigues, M.S.; Martin, N.; Epicier, T.; Vaz, F.; Steyer, P. Electron Tomography of Plasmonic Au Nanoparticles Dispersed in a TiO_2 Dielectric Matrix. *ACS Appl. Mater. Interfaces* **2018**, *10*, 42882–42890. [CrossRef]
71. Johnston, I.D.; McCluskey, D.K.; Tan, C.K.L.; Tracey, M.C. Mechanical characterization of bulk Sylgard 184 for microfluidics and microengineering. *J. Micromech. Microeng.* **2014**, *24*, 035017. [CrossRef]
72. Caputo, R.; Cataldi, U.; Bürgi, T.; Umeton, C. Plasmomechanics: A Colour-Changing Device Based on the Plasmonic Coupling of Gold Nanoparticles. *Mol. Cryst. Liq. Cryst.* **2015**, *614*, 20–29. [CrossRef]
73. Ferreira, A.; Correia, V.; Mendes, E.; Lopes, C.; Vaz, J.F.V.; Lanceros-Mendez, S. Piezoresistive Polymer-Based Materials for Real-Time Assessment of the Stump/Socket Interface Pressure in Lower Limb Amputees. *IEEE Sens. J.* **2017**, *17*, 2182–2190. [CrossRef]
74. Marin, B.C.; Liu, J.; Aklile, E.; Urbina, A.D.; Chiang, A.S.-C.; Lawrence, N.; Chen, S.; Lipomi, D.J. SERS-enhanced piezoplasmonic graphene composite for biological and structural strain mapping. *Nanoscale* **2017**, *9*, 1292–1298. [CrossRef] [PubMed]

sensors

MDPI

Article

Discrimination of Different Breast Cell Lines on Glass Substrate by Means of Fourier Transform Infrared Spectroscopy

Maria Lasalvia, Vito Capozzi and Giuseppe Perna *

Dipartimento di Medicina Clinica e Sperimentale, Università di Foggia, 71122 Foggia, Italy; maria.lasalvia@unifg.it (M.L.); vito.capozzi@unifg.it (V.C.)
* Correspondence: giuseppe.perna@unifg.it

Abstract: Fourier transform infrared (FTIR) micro-spectroscopy has been attracting the interest of many cytologists and histopathologists for several years. This is related to the possibility of FTIR translation in the clinical diagnostic field. In fact, FTIR spectra are able to detect changes in biochemical cellular components occurring when the cells pass to a pathological state. Recently, this interest has increased because it has been shown that FTIR spectra carried out just in the high wavenumber spectral range (2500–4000 cm^{-1}), where information mainly relating to lipids and proteins can be obtained, are able to discriminate cell lines related to different tissues. This possibility allows to perform IR absorption measurements of cellular samples deposited onto microscopy glass slides (widely used in the medical environment) which are transparent to IR radiation only for wavenumber values larger than 2000 cm^{-1}. For these reasons, we show that FTIR spectra in the 2800–3000 cm^{-1} spectral range can discriminate three different cell lines from breast tissue: a non-malignant cell line (MCF10A), a non-metastatic adenocarcinoma cell line (MCF7) and a metastatic adenocarcinoma cell line (MDA). All the cells were grown onto glass slides. The spectra were discriminated by means of a principal component analysis, according to the PC1 component, whose values have the opposite sign in the pairwise score plots. This result supports the wide studies that are being carried out to promote the translation of the FTIR technique in medical practice, as a complementary diagnostic tool.

Keywords: FTIR micro-spectroscopy; breast cell lines; principal component analysis

Citation: Lasalvia, M.; Capozzi, V.; Perna, G. Discrimination of Different Breast Cell Lines on Glass Substrate by Means of Fourier Transform Infrared Spectroscopy. *Sensors* **2021**, *21*, 6992. https://doi.org/10.3390/s21216992

Academic Editor: Longhua Tang

Received: 13 September 2021
Accepted: 19 October 2021
Published: 21 October 2021

1. Introduction

In recent years, Fourier transform infrared (FTIR) micro-spectroscopy has been held in increasing consideration as a versatile tool for biomedical and bioanalytical applications. The main reason for the interest in such a technique is that it provides biochemical information without employing reagents and has easy sample preparation. As a consequence, many works have proposed the FTIR technique as a complementary spectroscopic method to support diagnostic investigation of cytological samples [1–5]. In addition, FTIR has also been considered in clinical practice for biological tissues [6–8] and biofluids [9–11] analysis, as well as to monitor the response of cells and tissues to radiotherapy [12] and chemotherapy [13] treatments.

In addition to FTIR, other examples of spectroscopy techniques able to provide biochemical information in a label free away are fluorescence [14] and Raman spectroscopy [15]. In particular, the Raman technique has been successfully proposed for cancer diagnostic applications of cytological [16], tissue [17] and biofluids [18,19] samples. FTIR and Raman techniques are considered complimentary methods to provide biochemical information about biological samples. The former technique is based on the selective absorption of incident IR radiation, whereas the latter one analyzes the inelastic scattering of incident light. The Raman spectroscopy has a few limitations, such as a longer time for measuring a spectrum with respect to FTIR, the possible damage to the sample (because the excitation beam consists of a laser focused on the sample) and the fluorescence from the sample which can strongly interfere with the Raman signal. Although FTIR spectroscopy has several disadvantages compared with the Raman technique, such as the worst spatial resolution and the impossibility of measuring samples in aqueous environment, it is preferable for diagnostic purposes, especially for the capability of providing reliable information in a relatively short time.

One of the main obstacles to overcome before the adoption of the FTIR technique in cytological diagnostics is related to the substrate on which the cells must be deposited for the spectral measurements. In fact, the cell substrate most widely used by pathologists is a glass slide, with a thickness of about 1 mm. The advantage of these slides is that they are cheap, robust, and largely available. However, glass slides absorb IR radiation in the spectral range corresponding to the fingerprint region (1000–1800 cm^{-1}), so making impossible the FTIR spectrum acquisition in this important spectra range, characterized by peaks related to the most important cellular components (proteins, lipids and nucleic acids). Nonetheless, glass substrates are transparent to mid-IR radiation in the high wavenumber spectral range (2000–4000 cm^{-1}), where absorption peaks mainly related to lipid and protein components are located.

We recently showed that FTIR measurements in the high wavenumber spectral range are able to discriminate two cellular samples from different human cell lines, regardless of the growth substrate [20]. In particular, the FTIR spectra of the two cell lines were discriminated through the difference of the score values by performing a principal component analysis (PCA), which is a multivariate statistical technique widely used to discriminate spectral data [21]. Briefly, PCA transforms the N original variables, consisting in the absorption values for the N wavenumber values for the group of all the M measured samples, into a new set of N variables, called principal components (PCs), each one is a linear combination of the N original variables. Each original spectrum takes specific values in the set of PCs: such values are called scores. The criterion according to which the first PC is chosen is that it contains most of the variance of the scores, and each subsequent PC contains less variance. A score plot, reporting the score values of two different PCs for all the M samples, allows to visualize differences and similarities among the M samples, based on the original spectral characteristics.

Other authors successfully discriminated different types of cell lines, deposited on glass thick slides and thin coverslips, according to their FTIR spectra in the high wavenumber range and partially in the fingerprint range [22,23]. However, the cell lines investigated by these authors, as well as those measured by us in our previous paper [20], were very different among them. In particular, three types of cells were investigated by Rutter et al., i.e., cells from lung cancer (CALU-1 line), leukemia (K562 line), and peripheral blood (PBMC) [22,23]. Instead, we compared FTIR spectra of human neuroblastoma cells (SH-SY5Y line) with those of human normal keratynocytes cells (HUKE line).

In this paper, we investigate by FTIR micro-spectroscopy in the high wavenumber range three different samples of cells from breast tissue: non-malignant (MCF10A), malignant non-metastatic (MCF7), and metastatic (MDA). All the cellular samples were grown on conventional glass slides. We found that PCA technique is able to discriminate the spectra of the three samples according to the values of PC1 score. Although we found that the different relative amount of lipid component plays a fundamental role in the separation of score values of PC1, we remark that the principal component analysis was performed by considering the whole spectra of the samples (i.e., it was not limited to specific peaks). Therefore, full spectral data can be considered as spectral biomarkers of the cell type, without the need to search for any specific biochemical component. These results, obtained with cheap conventional glass slides, are promising for supporting the use of the FTIR technique as a complementary diagnostic tool in medical practice.

2. Materials and Methods

2.1. Cell Culture and Preparation

Three different breast cell lines, including non-malignant (MCF10A), malignant (MCF7), and metastatic (MDA-MB-231) cells were grown. MCF10A were cultured in DMEM/Ham's F-12 (Sigma-Aldrich, Milano, Italy) supplemented with 100 ng/mL cholera toxin, 20 ng/mL EGF epidermal growth factor, 0.01 mg/mL insulin, 500 ng/mL hydrocortisone, and 5% horse serum (Life Technologies, Monza, Italy). MCF7 and MDA-MB-231 cells were cultured in the DMEM medium (DMEM, Life Technologies, Monza, Italy) and supplemented with 2 mM L-glutamine, (Sigma-Aldrich, St. Louis, MO, USA), 10% heat-inactivated fetal bovine serum (FBS) (Thermo Scientific), 1% penicillin/streptomycin (Life Technologies, Monza, Italy) and 0.25 ug/mL amphotericin B.

Cells were grown on poly-lysine coated glass microscopy slides (Fisher Scientific, Rodano, Italy). The slides were located at the bottom of petri dishes incubated at 37 °C, and 5% CO_2. Before FTIR measurements, the cells were fixed by means of 3.7% PFA in PBS solution and preserved inside a desiccator.

2.2. FTIR Measurements and Spectral Analysis

A FTIR Microscope HYPERION 2000 (Bruker Optik GmbH, Ettlingen, Germany) was employed to perform FTIR spectra in the transmission mode. The IR radiation entering the microscope came from a Vertex 70 Bruker interferometer (Bruker Optik GmbH). The IR signal was measured by means of a mercury cadmium telluride (MCT) detector, cooled at liquid N_2 temperature. Each spectrum was estimated as an average of 64 scans in the 2500–4000 cm^{-1} range, with a resolution of 4 cm^{-1}. The IR radiation was focused with a 15× objective on a sampling area of about 80 × 80 µm size, including 3–4 cells of each type. The background signal was measured from a spatial region of the slide close to the sampling area, but without any cell. About 30 cells were measured for each type of cells. The data acquisition was performed by means of the Opus 6.5 software (Bruker Optik).

For each FTIR measurement, the spectral range between 2750 and 3700 cm^{-1} was analyzed, because it includes the main spectral peaks related to radiation absorption from the cellular lipid and protein components. The spectra were normalized using standard normal variate (SNV) which, for each absorption intensity value corresponding to each wavenumber value, subtracts the mean value and then divides by the standard deviation value. Such a normalization procedure reduces spectrum baseline shifts due to scattering effects resulting from the interaction between IR radiation and sample particles [24] and minimizes the contribution of the absorption from cells having a different thickness.

The R software was employed to achieve PCA by using the Chemospec package (version 3.4.1, R Core Team, Vienna, Austria, 2017) [25]. The t-test method was employed to evaluate the statistical differences between the groups of different cells, by means of the SigmaPlot software (version 12.5, Systat Software Inc., San Jose, CA, USA).

3. Results and Discussion

The comparison among the normalized spectra of the three cell types grown on glass slides is shown in Figure 1, where the mean and standard deviation signals are reported. The mean spectra are not very different from each other, and they are also similar to the FTIR spectra obtained by other authors for breast cell lines and tissues [26–28]. Each spectrum is characterized by a few absorption peaks, which can be attributed according to the published literature [29]. Specifically, the broadest band at about 3300 cm^{-1} is related to the amide A (N-H stretching mode of proteins amino acids and nucleic acids), although a contribution of O-H stretching mode from residual water inside cells cannot be excluded. In addition, the peaks at about 2958 and 2871 cm^{-1} are due to the asymmetric and symmetric stretching mode of the CH_3 groups of cellular proteins and lipids, respectively, whereas the peaks at 2924 and 2852 cm^{-1} can be attributed to the asymmetric and symmetric stretching of the CH_2 groups of lipids, respectively.

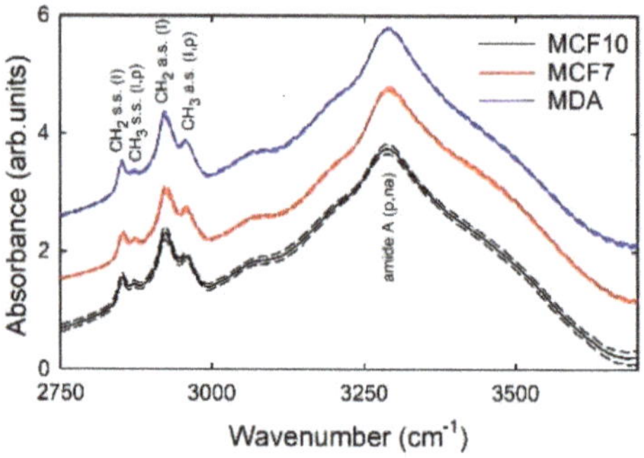

Figure 1. Mean FTIR spectra of MCF10A (continuous black line), MCF7 (continuous red line), and MDA (continuous blue line) cells after SNV normalization. Standard deviation spectra are also reported as dashed lines. The assignment of vibrational modes is also reported (c: carbohydrates; n.a.: nucleic acids; p: proteins; l: lipids). The spectra are vertically shifted for clarity of purpose.

Two main characteristics are evident in Figure 1. First, the standard deviation values suggest that the spectral intensities for the MCF10A cells spectra are more broadly distributed with respect to those of the other two cell lines, whose distributions of spectra are quite narrow. As the spectral variability is related to the biochemical content variability, we might deduce that non-malignant cells present larger differences of the relative content of cellular components with respect to the malignant and metastatic cancer cells. In addition, for each spectrum the FTIR signal between 2750 and 2850 cm^{-1} has a slightly larger intensity with respect to the signal at about 3700 cm^{-1}, although no specific absorption peaks are reported for both such spectral ranges: this means that a baseline signal is still present. Therefore, the SNV normalization fails to totally remove the scattering signal.

In order to decrease the effect of the unwanted baseline and improve the comparison among the three cellular samples, second derivative signals of the SNV normalized spectra were calculated and are reported in Figure 2 as far as the 2800–3000 cm^{-1} spectral range is concerned. The subsequent analysis was focused on this spectral range, for two reasons: first, because the most interesting signals related to protein and lipid components are located in such a spectral range. Secondly, because of the doubts about the attribution of large a band at about 3300 cm^{-1}, due to the uncertainty about the water contribution. Indeed, an important effect of the derivative process is that the signals of broad bands are suppressed relatively to those of sharp bands. So, since the scattering component resembles a very broad absorbance, using derivatives reduces the scattering effect. In particular, the second derivative was used because the most characteristic feature of a second-order derivative is a negative band with a minimum at the spectral position corresponding to the maximum on the zero-order band. An interesting feature visible in Figure 2 is that the intensity signals corresponding to the minima of the derivative spectra of the three cell lines show similar variability (as deduced by standard deviation values). By considering what was reported above for the comparison of the SNV spectra in Figure 1, it can be deduced that the largest variability of MCF10A cells is probably related to the distribution of the size of the scattering centers, which are the nuclei for cellular samples [30].

The PCA technique was used to assess whether FTIR is able to discriminate between the non-malignant MCF10A cells and the two types of mentioned cancer cells. Figure 3a shows the PCA score plots for the MCF10A and MCF7 cells: it is clearly visible that the two types of cells are separated according to the PC1 component. In fact, MCF10A scores (black dots) are characterized by negative values, whereas MCF7 scores (red dots) have mainly positive values. The average and standard deviation values of scores distributions are reported at the bottom of Figure 3a. The differences of PCA score values between the two types of cells were statistically significant, as confirmed by a *t*-test yielding a p-value < 0.001. The loading 1 plot, shown in Figure 3b (black line), is in good agreement with the difference plot (blue line) between the average spectra of the two types of cells (considering that spectral positions of positive and negative peaks are exchanged). In particular, the largest negative bands in the loading 1 plot are located at about 2855 and 2925 cm^{-1}, corresponding to the spectral positions of CH_2 stretching modes of lipid components in Figure 1. The loadings represent coefficients describing the influence of the variables (IR absorption at specific wavenumber values) on the score values for a given PC [31]; therefore, Figure 3b suggests that the spectral origin of the variations which differentiate the two types of cells is related to the different relative amount of lipid components. Such a result is in agreement with the results reported by C. Nieva et al. [32], by using the Raman micro-spectroscopy and immunocytochemistry methods. They found that the lipid content in MCF10A cells is larger than in MCF7 cells. Apart from this specific biochemical marker that differentiates the two types of cells, the results shown in Figure 3a suggests that the FTIR technique, associated with multivariate methods such as PCA, is able to discriminate the two cell types on the basis of a spectral marker, i.e., a combination of several peaks and bands that make up the whole spectrum in the measured wavenumber range.

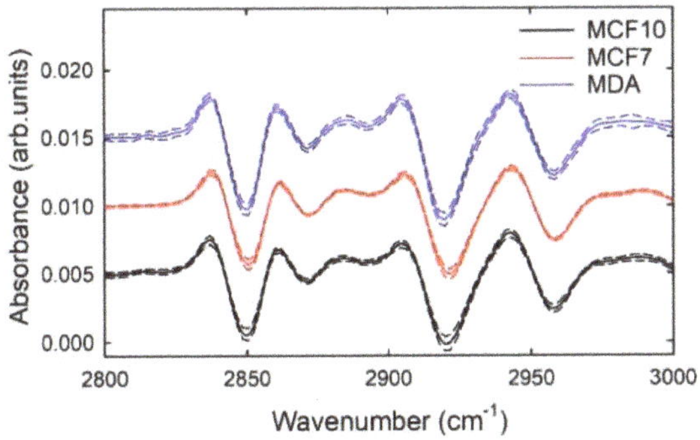

Figure 2. Mean second derivative signals of the SNV normalized FTIR spectra of MCF10A (continuous black line), MCF7 (continuous red line), and MDA (continuous blue line) cells. Standard deviation spectra are also reported as dashed lines. The spectra are vertically shifted for clarity of purpose.

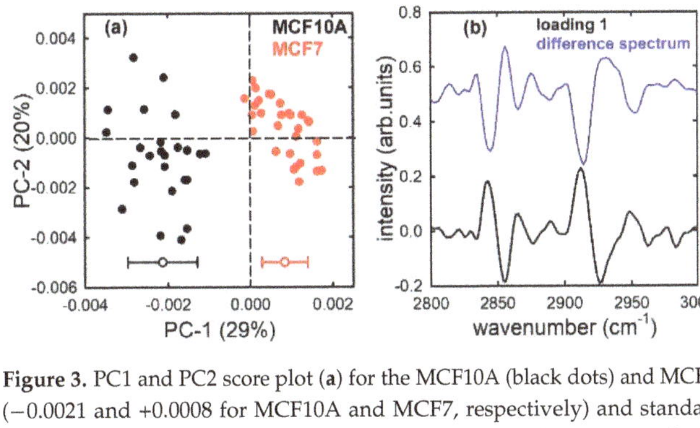

Figure 3. PC1 and PC2 score plot (**a**) for the MCF10A (black dots) and MCF7 (red dots) cells; average (−0.0021 and +0.0008 for MCF10A and MCF7, respectively) and standard deviation (0.0008 and 0.0005 for MCF10A and MCF7, respectively) of the PC 1 distribution values are shown at the bottom of (**a**). Loading 1 spectrum (black line), compared with the difference between average spectra (blue line), is reported in (**b**).

PC1 also discriminates MCF10A and MDA cells, as shown in the score plot Figure 4a, with the MCF10A cells corresponding mostly to positive score values and MDA cells mainly to negative values, although this separation is not as clear as in Figure 3a and some outliers are visible. Therefore, the FTIR spectrum in the high wavenumber range successfully can be considered as a spectral marker capable of separating the two types of cell without the need to investigate for any biochemical cellular component whose relative content is different for the two cell lines. However, the similarity of the loading plot to the difference of average spectra in Figure 4b suggests that the vibrational modes of lipids mainly contribute to the separation of the two cell lines. Small differences in the relative lipid content of MCF10A and MDA cells have been attributed to a large content of lipid droplets in MDA metastatic cells [32]. In fact, it is well-known that lipid metabolism is altered in metastatic cells, which use lipids as a source of energy to proliferate more and more [33,34].

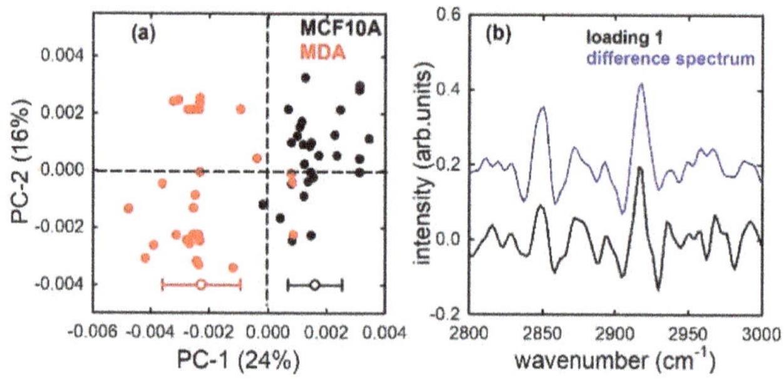

Figure 4. PC1 and PC2 score plot (**a**) for the MCF10A (black dots) and MDA (red dots) cells; average (+0.0016 and −0.0023 for MCF10A and MDA, respectively) and standard deviation (0.0009 and 0.0013 for MCF10A and MDA, respectively) of the PC 1 distribution values are shown at the bottom of (**a**). Loading 1 spectrum (black line), compared with the difference between average spectra (blue line), is reported in (**b**).

The PCA analysis performed for MCF7 and MDA cells is characterized by a large separation of the score values according to PC1, as shown in Figure 5a. In this case, from a biochemical point of view, the discrimination occurs as a consequence of the difference in the relative amount of lipid components, as suggested by the loading 1 and difference plots in Figure 5b. Previously reported findings also observed that MCF7 cells are characterized by a lower amount of fatty acids, both membrane phospholipids and cholesterol [32]. The lipidomic analysis performed by means of liquid chromatography mass spectrometry for these two different breast cancer cell lines showed that they are characterized by different mutual proportions of specific lipids [35]. However, we remark that Figure 5a shows that the FTIR spectra constitute a reliable marker for the spectral characterization of the cellular samples and for the differentiation of their different types.

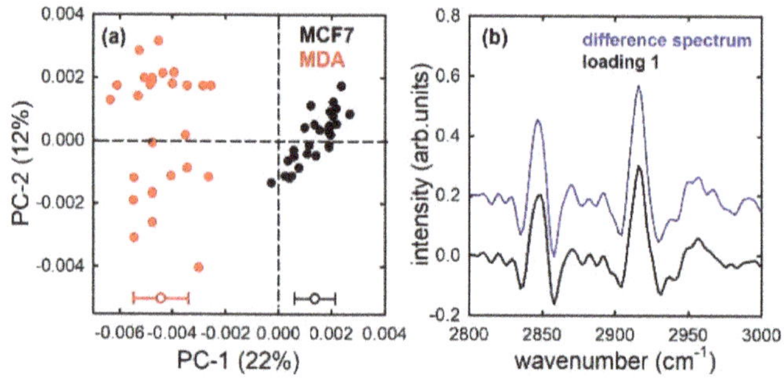

Figure 5. PC1 and PC2 score plot (**a**) for the MCF7 (black dots) and MDA (red dots) cells; average (+0.0014 and −0.0044 for MCF7 and MDA, respectively) and standard deviation (0.0008 and 0.0011 for MCF7 and MDA, respectively) of the PC 1 distribution values are shown at the bottom of (**a**). Loading 1 spectrum (black line), compared with the difference between average spectra (blue line), is reported in (**b**).

If we consider the difference of average PC1 score values as an estimate of the discrimination quality, Figures 3–5 suggest that non-malignant (MCF10A) cells can be better separated from metastatic (MDA) cells than from malignant non-metastatic (MCF7) cells. In particular, the difference of average PC1 score values is 0.0039 ± 0.0003 in the former case, and 0.0030 ± 0.0002 in the latter one. Moreover, the best separation occurs between MCF7 and MDA cells, whose difference of PC1 score values is 0.0058 ± 0.0002. This is probably related to the main role of lipid components in the discrimination among such cell lines: indeed, as described above, MCF7 cells are characterized by the smallest relative amount of lipid components [32]. Such different separation is also clearly visible in Figure 6, which shows the results of the PCA analysis of the three cell lines. Figure 6a confirms that PC1 score values are able to significantly discriminate ($p < 0.001$) the three types of cell line. In particular, PC1 scores are mainly positive for MCF7 cells and mostly negative for MDA and MCF10A cells, with the former cells having larger negative values than the latter ones. Figure 6b suggests that the separation between positive and negative values occurs because of the different lipid contents (the negative bands in Figure 6b are located at spectral positions corresponding quite to lipid peaks in the absorption spectra), as a consequence of the low lipid relative amount of MCF7 cells with respect to the other two types of cells.

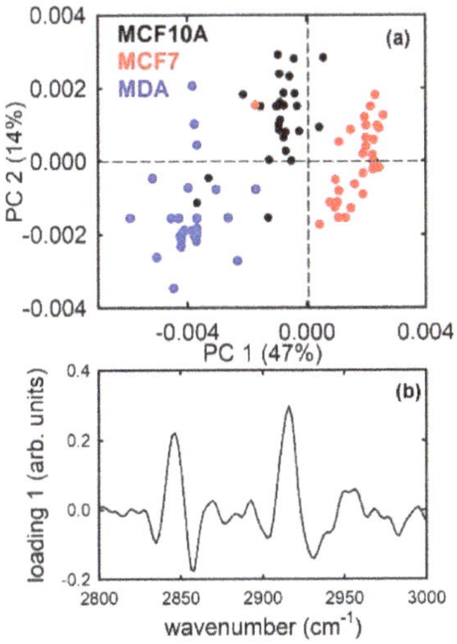

Figure 6. PC1 and PC2 score plot (**a**) for the MCF10A (black dots), MCF7 (red dots), and MDA (blue dots) cells and loading 1 spectrum (**b**).

Overall, the PCA analysis of FTIR spectra measured in the high wavenumber spectral range proved to be a reliable tool for discriminating non-malignant cells from malignant cells and from metastatic cells. The aspect to be remarked is that such separation can yield possible application in clinical diagnostics, even neglecting any discussion about the biochemical components.

4. Conclusions

The obtained result observe that FTIR spectra in the high wavenumber range and the principal components analysis technique are able to discriminate three cellular samples from different cell lines related to the same tissue. In particular, three breast cell lines were successfully separated: MCF10A, consisting of a non-malignant mammary epithelial cell line; MCF7, which is a human breast non-metastatic adenocarcinoma cell line; and MDA, which is a metastatic mammary adenocarcinoma cell line. The three cell samples, grown on conventional microscopy glass slides and measured by means of the FTIR technique in transmission mode, were discriminated according to PC1 score values. The separation occurs as a consequence of the different relative content of lipid components, as deduced by the features at 2924 and 2852 cm^{-1} in the loading plots, which are related to the asymmetric and symmetric stretching modes of the CH_2 groups of lipids.

Regardless of the biochemical point of view, a significant feature to remark is that FTIR spectrum in the 2800–3000 cm^{-1} range can be used as a spectroscopic biomarker for the discrimination of non-malignant, malignant, and metastatic breast cells. This means that the whole spectrum can be used as a marker of cell status, instead of specific spectral peaks (molecular markers). In addition, the use of the microscopy glass slide as cell growth substrates is worthy of attention because such slides are widely used in a clinical environment for cytological and histological diagnosis. Therefore, the obtained results suggest a possible implementation of the FTIR technique as a complementary diagnostic tool in medical practice.

However, it must be remarked that the PCA method, while displaying the discrimination between objects belonging to different classes, is not a classification method. Therefore, it does not allow to properly classify new objects (spectra, in our case) whose classification is unknown. For this purpose, appropriate statistical techniques must be used (e.g., k-nearest neighbors, soft independent modeling of class analogies, and support-vector machine). Hence, the PCA analysis is only a starting point for using FTIR in medical diagnostics.

In addition, there are also other limits and hindrances to be overcome before the use of FTIR in diagnostics can be accepted. First of all, an in-vivo analysis is difficult to perform. Being an in-vitro method, such a procedure is albeit minimally invasive, because it needs cellular samples to be extracted from patients. Further, it is necessary to also extract healthy cells from the patient if the aim is to discriminate healthy cells from pathological cells.

Other authors have shown the possibility to discriminate different types of cells grown onto a glass substrate by means of FTIR spectra in the high wavenumbers range [22,23]; however, they investigated cells from different tissues. Our findings point out that the discrimination is possible also for cells from the same type of tissue but at a different pathology stage. This is an interesting, but not decisive, step towards the introduction of this spectral investigation in the medical diagnostic field. Further steps are both to confirm the separation by using cell lines from other types of tissue than breast and particularly, to measure ex-vivo cells as well as tissues from patients.

Author Contributions: Conceptualization, V.C. and G.P.; methodology, G.P. and M.L.; formal analysis, G.P.; investigation, G.P. and M.L.; data curation, G.P.; writing—original draft preparation, G.P. and M.L.; writing—review and editing, V.C. and G.P.; supervision, V.C. All authors have read and agreed to the published version of the manuscript.

Funding: This research received no external funding.

Institutional Review Board Statement: Not applicable.

Informed Consent Statement: Not applicable.

Conflicts of Interest: The authors declare no conflict of interest.

References

1. Baker, M.J.; Trevisan, J.; Bassan, P.; Bhargava, R.; Butler, H.J.; Dorling, K.M.; Fielden, P.R.; Fogarty, S.W.; Fullwood, N.J.; Heys, K.A.; et al. Using Fourier transform IR spectroscopy to analyze biological materials. *Nat. Protoc.* **2014**, *9*, 1771–1791. [CrossRef]
2. Sheng, D.; Xu, F.; Yu, Q.; Fang, T.; Xia, J.; Li, S.; Wang, X. A study of structural differences between liver cancer cells and normal liver cells using FTIR spectroscopy. *J. Mol. Struct.* **2015**, *1099*, 18–23. [CrossRef]
3. Butler, H.; Cameron, J.M.; Jenkins, C.A.; Hithell, G.; Hume, S.; Hunt, N.T.; Baker, M.J. Shining a light on clinical spectroscopy: Translation of diagnostic IR, 2D-IR and Raman spectroscopy towards the clinic. *Clin. Spectrosc.* **2019**, *1*, 100003. [CrossRef]
4. Finlayson, D.; Rinaldi, C.; Baker, M.J. Is Infrared Spectroscopy Ready for the Clinic? *Anal. Chem.* **2019**, *91*, 12117–12128. [CrossRef] [PubMed]
5. Byrne, H.J.; Behl, I.; Calado, G.; Ibrahim, O.; Toner, M.; Galvin, S.; Healy, C.M.; Flint, S.; Lyng, F.M. Biomedical applications of vi-brational spectroscopy: Oral cancer diagnostics. *Spectrochim Acta A Mol. Biomol. Spectrosc.* **2021**, *252*, 119470. [CrossRef] [PubMed]

6. Elshemey, W.M.; Ismail, A.M.; Elbialy, N.S. Molecular-Level Characterization of Normal, Benign, and Malignant Breast Tissues Using FTIR Spectroscopy. *J. Med. Biol. Eng.* **2016**, *36*, 369–378. [CrossRef]
7. Talari, A.C.S.; Martinez, M.A.G.; Movasaghi, Z.; Rehman, S.; Rehman, I.U. Advances in Fourier transform infrared (FTIR) spectroscopy of biological tissues. *Appl. Spectrosc. Rev.* **2017**, *52*, 456–506. [CrossRef]
8. Depciuch, J.; Tołpa, B.; Witek, P.; Szmuc, K.; Kaznowska, E.; Osuchowski, M.; Król, P.; Cebulski, J. Raman and FTIR spec-troscopy in determining the chemical changes in healthy brain tissues and glioblastoma tumor tissues. *Spectrochim. Acta Part A Mol. Biomol. Spectrosc.* **2020**, *225*, 117526. [CrossRef] [PubMed]
9. Seredin, P.; Goloshchapov, D.; Ippolitov, Y.; Vongsvivut, J. Spectroscopic signature of the patho-logical processes of carious dentine based on FTIR investigations of the oral biological fluids. *Biomed. Opt. Express* **2019**, *10*, 4050–4058. [CrossRef]
10. Byrne, H.J.; Bonnier, F.; McIntyre, J.; Parachalil, D.R. Quantitative analysis of human blood serum using vibrational spectroscopy. *Clin. Spectrosc.* **2020**, *2*, 100004. [CrossRef]
11. Paraskevaidi, M.; Morais, C.L.; Lima, K.M.; Ashton, K.M.; Stringfellow, H.F.; Martin-Hirsch, P.L.; Martin, F.L. Potential of mid-infrared spectroscopy as a non-invasive diagnostic test in urine for endometrial or ovarian cancer. *Analyst* **2018**, *143*, 3156–3163. [CrossRef]
12. Meade, A.D.; Howe, O.; Unterreiner, V.; Sockalingum, G.D.; Byrne, H.J.; Lyng, F.M. Vibrational spectroscopy in sensing radiobiological effects: Analyses of targeted and non-targeted effects in human keratinocytes. *Faraday Discuss.* **2016**, *187*, 213–234. [CrossRef] [PubMed]
13. Depciuch, J.; Stanek-Widera, A.; Warchulska, M.; Lange, D.; Sarwa, K.; Koziorowska, A.; Kula, M.; Cebulski, J. Identification of chemical changes in healthy breast tissue caused by chemotherapy using Raman and FTIR spectroscopy: A preliminary study. *Infrared Phys. Technol.* **2019**, *102*, 102989. [CrossRef]
14. Monici, M. Cell and tissue autofluorescence research and diagnostic applications. *Biotechnol. Annu. Rev.* **2005**, *11*, 227–256. [CrossRef]
15. Butler, H.; Ashton, L.; Bird, B.; Cinque, G.; Curtis, K.; Dorney, J.; Esmonde-White, K.; Fullwood, N.J.; Gardner, B.; Martin-Hirsch, P.L.; et al. Using Raman spectroscopy to characterize biological materials. *Nat. Protoc.* **2016**, *11*, 664–687. [CrossRef]
16. Traynor, D.; Behl, I.; O'Dea, D.; Bonnier, F.; Nicholson, S.; O'Connell, F.; Maguire, A.; Flint, S.; Galvin, S.; Healy, C.M.; et al. Raman spectral cytopathology for cancer diagnostic applications. *Nat. Protoc.* **2021**, *16*, 1–20. [CrossRef]
17. Zhang, H.; Wang, X.; Ding, R.; Shen, L.; Gao, P.; Xu, H.; Xiu, C.; Zhang, H.; Song, D.; Han, B. Characterization and imaging of surgical specimens of invasive breast cancer and normal breast tissues with the application of raman spec-tral mapping: A feasibility study and comparison with randomized single-point detection method. *Oncol. Lett.* **2020**, *20*, 2969. [CrossRef] [PubMed]
18. Staritzbichler, R.; Hunold, P.; Estrela-Lopis, I.; Hildebrand, P.W.; Isermann, B.; Kaiser, T. Raman spectros-copy on blood serum samples of patients with end-stage liver disease. *PLoS ONE* **2021**, *16*, e0256045. [CrossRef]
19. Hardy, M.; Kelleher, L.; de Carvalho Gomes, P.; Buchan, E.; Chu, H.O.M.; Goldberg Oppenheimer, P. Hin On Martin Chu & Pola Gold-berg Oppenheimer Methods in Raman spectroscopy for saliva studies—A review. *Appl. Spectrosc. Rev.* **2021**, 1–57.
20. Perna, G.; Capozzi, V.; Lasalvia, M. A Comparison between FTIR Spectra from HUKE and SH-SY5Y Cell Lines Grown on Different Substrates. *Appl. Sci.* **2020**, *10*, 8825. [CrossRef]
21. Varmuza, K.; Filzmoser, P. *Introduction to Multivariate Statistical Analysis in Chemometrics*; CRC Press: Boca Raton, FL, USA, 2009.
22. Rutter, A.V.; Crees, J.; Wright, H.; van Pittius, D.G.; Yousef, I.; Sulé-Suso, J. Fourier transform infrared spectra of cells on glass coverslips. A further step in spectral pathology. *Analyst* **2018**, *143*, 5711–5717. [CrossRef]
23. Rutter, A.V.; Crees, J.; Wright, H.; Raseta, M.; Van Pittius, D.G.; Roach, P.; Sulé-Suso, J. Identification of a Glass Substrate to Study Cells Using Fourier Transform Infrared Spectroscopy: Are We Closer to Spectral Pathology? *Appl. Spectrosc.* **2020**, *74*, 178–186. [CrossRef] [PubMed]
24. Zeaiter, M.; Rutledge, D. Preprocessing methods. In *Comprehensive Chemometrics: Chemical and Biochemical Data Analysis Vol. 3*; Brown, S.D., Tauler, R., Walczak, B., Eds.; Elsevier: Amsterdam, The Netherlands, 2009; pp. 121–231.
25. Hanson, B.A. ChemoSpec: Exploratory Chemometrics for Spectroscopy. R Package Version 4.4.97. 2017. Available online: https://CRAN.R-project.org/package=ChemoSpec (accessed on 20 July 2021).
26. Kar, S.; Katti, D.; Katti, K.S. Fourier transform infrared spectroscopy based spectral biomarkers of metastasized breast cancer progression. *Spectrochim. Acta Part A Mol. Biomol. Spectrosc.* **2019**, *208*, 85–96. [CrossRef]
27. Verdonck, M.; Wald, N.; Janssis, J.; Yan, P.; Meyer, C.; Legat, A.; Speiser, D.E.; Desmedt, C.; Larsimont, D.; Sotiriou, C.; et al. Breast cancer and melanoma cell line identification by FTIR imaging after formalin-fixation and paraffin-embedding. *Analyst* **2013**, *138*, 4083–4091. [CrossRef] [PubMed]
28. Yang, W.; Xiao, X.; Tan, J.; Cai, Q. In situ evaluation of breast cancer cell growth with 3D ATR-FTIR spectroscopy. *Vib. Spectrosc.* **2009**, *49*, 64–67. [CrossRef]
29. Movasaghi, Z.; Rehman, S.; Rehman, I.U. Fourier Transform Infrared (FTIR) Spectroscopy of Biological Tissues. *Appl. Spectrosc. Rev.* **2008**, *43*, 134–179. [CrossRef]
30. Mohlenhoff, B.; Romeo, M.; Diem, M.; Wood, B.R. Mie-Type Scattering and Non-Beer-Lambert Absorption Behavior of Human Cells in Infrared Microspectroscopy. *Biophys. J.* **2005**, *88*, 3635–3640. [CrossRef] [PubMed]
31. Bonnier, F.; Byrne, H. Understanding the molecular information contained in principal component analysis of vibrational spectra of biological systems. *Analyst* **2011**, *137*, 322–332. [CrossRef]

32. Nieva, C.; Marro, M.; Codina, N.S.; Rao, S.; Petrov, D.; Sierra, A. The Lipid Phenotype of Breast Cancer Cells Characterized by Raman Microspectroscopy: Towards a Stratification of Malignancy. *PLoS ONE* **2012**, *7*, e46456. [CrossRef] [PubMed]
33. Beloribi-Djefaflia, S.; Vasseur, S.; Guillaumond, F. Lipid metabolic reprogramming in cancer cells. *Oncogene* **2016**, *5*, e189. [CrossRef]
34. Corbet, C.; Feron, O. Emerging roles of lipid metabolism in cancer progression. *Curr. Opin. Clin. Nutr. Metab. Care* **2017**, *20*, 254–260. [CrossRef] [PubMed]
35. Eiriksson, F.F.; Nøhr, M.K.; Costa, M.; Bödvarsdottir, S.K.; Ögmundsdottir, H.M.; Thorsteinsdottir, M. Lipidomic study of cell lines reveals differences between breast cancer subtypes. *PLoS ONE* **2020**, *15*, e0231289. [CrossRef] [PubMed]

sensors

MDPI

Article

Raman Natural Gas Analyzer: Effects of Composition on Measurement Precision

Dmitry V. Petrov [1,2,*], Ivan I. Matrosov [1], Alexey R. Zaripov [1] and Aleksandr S. Tanichev [1]

[1] Institute of Monitoring of Climatic and Ecological Systems, 634055 Tomsk, Russia; mii@imces.ru (I.I.M.); alexey-zaripov@rambler.ru (A.R.Z.); tanichev_aleksandr@mail.ru (A.S.T.)

[2] Department of Optics and Spectroscopy, Tomsk State University, 634050 Tomsk, Russia

* Correspondence: dpetrov@imces.ru

Abstract: Raman spectroscopy is a promising method for analyzing natural gas due to its high measurement speed and the potential to monitor all molecular components simultaneously. This paper discusses the features of measurements of samples whose composition varies over a wide range (0.005–100%). Analysis of the concentrations obtained during three weeks of experiments showed that their variation is within the error caused by spectral noise. This result confirms that Raman gas analyzers can operate without frequent calibrations, unlike gas chromatographs. It was found that a variation in the gas composition can change the widths of the spectral lines of methane. As a result, the measurement error of oxygen concentration can reach 200 ppm. It is also shown that neglecting the measurement of pentanes and n-hexane leads to an increase in the calculated concentrations of other alkanes and to errors in the density and heating value of natural gas.

Keywords: Raman spectroscopy; gas analysis; natural gas; methane; alkanes; isotopic composition; heating value

Citation: Petrov, D.V.; Matrosov, I.I.; Zaripov, A.R.; Tanichev, A.S. Raman Natural Gas Analyzer: Effects of Composition on Measurement Precision. *Sensors* **2022**, *22*, 3492. https://doi.org/10.3390/s22093492

Academic Editors: Maria Lepore and Ines Delfino

Received: 12 April 2022
Accepted: 1 May 2022
Published: 4 May 2022

1. Introduction

Natural gas (NG) is the most environmentally friendly of all fossil fuels and is also a raw material for the production of many chemicals, including hydrogen [1]. To date, the basic method for measuring its composition is gas chromatography. However, this method has some disadvantages. Among them are the need for consumables, frequent calibration checks, and a long analysis time. These features make real-time measurements impossible. Devices based on optical spectroscopy do not have such drawbacks. The application of infrared (IR) spectroscopy for the analysis of NG composition was demonstrated by Kireev et al. [2,3]. The measurement accuracy of hydrocarbons is close to the gas chromatography. However, it is impossible to measure the content of diatomic homonuclear molecules (such as N_2, O_2, H_2, etc.), using this method. Taking into account the ongoing development of energy technologies with minimal CO_2 emissions, the use of hydrogen-enriched natural gas will increase [4,5]. In this regard, IR spectroscopy is not an ideal method for measuring such gas mixtures. Raman spectroscopy is a promising alternative technique. It is possible to simultaneously control the content of all types of molecules using an instrument based on this effect. The capabilities of such gas analyzers were demonstrated in many studies [6–17]. It should be noted that many authors measure alkanes only up to C4. This is explained by the weakness of the Raman signals of gaseous components and the difficulty in deriving the concentrations of heavy alkanes from the Raman spectrum of NG due to the significant overlap of the spectra of various components [18]. According to ISO 6974-5 [19], the detection limit for C2–C6 alkanes is 0.005%. Thus, a Raman gas analyzer must measure NG composition with this accuracy to be competitive with gas chromatographs. In this work, we study the capabilities of the developed Raman gas analyzer using NG samples whose composition varies in ranges close to values indicated in ISO 6974-5 [19]. In addition, we investigate the influence of line broadening and the effect of ignoring the spectra of C5+ alkanes on measurement precision.

2. Materials and Methods

2.1. Raman Gas Analyzer

The Raman gas analyzer used in this work is an improved analog of the device as that used previously [6]. Its optical design is based on a 90-degree geometry of scattered light collection (see Figure 1) since spectra with a minimum background level can be recorded using this scheme. A solid-state continuous-wave laser with a power of 1.5 W at a wavelength of 532 nm was used as a source of exciting radiation. Two identical f/1.8-lenses were used for scattered light collection. An analysis of our previous results [6] and the Raman spectra of the main NG components [18] showed that it is necessary to improve the signal-to-noise ratio to improve the accuracy of measurements. In this regard, a new compact no-moving-parts f/1.8-spectrometer MKR-2m (Sibanalitpribor LLC, Tomsk, Russia) was used in this work. Its main difference from the previous spectrometer [6] is a higher spectral sensitivity (especially at the edges of the recorded range) due to the optimization of the optical scheme. The simultaneously recorded spectral range was 530–628 nm using the 1800 lines/mm grating. With an entrance slit of 40 μm, the half-width of instrumental function response was ~6 cm^{-1} at the center of this range. The signals were recorded using the charge-coupled device (CCD) sensor Hamamatsu S10141 (2048 × 256 pixels, 12 μm in size) with thermoelectric cooling down to −10°C. About 10-fold amplification of the Raman signals was obtained in the range of 300–1000 cm^{-1}, where the characteristic peaks of C2+ alkanes are located, using this spectrometer (in comparison with Ref. [6]).

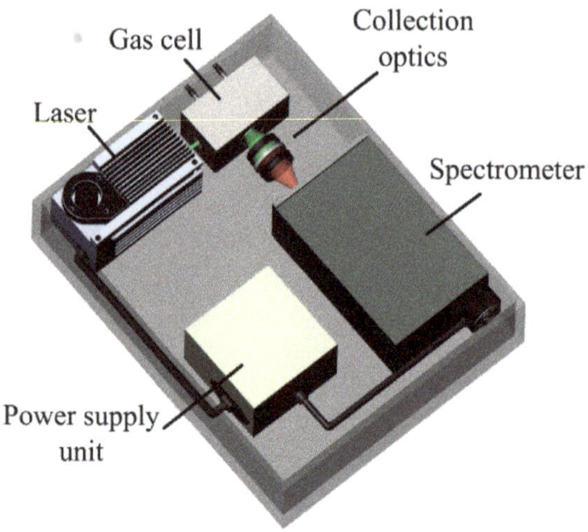

Figure 1. Schematic of Raman gas analyzer.

2.2. Concentration Measurement Method

The contour fit method was used to derive the concentrations due to the significant overlap in the spectra of NG species [18]. Its essence is as follows. The NG spectrum $I_{mix}(\nu)$ at each wavenumber ν can be represented as the sum of the spectra of its components $I_i(\nu)$:

$$I_{mix}(\nu) = \sum_{i=1}^{m} a_i I_i(\nu), \quad (1)$$

where a_i is the contribution of the spectrum of the ith component to the spectrum of the mixture [0..1], and m is the number of measured components.

Taking into account the number of CCD sensor columns, a system of 2048 equations can be obtained. Its solution (contributions a_i) can be found using the least-squares method. The required relative concentrations (N_i) can be found using Equation (2).

$$N_i = \frac{n_i a_i}{\sum\limits_{j=1}^{m} n_j a_j} \cdot 100\%, \quad (2)$$

where n_i is the absolute concentration of the ith component in the reference spectrum $I_i(\nu)$.

According to Ref. [20], the spectral characteristics (peak positions and half-widths) of the reference spectra and the spectra of the mixture should be equivalent to obtain the most accurate results. First of all, to ensure this condition, all measurements of mixtures were carried out at a pressure of 25 atm and a temperature of 300 K. Reference spectra of pure methane, ethane, nitrogen, carbon dioxide, hydrogen, and oxygen were also obtained at these parameters. The spectra of heavier alkanes (propane, n-butane, isobutane, n-pentane, iso-pentane, neo-pentane, and n-hexane) liquefy under the above conditions. For this reason, they were obtained at saturated vapor pressure. The exposure time for each reference spectrum was 1000 s.

2.3. Experiment

Three samples of synthetic NG with significantly different compositions were used for research (see Table 1). These samples are the reference gas mixtures with low uncertainties that were purchased from Monitoring LLC (Saint Petersburg, Russia). Measurements were carried out for three weeks, once a week, to assess the long-term stability of the results. The sequence of analysis of mixtures is presented in Table 2. A series of five measurements were performed for each mixture with the replacement of the sample in the cell. The time of one analysis was 30 s. Note that the set of reference spectra of pure components was obtained once before the measurement procedure was started. Additional calibration procedures were not performed during all measurements.

Table 1. Composition of natural gas samples used.

Component	Concentration (%)		
	Sample 1	Sample 2	Sample 3
CH_4	99.9403	95.998	49.0379
C_2H_6	0.00496	0.997	15.1
C_3H_8	0.00474	0.509	6.05
n-C_4H_{10}	0.00493	0.105	0.709
iso-C_4H_{10}	0.00497	0.102	0.816
n-C_5H_{12}	0.00503	0.0474	0.205
iso-C_5H_{12}	0.00522	0.0472	0.19
neo-C_5H_{12}	0.0048	0.01	0.0511
n-C_6H_{14}	0.00445	0.0236	0.131
CO_2	0.0047	1	10.1
N_2	0.0054	1.039	15.1
H_2	0.00559	0.102	0.5
O_2	0.0048	0.0198	2.01

Table 2. Program of measurements.

Day	Sequence of Sample Analysis
1st	#1–#2–#3–#2–#1–#3–#2
2nd	#2–#1–#2–#1–#2–#1–#2–#3
3rd	#1–#3–#1–#3–#1–#3–#1–#2

3. Results and Discussion

3.1. Mixture Measurements

Figures 2 and 3 show the obtained Raman spectra of the samples of NG. Despite mutual overlaps, the characteristic peaks of most components are distinguishable at the resolution of the spectrometer used. The achieved sensitivity makes it possible to see the lines of the ν_4 band of methane down to ~800 cm^{-1}. In addition, a wide unresolved band is observed in the methane spectrum in the region of 300–600 cm^{-1}. We suppose this is a collision-induced rotational band [21,22], which is attenuated up to ~350 cm^{-1} by the notch filter. Bands of C–C–C deformation vibrations of C3+ hydrocarbons are also located in the region of 300–500 cm^{-1} (see Figure 4). The accuracy of concentration measurements can be improved using this range due to intense peaks of n-butane (429 cm^{-1}), n-pentane (398 cm^{-1}), and iso-pentane (459 cm^{-1}), the overlap of which is not as significant as in the region of 700–1000 cm^{-1}. Thus, to measure low concentrations, it is necessary to take into account the contribution of the methane spectrum to the spectrum of NG not only in the region of >990 cm^{-1} (as indicated in Ref. [18]) but also in the region of lower wavenumbers. The inset in Figure 2 shows the vibrational band of nitrogen (2330 cm^{-1}), whose concentration in sample 1 is 54 ppm, despite its significant overlap with the lines of the $2\nu_4$ and ν_3 bands of methane, is also well observed. Hence, concentrations with a sensitivity of <50 ppm can be measured due to the achieved signal-to-noise ratio. The limits of detection will be estimated below.

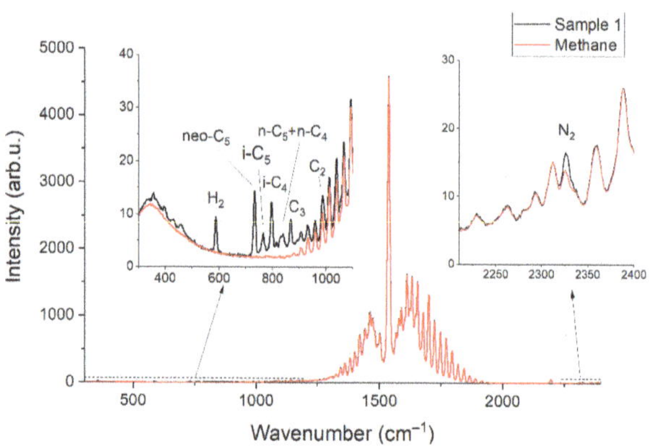

Figure 2. Raman spectra of pure methane and sample 1.

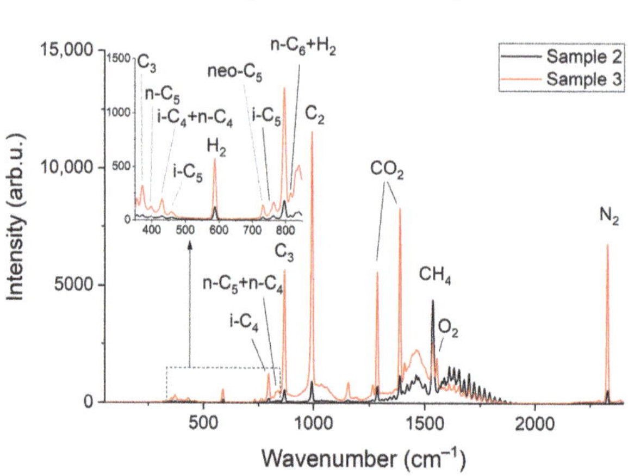

Figure 3. Raman spectra of sample 2 and sample 3.

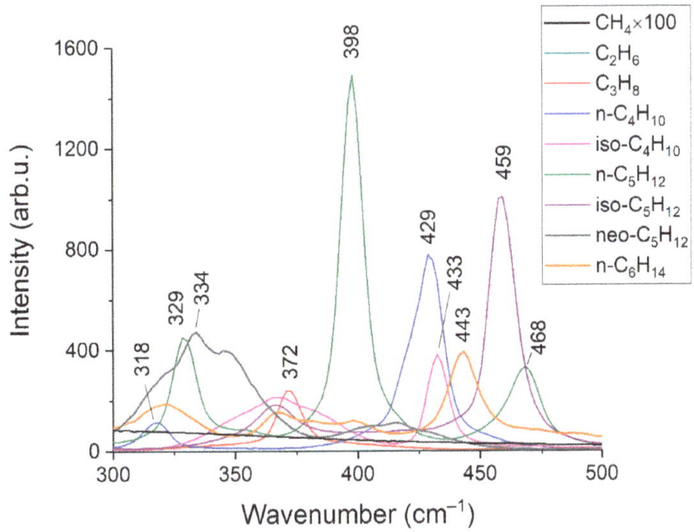

Figure 4. Raman spectra of C1–C6 alkanes in the range of 300–500 cm^{-1}. The intensities correspond to the equivalent pressure.

The range of 300–2400 cm^{-1} was used to determine the composition of mixtures. All measured concentrations during one day for each mixture were averaged. The concentrations (C) and their standard deviations (σ) are presented in Tables 3–5. It can be seen that the measured and reference concentrations are in good agreement taking into account the uncertainties. The only exception is data of n-hexane in samples 2 and 3. For most components, the variation in measured concentrations over all days is within their mean standard deviation. It indicates these variations are due to noise in the spectra. Thus, the presented data confirm that Raman gas analyzers can operate for a long time without calibration, unlike gas chromatographs.

Table 3. Measurement results for sample 1.

Component	Reference Data		Data Obtained					
			1st Day		2nd Day		3rd Day	
	C (%)	σ (%)	C (%)	σ (%)	C (%)	σ (%)	C (%)	σ (%)
CH_4	99.9403	0.0023	99.94	0.0023	99.938	0.0022	99.9401	0.0046
C_2H_6	0.00496	0.00018	0.00479	0.00027	0.00508	0.00025	0.00526	0.00047
C_3H_8	0.00474	0.00022	0.00496	0.00011	0.0052	0.00024	0.0052	0.00031
n-C_4H_{10}	0.00493	0.00023	0.00453	0.00025	0.00501	0.00027	0.00466	0.00030
iso-C_4H_{10}	0.00497	0.00023	0.00492	0.00006	0.0049	0.00007	0.00486	0.00011
n-C_5H_{12}	0.00503	0.00023	0.00545	0.00019	0.00549	0.00018	0.00514	0.00032
iso-C_5H_{12}	0.00522	0.00024	0.00496	0.00019	0.00517	0.00015	0.00508	0.00016
neo-C_5H_{12}	0.0048	0.00023	0.00492	0.00004	0.00493	0.00005	0.00494	0.00005
n-C_6H_{14}	0.00445	0.00021	0.00505	0.00064	0.00524	0.00072	0.00429	0.00098
CO_2	0.0047	0.0005	0.00527	0.00071	0.00509	0.00031	0.00504	0.0011
N_2	0.0054	0.0005	0.00539	0.00035	0.0048	0.00027	0.00584	0.0006
O_2	0.0048	0.0005	0.00457	0.0010	0.00595	0.0011	0.00428	0.0014
H_2	0.00559	0.00025	0.0051	0.00008	0.00508	0.00008	0.0052	0.00009

Table 4. Measurement results for sample 2.

Component	Reference Data		Data Obtained					
			1st Day		2nd Day		3rd Day	
	C (%)	σ (%)	C (%)	σ (%)	C (%)	σ (%)	C (%)	σ (%)
CH_4	95.998	0.09	95.9512	0.0042	95.9509	0.0046	95.9503	0.0029
C_2H_6	0.997	0.02	1.0172	0.0010	1.0181	0.0011	1.0179	0.0009
C_3H_8	0.509	0.015	0.5166	0.0006	0.5168	0.0008	0.5173	0.0005
n-C_4H_{10}	0.105	0.003	0.1038	0.0004	0.1035	0.0005	0.1042	0.0003
iso-C_4H_{10}	0.102	0.003	0.1018	0.0002	0.1018	0.0002	0.1019	0.0002
n-C_5H_{12}	0.0474	0.0015	0.0455	0.0003	0.0446	0.0003	0.0446	0.0003
iso-C_5H_{12}	0.0472	0.0015	0.0479	0.0002	0.0481	0.0004	0.0482	0.0003
neo-C_5H_{12}	0.01	0.0004	0.0096	0.00005	0.0096	0.00006	0.0096	0.00004
n-C_6H_{14}	0.0236	0.0008	0.0184	0.0007	0.0183	0.0006	0.0186	0.0006
CO_2	1	0.03	1.0238	0.0012	1.0234	0.0010	1.0228	0.0006
N_2	1.039	0.021	1.0447	0.0014	1.0451	0.0015	1.0432	0.0005
O_2	0.0198	0.001	0.0206	0.0015	0.0205	0.0017	0.0221	0.0008
H_2	0.102	0.003	0.0989	0.0002	0.0988	0.0002	0.099	0.0001

Table 5. Measurement results for sample 3.

Component	Reference Data		Data Obtained					
			1st Day		2nd Day		3rd Day	
	C (%)	σ (%)	C (%)	σ (%)	C (%)	σ (%)	C (%)	σ (%)
CH_4	49.038	1.12	49.499	0.0285	49.517	0.0049	49.518	0.0071
C_2H_6	15.1	0.3	14.908	0.0079	14.913	0.0081	14.905	0.0103
C_3H_8	6.05	0.18	6.0128	0.0036	6.0138	0.0021	6.0091	0.0043
n-C_4H_{10}	0.709	0.021	0.6987	0.0024	0.6985	0.0019	0.698	0.0017
iso-C_4H_{10}	0.816	0.025	0.8177	0.0005	0.8175	0.0006	0.817	0.0007
n-C_5H_{12}	0.205	0.006	0.204	0.0015	0.209	0.0017	0.2089	0.0022
iso-C_5H_{12}	0.19	0.006	0.1832	0.001	0.1829	0.0009	0.1828	0.0009
neo-C_5H_{12}	0.0511	0.0016	0.0502	0.0001	0.0502	0.0001	0.0508	0.0001
n-C_6H_{14}	0.131	0.004	0.1444	0.0033	0.1564	0.0044	0.1566	0.0049
CO_2	10.1	0.3	9.9551	0.0151	9.931	0.0106	9.9319	0.0102
N_2	15.1	0.3	15.035	0.015	15.02	0.0078	15.032	0.0134
O_2	2.01	0.06	1.978	0.0017	1.9772	0.0007	1.9766	0.0012
H_2	0.5	0.015	0.5141	0.0008	0.5134	0.0006	0.5125	0.001

The relative measurement errors of each component were obtained using the mean standard deviations (see Figure 5). It can be seen that these values depend both on the concentration and the type of molecule (due to different scattering cross-sections and the level of overlap of the spectral bands). Taking into account that the measurement errors of gas chromatographs are close to 5%, it can be concluded that the accuracy of the presented Raman gas analyzer is higher for species with a concentration of more than ~100 ppm.

3.2. Limits of Detection

Limits of detection (LOD_i) were estimated using Equation (3). Here, we defined the concentrations at which the signal of ith component is three times the standard deviation of the noise. The spectrum of sample 1 was used to obtain these data. Peak intensities of each component (S_i) were estimated, taking into account their contribution to the spectrum of the mixture (see Figure 6). The difference between two successive spectra of sample 1 was obtained to estimate the magnitude of the noise (see Figure 7). It can be seen that the noise in the region of 500–1000 cm^{-1}, where the characteristic bands of C2+ alkanes are located, is less than in the region of intense lines of the ν_2 band of methane (1200–1700 cm^{-1}). This feature is related to the effect of photon shot noise, which is proportional to the square root of the signal intensity. In this regard, the noises that affect measurement errors and *LOD*s are higher for CO_2 and O_2 than for all other components. The standard deviations of noise (N_i) were calculated using the intensities in the spectrum shown in Figure 7 in the following regions: 1540–1580 cm^{-1} (for O_2), 1280–1380 cm^{-1} (for CO_2), and 700–900 cm^{-1} (for other components). Concentrations of components (C_i) in sample 1 for calculations were taken from Table 1. The results obtained are presented in Table 6. It can be seen that the *LOD* values are within the range of 2–35 ppm. Thus, the achieved sensitivity of the Raman analyzer meets the requirements of ISO 6974-5 [19].

$$LOD_i = 3\frac{C_i}{S_i/N_i}, \quad (3)$$

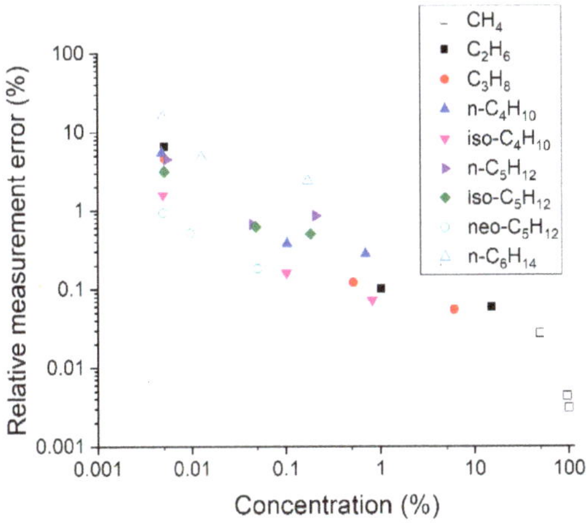

Figure 5. Relative measurement errors of alkanes at different concentrations.

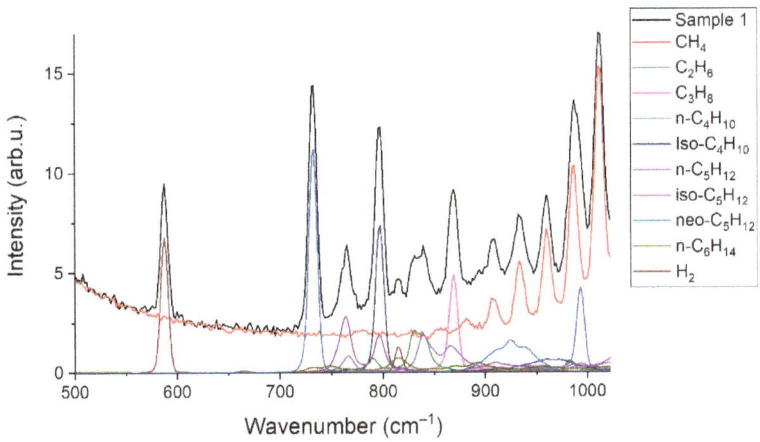

Figure 6. Contributions of the species to the spectrum of sample 1.

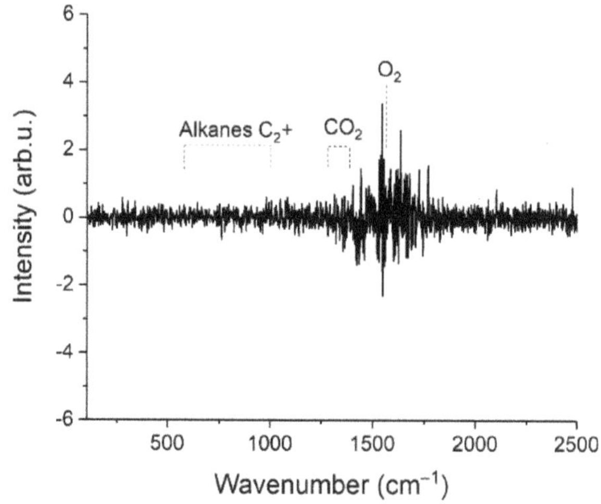

Figure 7. Difference between two successive spectra of sample 1.

Table 6. Parameters for Equation (3) and limits of detection of the Raman natural gas analyzer.

Component	*S* (arb.u.)	*N* (arb.u.)	*LOD* (ppm)
C_2H_6	4.3	0.017	5.9
C_3H_8	4.94	0.017	4.9
n-C_4H_{10}	2.15	0.017	11.7
iso-C_4H_{10}	7.39	0.017	3.4
n-C_5H_{12}	1.94	0.017	13.2
iso-C_5H_{12}	2.87	0.017	9.3
neo-C_5H_{12}	11.27	0.017	2.1
n-C_6H_{14}	0.73	0.017	31.1
CO_2	5.5	0.036	9.2
N_2	2.8	0.017	9.8
O_2	2.8	0.068	35.1
H_2	6.8	0.017	4.2

3.3. *Influence of Line Broadening on Measurements*

Let us consider the features of O_2 measurement. It has one fundamental vibrational band with the position of the maximum at 1555 cm^{-1}, which is overlapped by the ν_2 band of methane (see Figure 3). Hence, the measurement accuracy is affected by the broadening of the spectral lines of methane [20] besides the signal-to-noise ratio. Pressure [23] and molecular environment [24,25] influence the half-widths of the lines. The line at 1793 cm^{-1} was analyzed to assess the influence of the composition on the line half-widths of the ν_2 band of methane. This line was chosen since it is not overlapped by the spectra of other species and, therefore, the measurement error of its half-width in mixtures is eliminated. The data obtained and the half-width of this line as a function of pure methane pressure are shown in Figure 8. It can be seen that the half-width increases with a decrease in the fraction of methane in the mixtures. This broadening is related to an increase in the concentration of heavy hydrocarbons in the mixture since the methane-methane broadening coefficients are less than the broadening coefficients of methane-ethane, methane-propane, etc. [25].

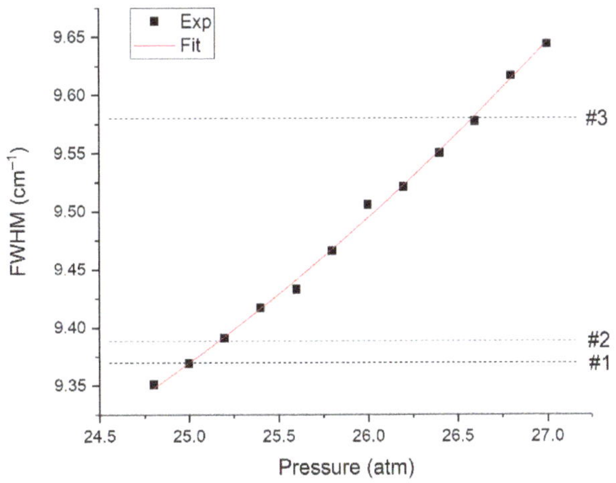

Figure 8. Half-width of the methane line at 1793 cm^{-1} in pure methane at various pressures and in analyzed samples at 25 atm.

According to Figure 8, an increase in the pressure of pure methane to 26.6 atm leads to the same broadening as in the spectrum of sample 3 at a pressure of 25 atm. Thus, in our case, we can use the spectra of pure methane at pressures of 25.0 and 26.6 atm to estimate the error in oxygen measurements due to the broadening of methane lines. The spectrum at a pressure of 26.6 atm was multiplied by the 25/26.6 value to ensure equal integral intensities of these spectra. Figure 9 shows the difference between these methane spectra in the region of 1555 cm^{-1}, denoted as R. According to Equation (4), this effect leads to an oxygen measurement error (Δ) close to 200 ppm.

$$\Delta = \frac{R \cdot 100\%}{I_{MAX}}, \quad (4)$$

where I_{MAX} is the peak intensity of the spectrum of pure oxygen at 25 atm. Taking into account the concentration ranges of C2+ alkanes in NG [19], it can be concluded that the systematic error in oxygen measurement can reach 200 ppm (depending on the composition). This error is less than the uncertainty of the reference O_2 concentration in sample 3. However, in the case of an O_2 concentration in such a mixture below 200 ppm, this is a sufficiently large value that cannot be ignored. Calibration coefficients or a reference spectrum of pure methane at a pressure that results in the required line broadening can be used to obtain reliable data.

We believe that the deviations of the measured hexane concentrations from the reference values are due to similar effects. Although hexane has several bands in the region of 700–900 cm^{-1}, their peak intensity is relatively low (see Figure 6), and all of them are overlapped by the spectra of other molecules [18]. Thus, a change in the spectral characteristics of alkanes in a mixture compared to a pure substance can lead to errors in measurements of the hexane concentration. We plan to study these features in more detail in the future.

3.4. Estimation of Errors in the Case of Ignoring C5+ Spectra

We decided to estimate the errors in the case of neglecting pentanes and hexane since many authors analyze the composition of mixtures only up to C4 [7,9–15]. All spectra of mixtures obtained during the first day of experiments were used. The spectra of pentanes and hexane were excluded from the set of reference spectra of pure components to calculate the concentrations. The results obtained are presented in Table 7. It can be seen that ignoring these components leads to an increase in the measured concentrations of ethane, propane, and butanes. Taking into account that this effect is due to the overlap of their spectra, the errors depend on the composition of the mixture and cannot be eliminated using calibration coefficients. In addition to these data, the characteristic parameters [26], which are required for power plant operators, were calculated. To this end, the concentrations shown in Tables 3 and 7, Tables 4 and 5 (1st day) were used. As shown in Table 8, these characteristics correspond to the reference data when all components are measured. In turn, only the heating value of sample 1 corresponds to the reference value in the case of ignoring the measurement of pentanes and hexane. Despite the increase in the measured concentrations of other alkanes, other characteristics are significantly less than the reference ones. Thus, reliable characteristic parameters of NG cannot be obtained by measuring alkanes only up to C4.

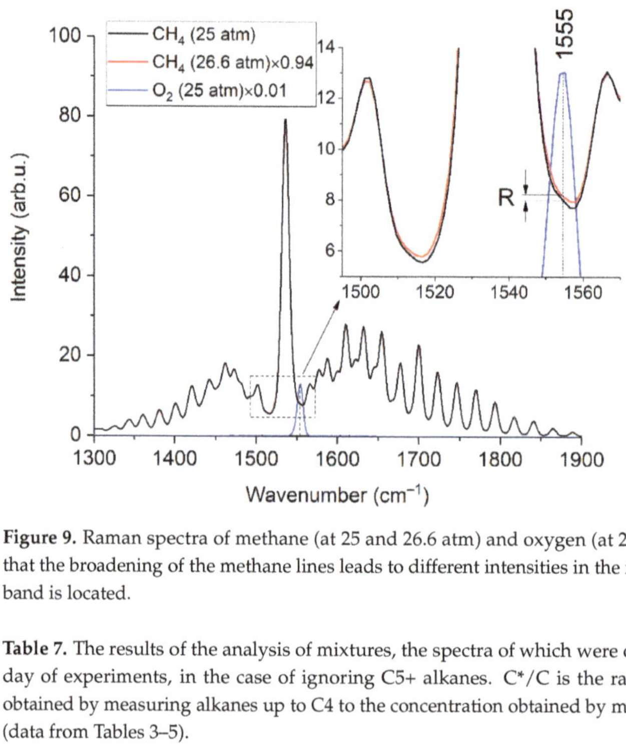

Figure 9. Raman spectra of methane (at 25 and 26.6 atm) and oxygen (at 25 atm). The inset shows that the broadening of the methane lines leads to different intensities in the region where the oxygen band is located.

Table 7. The results of the analysis of mixtures, the spectra of which were obtained during the first day of experiments, in the case of ignoring C5+ alkanes. C*/C is the ratio of the concentration obtained by measuring alkanes up to C4 to the concentration obtained by measuring all components (data from Tables 3–5).

Component	Sample 1		Sample 2		Sample 3	
	C* (%)	C*/C	C* (%)	C*/C	C* (%)	C*/C
CH_4	99.9419	1.000	95.960	1.000	49.779	1.006
C_2H_6	0.00644	1.353	1.0280	1.011	14.872	0.998
C_3H_8	0.0088	1.774	0.5405	1.046	6.1022	1.015
n-C_4H_{10}	0.01348	2.975	0.1588	1.530	0.9894	1.416
iso-C_4H_{10}	0.00746	1.516	0.1194	1.173	0.8907	1.089
n-C_5H_{12}	–	–	–	–	–	–
iso-C_5H_{12}	–	–	–	–	–	–
neo-C_5H_{12}	–	–	–	–	–	–
n-C_6H_{14}	–	–	–	–	–	–
CO_2	0.00629	1.194	1.0273	1.003	9.9140	0.996
N_2	0.0055	1.020	1.0443	0.999	14.963	0.995
O_2	0.00447	0.978	0.0199	0.966	1.9616	0.992
H_2	0.00561	1.100	0.1016	1.027	0.5285	1.028

Table 8. Comparison of characteristics of natural gas samples.

Sample	Parameter	Reference Data	Data Obtained	
			All Species Were Measured	C5 and C6 Were Ignored
1	Lower heating value (MJ/kg)	33.45 ± 0.03	33.45	33.44
	Relative density	0.55545 ± 0.00004	0.55547	0.55528
2	Lower heating value (MJ/kg)	33.54 ± 0.04	33.53	33.47
	Relative density	0.5838 ± 0.0004	0.5841	0.5830
3	Lower heating value (MJ/kg)	33.12 ± 0.19	33.13	32.86
	Relative density	0.8908 ± 0.0040	0.8876	0.8807

3.5. Variation in the Isotopic Composition of Methane

We noticed the different intensity of the peak with a wavenumber of 2196 cm^{-1} between the spectra of pure methane and sample 1 during the experiments. The ν_2 band of the CH_3D methane isotopologue is located in this region (see Figure 10). This discrepancy may be due to the different nature of the origin of the pure methane and methane in the mixtures. The difference in the peak intensity is ~0.4% and agrees with possible CH_3D/CH_4 variations in NG [27]. We did not find signs of $^{13}CH_4/^{12}CH_4$ variation in our samples since there is a small shift in their lines relative to each other in the ν_2 region [28]. It is worth noting that knowledge of the isotopic composition of methane is also useful. It is possible to determine the type of reservoir (gas, gas condensate, or oil), as well as the origin of natural gas (biogenic or thermogenic) based on this information [29]. Raman gas analyzer can also measure the content of $^{13}CH_4$ by registration of spectra up to 3100 cm^{-1} [30,31]. Note that when using the contour fit method, the discrepancy in the isotopic composition of methane in comparison to the reference methane can lead to a difference between their spectra and, consequently, to errors in the measurement of other components. In this case, the simulation of spectra can be used to improve the reliability of measurements [32]. The effects of pressure, molecular environment, and the contributions of all isotopologues can be taken into account to obtain a spectrum using this approach.

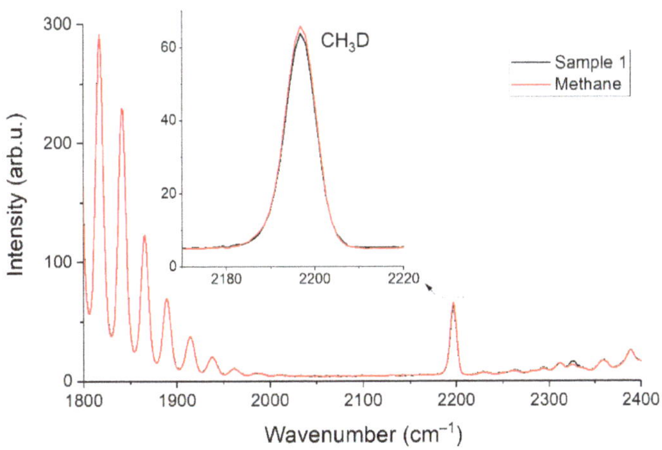

Figure 10. Raman spectra of pure methane and sample 1 in the range of 1800–2400 cm^{-1}.

4. Conclusions

This study presents the features of natural gas analysis using Raman spectroscopy. The use of the contour fit method to derive concentrations from the spectra of mixtures makes it possible to obtain reliable results even with a significant change in the composition of the samples. However, in the case of measuring low concentrations of components whose characteristic peaks are overlapped by intense bands of other molecules, it is necessary to take into account the change in spectral characteristics due to changes in the molecular environment to increase the accuracy. The data obtained confirmed that such devices can operate for a long time without calibration. This is a very important advantage of Raman gas analyzers over analogs. The achieved detection limits of the developed compact Raman gas analyzer are 2–35 ppm at a pressure of 25 atm and an analysis time of 30 s. This level of sensitivity makes it possible to monitor the isotopic composition of methane. In turn, it is possible to reduce the analysis time or improve the accuracy by using a more powerful laser and/or a photodetector with a lower noise level. Taking into account the advantages of Raman gas analyzers, we believe that they have great potential in natural gas analysis and can replace conventional gas chromatographs.

Author Contributions: Conceptualization, methodology, writing—original draft, D.V.P.; investigation, I.I.M.; resources, A.R.Z.; visualization, writing—review and editing, A.S.T. All authors have read and agreed to the published version of the manuscript.

Funding: This work was supported by the Russian Science Foundation, grant no. 19-77-10046.

Institutional Review Board Statement: Not applicable.

Informed Consent Statement: Not applicable.

Data Availability Statement: Data are contained within the article.

Conflicts of Interest: The authors declare no conflict of interest.

References

1. Chen, L.; Qi, Z.; Zhang, S.; Su, J.; Somorjai, G.A. Catalytic Hydrogen Production from Methane: A Review on Recent Progress and Prospect. *Catalysts* **2020**, *10*, 858. [CrossRef]
2. Kireev, S.V.; Podolyako, E.M.; Symanovsky, I.G.; Shnyrev, S.L. Optical absorption method for the real-time component analysis of natural gas: Part 1. Analysis of mixtures enriched with ethane and propane. *Laser Phys.* **2011**, *21*, 250–257. [CrossRef]
3. Kireev, S.V.; Podolyako, E.M.; Simanovsky, I.G.; Shnyrev, S.L. Optical absorption method of natural gas component analysis in real time. Part II. Analysis of mixtures of arbitrary composition. *Laser Phys.* **2012**, *22*, 1495–1501. [CrossRef]
4. Gondal, I.A. Hydrogen integration in power-to-gas networks. *Int. J. Hydrog. Energy* **2019**, *44*, 1803–1815. [CrossRef]
5. Deng, Y.; Dewil, R.; Appels, L.; Van Tulden, F.; Li, S.; Yang, M.; Baeyens, J. Hydrogen-enriched natural gas in a decarbonization perspective. *Fuel* **2022**, *318*, 123680. [CrossRef]
6. Petrov, D.V.; Matrosov, I.I. Raman Gas Analyzer (RGA): Natural Gas Measurements. *Appl. Spectrosc.* **2016**, *70*, 1770–1776. [CrossRef]
7. Kiefer, J.; Seeger, T.; Steuer, S.; Schorsch, S.; Weikl, M.C.; Leipertz, A. Design and characterization of a Raman-scattering-based sensor system for temporally resolved gas analysis and its application in a gas turbine power plant. *Meas. Sci. Technol.* **2008**, *19*, 1–9. [CrossRef]
8. Hippler, M. Cavity-Enhanced Raman Spectroscopy of Natural Gas with Optical Feedback cw-Diode Lasers. *Anal. Chem.* **2015**, *87*, 7803–7809. [CrossRef]
9. Dąbrowski, K.M.; Kuczyński, S.; Barbacki, J.; Włodek, T.; Smulski, R.; Nagy, S. Downhole measurements and determination of natural gas composition using Raman spectroscopy. *J. Nat. Gas Sci. Eng.* **2019**, *65*, 25–31. [CrossRef]
10. Sharma, R.; Poonacha, S.; Bekal, A.; Vartak, S.; Weling, A.; Tilak, V.; Mitra, C. Raman analyzer for sensitive natural gas composition analysis. *Opt. Eng.* **2016**, *55*, 104103. [CrossRef]
11. Gao, Y.; Dai, L.-K.; Zhu, H.-D.; Chen, Y.-L.; Zhou, L. Quantitative Analysis of Main Components of Natural Gas Based on Raman Spectroscopy. *Chin. J. Anal. Chem.* **2019**, *47*, 67–76. [CrossRef]
12. Zhu, H.; Zhou, L.; Chang, H.; Sun, X. Study on Standardization of Natural Gas Composition Analysis by Laser Raman Spectroscopy. *Spectrosc. Spectr. Anal.* **2018**, *38*, 3286–3294.
13. Sieburg, A.; Knebl, A.; Jacob, J.M.; Frosch, T. Characterization of fuel gases with fiber-enhanced Raman spectroscopy. *Anal. Bioanal. Chem.* **2019**, *411*, 7399–7408. [CrossRef] [PubMed]
14. Knebl, A.; Domes, C.; Domes, R.; Wolf, S.; Popp, J.; Frosch, T. Hydrogen and C2–C6 Alkane Sensing in Complex Fuel Gas Mixtures with Fiber-Enhanced Raman Spectroscopy. *Anal. Chem.* **2021**, *93*, 10546–10552. [CrossRef]
15. Khannanov, M.N.; Van'kov, A.B.; Novikov, A.A.; Semenov, A.P.; Gushchin, P.A.; Gubarev, S.I.; Kirpichev, V.E.; Morozova, E.N.; Kulik, L.V.; Kukushkin, I.V. Analysis of Natural Gas Using a Portable Hollow-Core Photonic Crystal Coupled Raman Spectrometer. *Appl. Spectrosc.* **2020**, *74*, 1496–1504. [CrossRef]
16. Chibirev, I.; Mazzoleni, C.; van der Voort, D.D.; Borysow, J.; Fink, M. Raman spectrometer for field determination of H_2O in natural gas pipelines. *J. Nat. Gas Sci. Eng.* **2018**, *55*, 426–430. [CrossRef]

17. Sandfort, V.; Trabold, B.M.; Abdolvand, A.; Bolwien, C.; Russell, P.S.J.; Wöllenstein, J.; Palzer, S. Monitoring the Wobbe Index of Natural Gas Using Fiber-Enhanced Raman Spectroscopy. *Sensors* **2017**, *17*, 2714. [CrossRef]
18. Petrov, D. Comment on Hydrogen and C2-C6 Alkane Sensing in Complex Fuel Gas Mixtures with Fiber-Enhanced Raman Spectroscopy. *Anal. Chem.* **2021**, *93*, 16282–16284. [CrossRef]
19. *ISO 6974-5:2014*; Natural Gas—Determination of Composition and Associated Uncertainty by Gas Chromatography—Part 5: Isothermal Method for Nitrogen, Carbon Dioxide, C1 to C5 Hydrocarbons and C6+ Hydrocarbons; International Organization for Standardization: Geneva, Switzerland, 2014.
20. Petrov, D.V.; Matrosov, I.I.; Zaripov, A.R.; Maznoy, A.S. Application of Raman Spectroscopy for Determination of Syngas Composition. *Appl. Spectrosc.* **2020**, *74*, 948–953. [CrossRef]
21. Penner, A.R.; Meinander, N.; Tabisz, G.C. The spectral intensity of the collision-induced rotational raman scattering by gaseous CH_4 and CH_4-inert gas mixtures. *Mol. Phys.* **1985**, *54*, 479–492. [CrossRef]
22. Meinander, N.; Tabisz, G.C.; Barocchi, F.; Zoppi, M. Revisiting the collision-induced light scattering spectrum of gaseous CH_4. *Mol. Phys.* **1996**, *89*, 521–531. [CrossRef]
23. Petrov, D.V. Pressure dependence of peak positions, half widths, and peak intensities of methane Raman bands (ν_2, $2\nu_4$, ν_1, ν_3, and $2\nu_2$). *J. Raman Spectrosc.* **2017**, *48*, 1426–1431. [CrossRef]
24. Petrov, D.V.; Matrosov, I.I.; Zaripov, A.R.; Maznoy, A.S. Effects of pressure and composition on Raman spectra of CO-H_2-CO_2-CH_4 mixtures. *Spectrochim. Acta Part A Mol. Biomol. Spectrosc.* **2019**, *215*, 363–370. [CrossRef] [PubMed]
25. Petrov, D.V. Raman spectrum of methane in nitrogen, carbon dioxide, hydrogen, ethane, and propane environments. *Spectrochim. Acta-Part A Mol. Biomol. Spectrosc.* **2018**, *191*, 573–578. [CrossRef]
26. *ISO 6976:2016*; Natural Gas—Calculation of Calorific Values, Density, Relative Density and Wobbe Indices from Composition; International Organization for Standardization: Geneva, Switzerland, 2016.
27. Quay, P.; Stutsman, J.; Wilbur, D.; Snover, A.; Dlugokencky, E.; Brown, T. The isotopic composition of atmospheric methane. *Global Biogeochem. Cycles* **1999**, *13*, 445–461. [CrossRef]
28. Nikitin, A.V.; Mikhailenko, S.; Morino, I.; Yokota, T.; Kumazawa, R.; Watanabe, T. Isotopic substitution shifts in methane and vibrational band assignment in the 5560-6200 cm^{-1} region. *J. Quant. Spectrosc. Radiat. Transf.* **2009**, *110*, 964–973. [CrossRef]
29. Stolper, D.A.; Lawson, M.; Formolo, M.J.; Davis, C.L.; Douglas, P.M.J.; Eiler, J.M. The utility of methane clumped isotopes to constrain the origins of methane in natural gas accumulations. *Geol. Soc. Lond. Spec. Publ.* **2017**, *468*, 23–52. [CrossRef]
30. Vitkin, V.; Polishchuk, A.; Chubchenko, I.; Popov, E.; Grigorenko, K.; Kharitonov, A.; Davtian, A.; Kovalev, A.; Kurikova, V.; Camy, P.; et al. Raman Laser Spectrometer: Application to $^{12}C/^{13}C$ Isotope Identification in CH_4 and CO_2 Greenhouse Gases. *Appl. Sci.* **2020**, *10*, 7473. [CrossRef]
31. Uda, T.; Okuno, K.; O'hira, S.; Naruse, Y. Preliminary Study of Isotopic Methanes Analysis by Laser Raman Spectroscopy for *In-Situ* Measurement at Fusion Fuel Gas Processing. *J. Nucl. Sci. Technol.* **1991**, *28*, 618–626. [CrossRef]
32. Tanichev, A.S.; Petrov, D.V. Simulation of ν_2 Raman band of methane as a function of pressure. *J. Raman Spectrosc.* **2022**, *53*, 654–663. [CrossRef]

sensors

Article

A New Pattern Quality Assessment Criterion and Defocusing Degree Determination of Laser Speckle Correlation Method

Wenxin Hu [1], Zhipeng Sheng [1], Keyu Yan [1], Hong Miao [2] and Yu Fu [1,*]

[1] College of Physics and Optoelectronic Engineering, Shenzhen University, 3688 Nanhai Avenue, Shenzhen 518060, China; huwenxin@szu.edu.cn (W.H.); shengzhipeng1993@163.com (Z.S.); yankeyu@szu.edu.cn (K.Y.)
[2] CAS Key Laboratory of Mechanical Behavior and Design of Materials, Department of Modern Mechanics, University of Science and Technology of China, Hefei 230027, China; miaohong@ustc.edu.cn
* Correspondence: fuyuoptics@gmail.com

Abstract: The laser speckle correlation method has found widespread application for obtaining information from vibrating objects. However, the resolution and accuracy of the laser speckle correlation method as they relate to the defocusing degree have not been analyzed sufficiently. Furthermore, the possible methods for speckle pattern quality assessment and enhancement have not been studied. In this study, the resolution and accuracy of the laser speckle correlation method are analyzed, and it is found that they are affected by the defocusing degree and speckle pattern quality, respectively. A new speckle pattern quality criterion combining the mean intensity gradient and frequency spectrum was proposed, called CMZ. The quality of the speckle pattern is higher when the CMZ is closer to zero. The proposed criterion was verified by simulated speckle patterns and real speckle patterns with different speckle sizes, densities, and gray contrasts. In the experimental setup stage, a suitable defocusing degree can be selected based on the resolution requirement and optimal speckle size, and other experimental parameters can be determined according to the CMZ criterion. Rotation and vibration experiments verified the effectiveness of the laser speckle correlation method and confirmed the reliability of the experiment preparation based on proposed CMZ criterion.

Keywords: laser speckle correlation; defocusing degree; speckle pattern quality criterion; vibration measurement; rotation measurement

Citation: Hu, W.; Sheng, Z.; Yan, K.; Miao, H.; Fu, Y. A New Pattern Quality Assessment Criterion and Defocusing Degree Determination of Laser Speckle Correlation Method. *Sensors* **2021**, *21*, 4728. https://doi.org/10.3390/s21144728

Academic Editors: Maria Lepore and Ines Delfino

Received: 7 May 2021
Accepted: 8 July 2021
Published: 10 July 2021

1. Introduction

Optical dynamic measurements have been widely used to detect noncontact vibrations, continuous deformation, or movement of objects in various research and industrial applications. The common methods are divided into interferometric and imaging-based methods. Interferometric methods include electric speckle pattern interferometry (ESPI) [1–4], shear interferometry [5,6], and holographic interferometry [7–9], and these methods generally produce subwavelength accuracy. Adopting high-efficiency phase extraction methods, such as the temporal phase-shifting method [10,11], spatial carrier phase-shifting method [12,13], and Fourier transform method [14,15], nanometer accuracy can be achieved under laboratory conditions. However, its applications are limited by the sampling rate of camera and environmental requirements. The more powerful laser Doppler vibrometry (LDV) technique [16–18] can provide single-point high-speed dynamic measurements using a photoelectric detector, but it is still essentially an interferometer that is sensitive to environmental fluctuations. Furthermore, the digital image correlation (DIC) method [19–21] is sensitive to object surface displacement, especially in-plane displacement. In DIC, artificial speckle or laser speckle is as a carrier of deformation information and deforms together with the specimen surface. Artificial speckle is most commonly used and is usually prepared by spraying paints on the sample surface [22] or transferring speckle patterns to sample surface using the water transfer printing technique [23]. However, laser speckles, formed by the reflection and scattering of laser irradiation onto a rough surface, are desirable in some situations. For example, in a high temperature environment, artificial speckle will inevitably fall off and discoloration will occur. Song et al. [24] and Zheng et al. [25] have applied laser speckle correlation method to a high temperature measurement field successfully.

Gregory et al. [26] proposed defocused speckle photography and described how to separate tilt (differential) topology variations on a scatter surface from linear displacements. Horváth et al. [27] deduced the relationship between the small-deformation tensor and the speckle field displacement in detail. Based on these analyses, another simple and prospective application of the laser speckle correlation method has emerged. Jo et al. [28] proposes to observe the movement of the secondary speckle patterns that are generated on top of a target when it is illuminated by a laser beam spot. Through proper defocusing, the movement of the object creates a scenario in which the same speckle pattern moves or vibrates in the transverse plane, instead of the speckle pattern constantly changing. Gradually, this method has achieved widespread application for obtaining information from vibrating objects. For instance, Zeev et al. [29], Lin et al. [30], and Yevgeny et al. [31] applied this method for the simultaneous remote extraction of multiple speech sources, vibration measurements, and blood pulse pressure measurements, respectively. Furthermore, Wu et al. [32] introduced a high-speed optical flow algorithm to tracking laser speckle images to realize real-time audio detection and regeneration of a moving sound source. However, the influence of the laser speckle quality and possible ways to achieve quality assessment and enhancement have not received enough attention in vibration measurements. Furthermore, the key parameter defocusing degree can be selected combining optimal speckle pattern quality and resolution requirement.

In the DIC field, many quality assessment criteria aimed at sprayed speckle patterns have been developed gradually. Subset entropy [33] and the sum of square subset intensity gradients (SSSIG) [34] are suitable for subset optimization. In order to evaluate the quality of the whole speckle pattern, Lecompte et al. [35] first proposed the mean speckle size based on the image morphology, and then Grammond et al. [36] applied edge detection to determine the speckle size and density. These methods based on speckle morphology lack the ability to evaluate gray information, such as the contrast influence on the speckle pattern quality. To overcome this deficiency, the mean intensity gradient (MIG) [37], the mean intensity of the second derivative (MIOSD) [38] and standard deviation of gray intensities within each speckle (SDGIS) [39] are proposed successively. Another trend is to consider the primary and secondary peaks of the autocorrelation functions [40–42]. However, compared with the sprayed speckle, the laser speckle has a more uniform distribution of speckle particles, a smaller difference of the gray standard deviations between individual speckles, and non-obvious secondary auto-correlation peaks. Thus, the assessment criteria described above cannot be used directly. Song et al. proposed a new index, the multi-factor fusion index (MFFI) [43], which took the inhomogeneity of the gray contribution, the mean square deviation of the gray contribution, and the standard deviation of the speckle particles size into consideration.

In this study, the resolution of the laser speckle correlation method is analyzed, and its main influence factors are distance relationships between the measurement planes, which depend on the defocusing degree. The defocusing degree also affects the speckle pattern quality, further influencing the accuracy of the laser speckle correlation method. To ensure a high quality of the speckle pattern, a new speckle pattern quality criterion combining the MIG and frequency spectrum was proposed, called CMZ, which accounts for both random error and the interpolation bias. A simple rule is presented based on the balance of random error and interpolation bias, and it was verified that the quality of the speckle pattern is higher when CMZ is closer to zero. Furthermore, the particular characteristics of the laser speckle have been used, which were distinguished using traditional indices, such as MIG and MIOSD. The proposed criterion was demonstrated by simulated speckle patterns with different speckle sizes and densities. Experimental speckle patterns at different defocusing degrees, exposure times, and measurement locations and the corresponding translation patterns were then used to validate the proposed criterion. Based on the sufficient resolution requirement and optimal speckle size, a suitable defocusing degree (such that the distance relationships between the measurement planes can be determined) and other experimental parameters can be determined according to the CMZ criterion during the experimental setup. Rotation experiments were used to illustrate the relationship between the resolution and the defocusing degree, which further verified the CMZ criterion. Vibration experiments simultaneously verified the effectiveness of the laser speckle correlation method and the reliability of the experimental setup based on the proposed CMZ criterion.

2. Defocusing Degree Determination of Laser Speckle Correlation Method

By illuminating an object with a laser beam spot, a speckle pattern can be generated due to the roughness of the object surface. When a spatially coherent beam is reflected from the object whose roughness generates a random phase distribution, we may obtain the self interfering speckle pattern in the far field.

As shown in Figure 1a,b the camera is focused on the plane behind or in front of the object such that the object itself is defocused, respectively, where the focal plane is at a distance of Z_1.

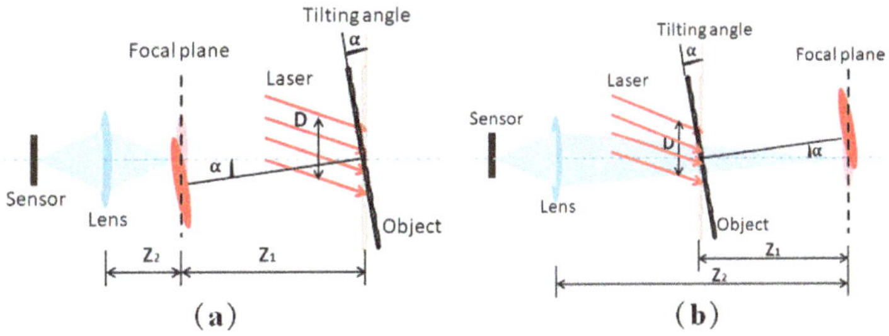

Figure 1. Schematic of the system: camera is focused on the plane (**a**) behind or (**b**) in front of the object.

According to the analysis conducted by Zeeval et al. [29], this system was sensitive to the tilt, and the effect caused by transversal and axial movement is negligible. When slightly defocusing, object tilt creates a situation in which the same speckle pattern only moves or vibrates in the transverse plane instead of constantly changing the speckle pattern. Thus, shifts of the speckle pattern due to tilt can be easily detected by spatial pattern correlation. According to the geometric relation, the tilt angle α can then be determined as follows:

$$\alpha = \frac{Z_1 U}{M} \quad (1)$$

where U represents the displacement of the pattern on the camera. M is the imaging system magnification. Once tilt angles along the time axis are obtained during vibration, vibration information including frequency and strain can be calculated. Thus, accurate correlation tracking is a prerequisite to tilt angle calculation, even to vibration analysis.

Valid correlation calculations require a suitable speckle size to be imaged to the sensor plane. In the case of an objective laser speckle, the speckle size S is described as follows:

$$S \approx \frac{\lambda Z_1}{D} \quad (2)$$

where λ and D are the optical wavelength and the dimension of the illuminated spot, respectively. The size S'. of the speckle imaged to the sensor plane, which is obtained at the Z_1 plane, is expressed as follows:

$$S' = \frac{\lambda Z_1}{DM} \tag{3}$$

To ensure that every speckle in the sensor plane equals K pixels, the condition is described as follows:

$$\frac{\lambda Z_1}{DM} = K \times L_x \tag{4}$$

where L_x is the physical size of one pixel in the CCD sensor.

According to formula (1), a greater distance Z_1 and a smaller magnification factor M correspond to a higher tilt angle resolution U. When the camera magnification is fixed, the larger Z_1 corresponds to larger angle resolution U in situations (a) and (b). However, usually the distance between the sensor plane and object is fixed, but the defocusing degree can be adjusted by changing the focal length. When the adjusted parameter is the defocusing degree, the rule is different in the two situations. If the camera is focused on the plane behind the object, as in situation (a), a greater distance Z_1 means a shorter object distance Z_2, that is, a smaller magnification factor M and a higher angle resolution U. For situation (b), a greater distance Z_1 corresponds to a greater magnification factor M, so the variation of $\frac{Z_1}{M}$ cannot be judge directly, causing non-determinacy of the angle resolution change. Thus, a reasonable defocusing degree that determines the relative distances between the object, the focal plane, and the sensor plane can be obtained according to speckle size requirement imaged to the sensor plane and a sufficient tilt angle resolution requirement.

3. Laser Speckle Pattern Quality Assessment

An effective speckle pattern quality assessment criterion is a prerequisite to ensure correlation tracking. In general, correlation calculation error consists of random error and interpolation bias. Random error highly depends on image noise, which is related to the gray scale of the image. For zero- and first-order shape functions, Pan et al. [34] pointed out that the random error Std is defined as follows:

$$\mathrm{Std} = \frac{\sigma}{\sqrt{\sum_{i=-N}^{N}\sum_{j=-N}^{N}\frac{(g_x[i,j])^2+(g_y[i,j])^2}{2}}} \tag{5}$$

where N is half of the size of the subset, σ is the standard deviation of the image noise. $g_x[i,j]$ and $g_y[i,j]$ are the x- and y-directional gray derivatives at point $[i,j]$, respectively. Su et al. [44] introduced interpolation bias kernel to characterize the frequency response of the interpolation bias, and interpolation bias kernel was defined as

$$E(v_x, v_y) = (v_x - 1)\varphi(v_x - 1, v_y) - (v_x + 1)\varphi(v_x + 1, v_y) + \varphi(v_x, v_y)(v_x + 1, v_y) + \varphi(v_x, v_y)(v_x - 1, v_y) \tag{6}$$

where $\varphi(v_x, v_y)$ represents interpolation function (cubic BSpline) at frequency (v_x, v_y). The result curve of interpolation bias kernel verified that high-frequency components are the major source of interpolation bias.

As for sprayed speckle patterns assessment, MIG and MIOSD are the most commonly used. MIG is defined as

$$\mathrm{MIG} = \frac{\sum_{i=1}^{W}\sum_{j=1}^{H}\sqrt{g_x(x)_{ij}^2 + g_y(x)_{ij}^2}}{W \times H} \tag{7}$$

MIOSD is defined as:

$$\text{MIOSD} = \frac{\sum_{i=1}^{W}\sum_{j=1}^{H}\sqrt{g_{xx}(x)_{ij}^{2} + g_{yy}(x)_{ij}^{2}}}{W \times H} \tag{8}$$

where $g_x(x)_{ij}$ and $g_y(x)_{ij}$ are the x- and y-directional gray derivatives at position x_{ij}, respectively. $g_{xx}(x)_{ij}$ and $g_{yy}(x)_{ij}$ are the x- and y-directional intensity of the second derivatives at position x_{ij}, respectively. W and H represent the pixel width and pixel height, respectively.

The MIG and MIOSD are defined according to gray gradient, which are supposed to assess random error sufficiently. However, the other component interpolation bias is not only related to the gray gradient. Typically, MIG should be large and continue increasing as the speckle particle size decreases, but this does not mean the smallest speckle particle size is optimal. Based on the rich research in the DIC field, the optimal speckle size is 3–5 pixels [41–44]. If the speckle size is too small, it leads to image under-sampling, which causes a large interpolation bias. If the speckle size is too large, the details of the image are not rich enough, and the contrast is poor, resulting in a large random error.

Considering high-frequency components are the major source of interpolation bias, we present a new concept called the zero spectrum ratio (ZSR) to quantize frequency spectrum component. The ZSR is defined as

$$\text{ZSR} = \frac{\max[\text{FFT}(g)]}{\text{sum}[\text{FFT}(g)]} \tag{9}$$

where $\text{FFT}(f)$ represents the Fourier transform of the speckle pattern, and $\max[\text{FFT}(g)]$ and $\text{sum}[\text{FFT}(g)]$ are the maximum and sum of the frequency spectrum, respectively. The value of ZSR represents the proportion of the zero-order spectrum, which is designed to be related to interpolation bias.

To take both the interpolation error and the random error into account, we propose to combine these two indexes. Because laser speckles have a more uniform distribution of speckle particles and a smaller difference of the gray standard deviation between individual speckle particles, we first analyze the speckle particle size and speckle density effects on the MIG and the ZSR. A speckle density of 100% means that adjacent speckle particles are in contact with each other.

As shown in Figure 2, we simulated two series of speckle patterns. In series (a), the speckle size increased from 2 to 14 pixels, and the speckle density remained at 50%. In series (b), the speckle size was unchanged, but the speckle density decreased from 80% to 20%. Although real speckle patterns for surfaces could have significantly different appearances [45], this simple model was used to illustrate the proposed quality assessment criterion, then real experiment speckle patterns were applied to ensure its validation.

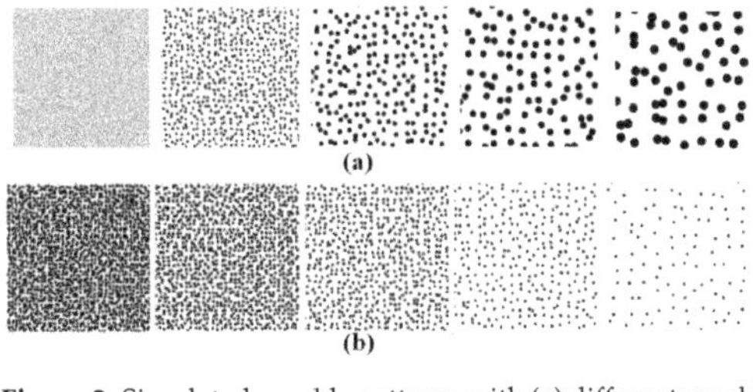

Figure 2. Simulated speckle patterns with (**a**) different speckle sizes and (**b**) different speckle densities.

Figure 3a,b give the results of the speckle pattern series (a) and (b), respectively. We found that the values of MIG and ZSR undergo opposite changes with the speckle size increasing or the speckle density decreasing. We propose a new assessment criterion: the speckle pattern quality is higher when the normalized MIG and normalized ZSR are closer to the same value, which can also be understood as finding the intersection of the MIG and ZSR curves. Thus, neither MIG nor ZSR is too big, balancing the interpolation bias and random error. Based on this idea, a new parameter combining the MIG and the ZSR, named CMZ, is defined as

$$CMZ = |MIG' - ZSR'| \tag{10}$$

$$MIG' = \begin{cases} 1 \text{ If } MIG > 35 \\ \frac{MIG-5}{30} \text{ If } 5 < MIG < 35 \\ 0 \text{ If } MIG < 5 \end{cases} \tag{11}$$

$$ZSR' = \begin{cases} 1 \text{ If } ZSR > 0.08 \\ \frac{ZSR-0.01}{0.07} \text{ If } 0.01 < ZSR < 0.08 \\ 0 \text{ If } ZSR < 0.01 \end{cases} \tag{12}$$

where MIG′ and ZSR′ are the normalized MIG and normalized ZSR, respectively. The ranges of MIG and ZSR need to be determined in advance. Considering the limiting case, the density and the size of speckle were set to be 100% and 1 pixel, respectively, and the corresponding MIG was 35. Thus, we set the MIG range to be 0–35, ignoring the exceeding part. Similarly, in the limiting case where the densities were set to be 20% and 85%, the corresponding ZSR values were 0.0817 and 0.011, respectively, where the size of speckle was 2 pixels. Thus, we set the ZSR range to be 0.01–0.08. We concluded that the quality of the speckle pattern was higher when CMZ was closer to zero.

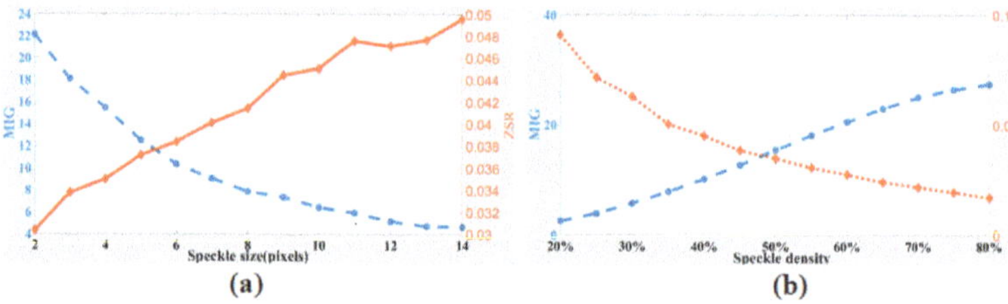

Figure 3. Assessment results of speckle patterns with (**a**) different speckle sizes and (**b**) different speckle densities.

4. Experimental Verification

4.1. CMZ Assessment Criterion

To verify the effectiveness of the proposed assessment criterion based on the CMZ, experimental speckle patterns at different defocusing degrees, exposure times, and surface roughness were evaluated, and results were compared with the MIG and the MIOSD. As shown in Figure 4a, the object was irradiated by a laser beam, and then the surface of the object formed bright spots and dark spots due to coherent subwaves interference. The defocused speckle pattern was captured by a charge-coupled device (CCD) camera and analyzed by the computer. Figure 4b shows the experimental setup. The laser power of the He-Ne laser was 20 mW, and the wavelength was 632.8 nm. Defocusing degree related to speckle particle size was controlled by focal length. Exposure time related to gray contrast was adjusted through camera software. In order to ensure different roughness at different measurement locations, uniform white paint was sprayed on the surface of the specimen, and then sandpaper was used to polish different parts in different degrees.

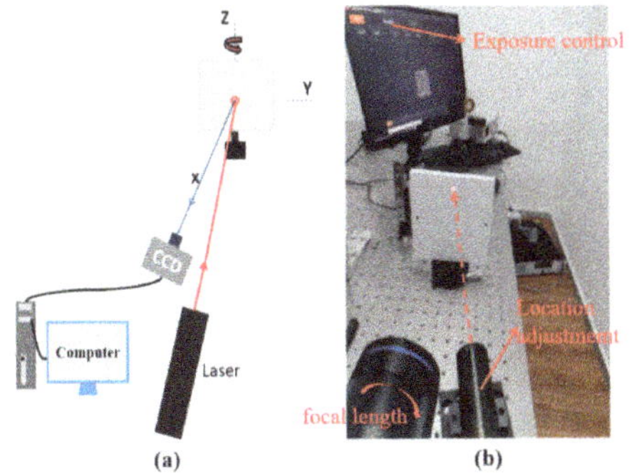

Figure 4. Experimental setup: (**a**) schematic of the experimental setup and (**b**) physical diagram of experimental setup.

The quality of the speckle pattern can be affected by factors: the defocusing degree, the camera exposure time, and the measurement location. The defocusing degree and the camera exposure time are related to speckle particle size and gray contrast, respectively, and the different measurement locations are due to the roughness difference.

During speckle pattern acquisition, the defocusing degree was adjusted gradually, and the exposure time of the camera was then slightly adjusted around the value 8000 to retain the gray contrast. Twenty-eight speckle patterns (Group A: A1–A28, size: 256 × 256 pixels) with increasing speckle sizes were obtained, which are partially shown in Figure 5. In order to obtain deformed images with 0–1 pixel translation, sub-pixel shifted operation along the x-direction in the Fourier domain [34] was done. The step of translation was set to be 0.1 pixels. Decorrelation caused by big displacement or tilt should be avoided to ensure the validation of correlation calculation. Based on the DIC algorithm described in [46], the displacements of 81 points of each deformed image were calculated, and then the curves of the mean bias errors and the standard deviations of the displacements were obtained.

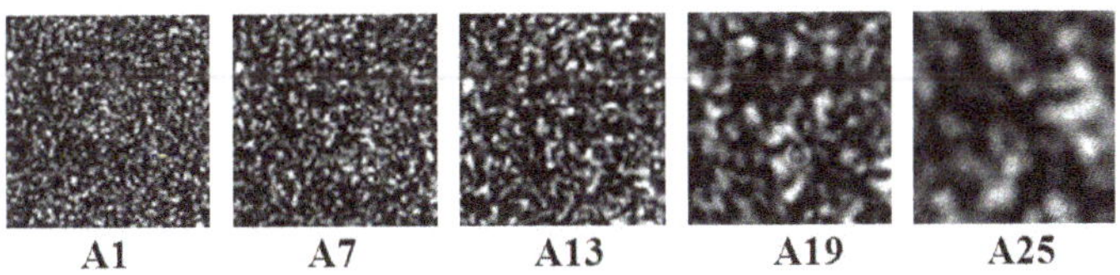

Figure 5. Speckle patterns collected at different defocusing degrees.

The average speckle size curve of the speckle pattern in group A is shown in Figure 6a. Results curves of MIG, MIOSD, ZSR, and CMZ assessment criteria are shown in Figure 6b. When sub-pixel translation is imposed to be 0.3 pixels, mean bias error curve of calculated displacement with different CMZ values is shown in Figure 6c, and standard deviation curve of calculated displacement with different CMZ values is shown in Figure 6d. The MIG and MIOSD values decreased as the speckle size increased. Conversely, the ZSR values increased as the speckle size increased. According to Figure 6b, the values of the proposed CMZ assessment criterion decreased first and then increased. Based on our research, the smaller the value of CMZ was, the higher the speckle pattern quality was. Thus, speckle patterns A15–A20, whose CMZ values were less than 0.3, were superior to the other patterns. As shown in Figure 6a, the speckle sizes of speckle patterns A15–A20 were between 3 and 5 pixels, which are consistent with the optimal speckle size during the DIC calculation. Results of Figure 6c,d verified that quality of the speckle pattern was higher when CMZ was closer to zero. Furthermore, standard deviation curve was relatively flat when CMZ was small.

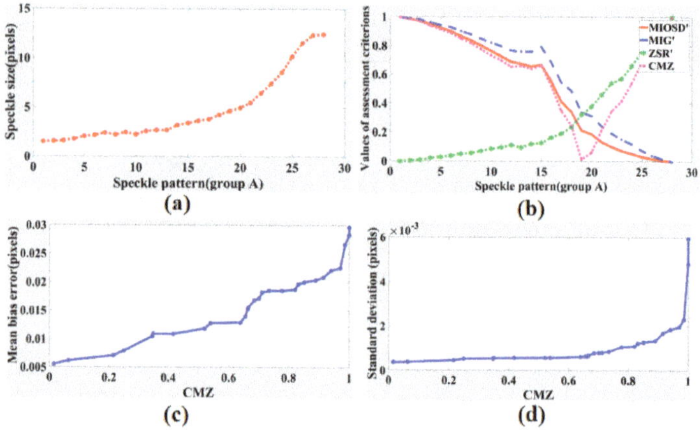

Figure 6. Results of speckle patterns in group A: (**a**) speckle size curve, (**b**) assessment criteria, (**c**) mean bias errors with different CMZ values, and (**d**) standard deviations with different CMZ values.

The curves of the mean bias errors and standard deviations calculated with the subset of 31 × 31 pixels are shown in Figure 7a,c, respectively. The curves of the mean bias errors and standard deviations calculated with the subset of 61 × 61 pixels are shown in Figure 7b,d, respectively. The mean bias errors and standard deviations both decreased first and then increased, and the displacement calculation errors of pattern A19 were the lowest. The results were consistent with the results predicted by the proposed CMZ criterion. Furthermore, calculation with the subset of 61 × 61 pixels performed better. Therefore, choosing a larger subset can improve the calculation accuracy.

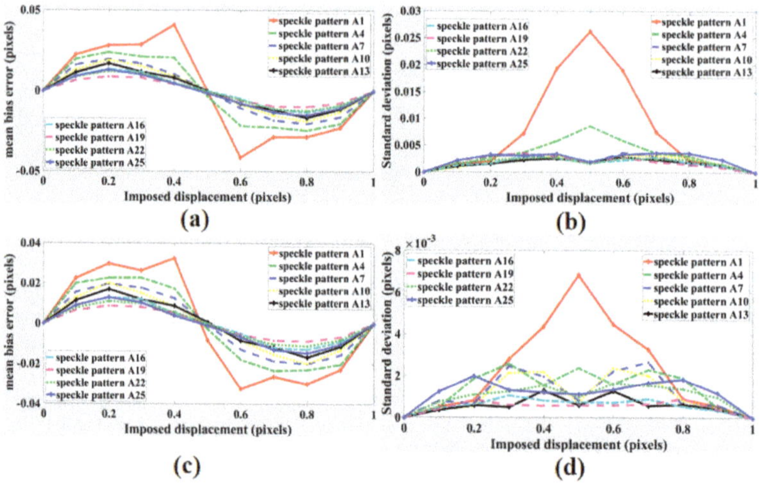

Figure 7. (**a**) Mean bias errors and (**b**) standard deviations of speckle patterns displacements in group A calculated with the subset of 31 × 31 pixels. (**c**) Mean bias errors and (**d**) standard deviations of speckle patterns displacements in group A calculated with the subset of 61 × 61 pixels.

To obtain speckle patterns with different gray contrasts, the defocusing degree was unchanged, and the exposure time of the camera was adjusted from 1000 to 16,000 gradually. Sixteen speckle patterns (Group B: B1–B16) were obtained, which are partly shown in

Figure 8, and then the same translation operation and displacement error calculation were performed.

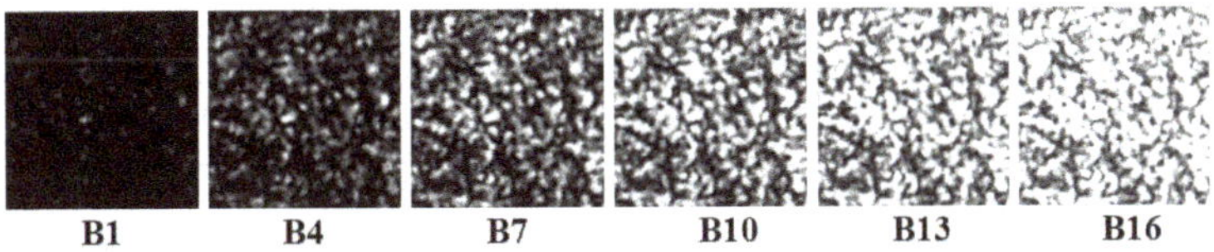

Figure 8. Speckle patterns collected at different camera exposure times.

The result curves of different assessment criteria are shown in Figure 9a. The MIG value increased and then decreased as the exposure time increased. According to the greater MIG principle, speckle patterns B13–B15 had stronger abilities to resist noise, which are supposed to have higher calculation accuracies. However, the proposed CMZ principle considers that the qualities of speckle patterns B6–B10 were superior. The trend of MIG and CMZ both indicate that the speckle pattern quality will decline when overexposure or underexposure because of decline of gray contrast. Mean bias error curve and standard deviation curve of calculated displacement with different CMZ values are drawn in Figure 9b,c, respectively. Mean bias error increased when CMZ value increased. Standard deviation and CMZ value also showed synchronous growth.

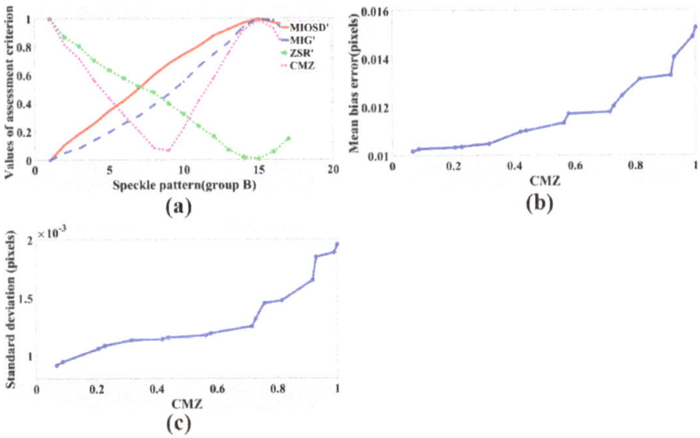

Figure 9. (**a**) Results of speckle patterns in group B: (**a**) assessment criteria, (**b**) mean bias errors with different CMZ values, and (**c**) standard deviations with different CMZ values.

The curves of the mean bias errors and standard deviations calculated with the subset of 31 × 31 pixels are shown in Figure 10a,c, respectively. The curves of the mean bias errors and standard deviations calculated with the subset of 61 × 61 pixels are shown in Figure 10b,d, respectively. The mean bias errors and standard deviation both decreased first and then increased, which verified that the contrast of speckle patterns will be reduced due to overexposure or underexposure, causing decline of speckle pattern quality. According to displacement calculation errors, pattern B6–B10 perform better, which was closer to the result predicted by the proposed criterion CMZ. Similarly, calculation with the subset of 61 × 61 pixels performed better.

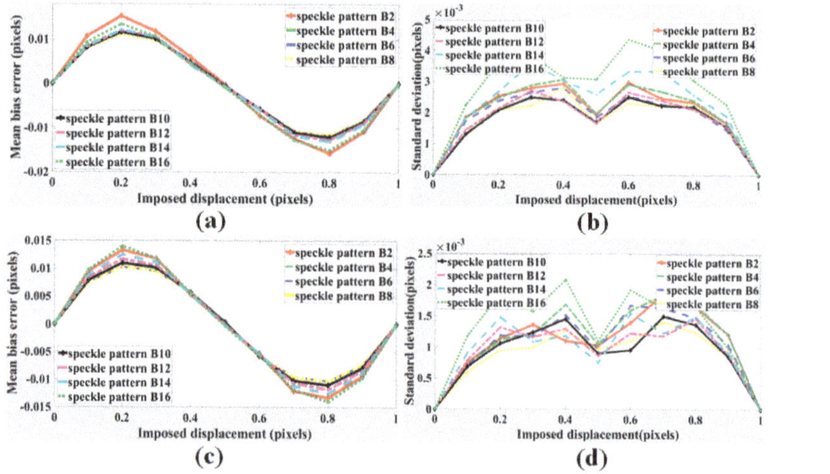

Figure 10. (**a**) Mean bias errors and (**b**) standard deviations of of speckle patterns displacements in group B calculated with the subset of 31 × 31 pixels. (**c**) Mean bias errors and (**d**) standard deviations of speckle patterns displacements in group B calculated with the subset of 61 × 61 pixels.

To further evaluate the proposed CMZ assessment criterion, we changed the measurement locations to acquire speckle patterns C1–C25 with nearly the same gray contrasts and speckle sizes but different speckle particle location distributions due to the object surface roughness, as partially shown in Figure 11. The mean bias error and standard deviation curves calculated with the subset of 31 × 31 pixels are shown in Figure 12a,b, respectively, and those calculated with the subset of 61 × 61 pixels are shown in Figure 12c,d, respectively. The results of different speckle patterns were similar, and no trend was evident.

The results for different assessment criteria are shown in Figure 13. The MIG, MIOSD, ZSR, and CMZ values all remained stable, which is consistent with the displacement calculation results presented in Figure 12. Thus, the surface roughness of the object has little effect on its speckle pattern quality. High laser speckle pattern quality depends on suitable speckle size and gray contrast, which can be controlled by defocus degree and exposure time and so on. When the gray contrast is unchanged and the speckle size keeps within a suitable range, the displacement calculation accuracy of the laser speckle pattern remains stable with different speckle particle distributions.

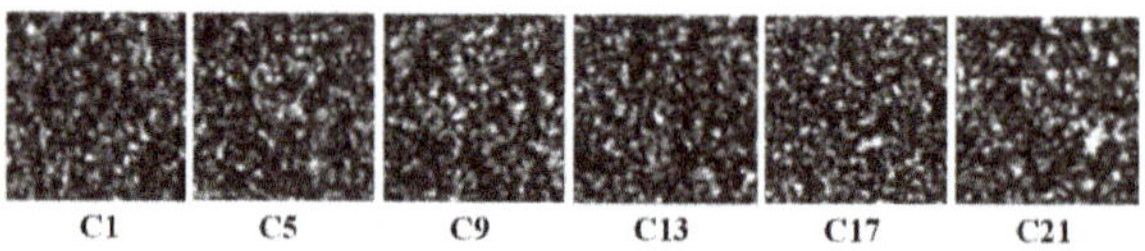

Figure 11. Speckle patterns collected at different measurement locations.

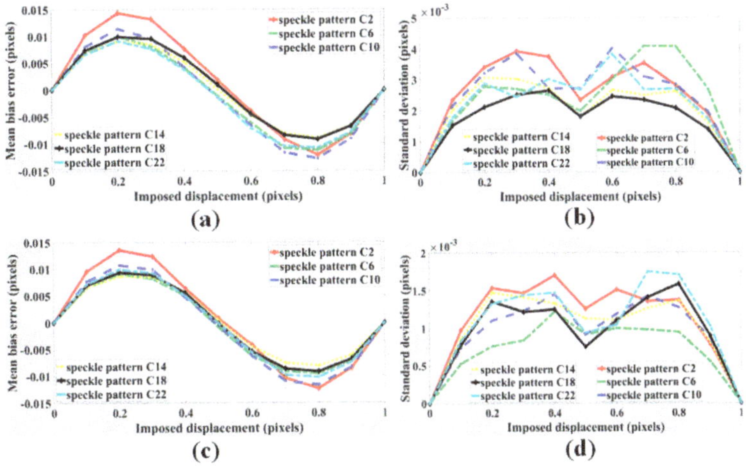

Figure 12. (**a**) Mean bias errors and (**b**) standard deviations of speckle patterns displacements in group C calculated with the subset of 31 × 31 pixels. (**c**) Mean bias errors and (**d**) standard deviations of speckle patterns displacements in group C calculated with the subset of 61 × 61 pixels.

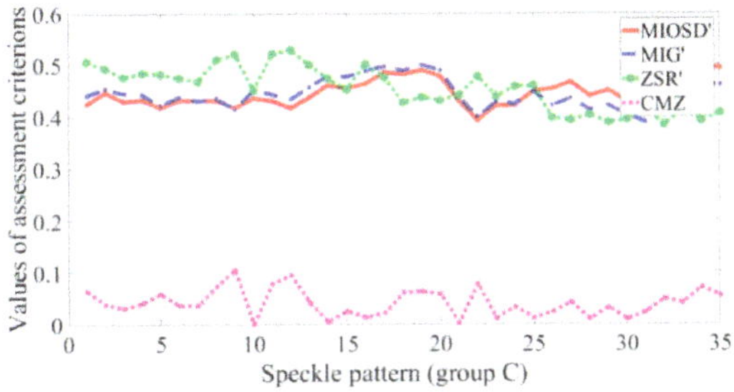

Figure 13. Comparison of different assessment criteria.

4.2. Rotation Experiment

Considering the vibration information is obtained from tilt angle, rotation experiments were first designed to test angle calculation accuracy and further verify the proposed quality assessment. Besides, different relative distances between the object, the focal plane, and the sensor plane were adopted, which corresponded to different defocusing degree, so as to illustrate the relationship between defocusing degree and the angle resolution. The schematic diagram of the rotation experiment is shown in Figure 4a. A square measured plate was placed on a rotary platform, and the position accuracy was $2'$. The plate was irradiated by a laser beam, and then the defocused speckle pattern was acquired by a CCD camera. The laser power of He-Ne laser was 20 mW, and the wavelength was 632.8 nm. We conducted four experiments. In two, the measured object was placed behind the focal plane, and in the other two, the measured object was placed in front of the focal plane, as shown in Figure 14a,c. In the two experiments with the object in front (shown in Figure 14b we changed the focal plane by adjusting the focal length slightly and keeping the other parameters fixed. In the two experiments with the object behind (shown in Figure 14d, only the location of the object changed. Based on the optimal speckle size determined by experience, the relative distances were first determined according to Equation (3) to ensure that the speckle sizes were all around 5 pixels. Next, the exposure time of the camera was adjusted to ensure a low CMZ value, because a low CMZ value corresponds to a high quality of the speckle pattern. The four speckle patterns given in Figure 15 are from the four different experiments. The speckle sizes and CMZ values of these speckle patterns were also calculated and are shown. The CMZ value of the third experiment was highest due to its low gray contrast. The CCD camera was a CP70 1HS M/C (Optronis, Germany). The physical size of the CCD sensor was 17.536 × 11.782 mm, and the size of the acquired speckle pattern was 1280 × 860 pixels.

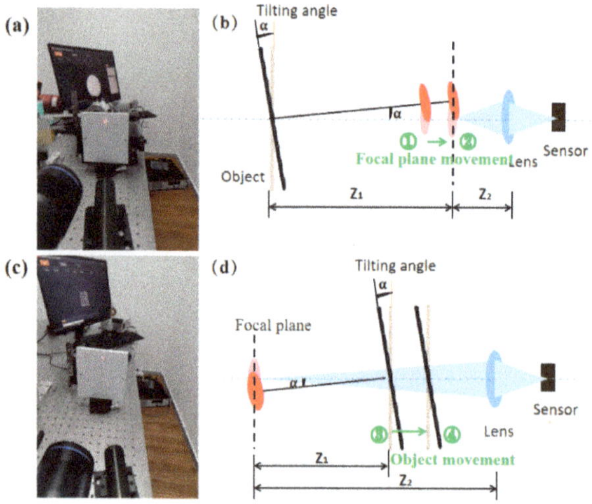

Figure 14. (**a**) Physical and (**b**) schematic diagrams of the first two experiments. (**c**) Physical and (**d**) schematic diagrams of the last two experiments.

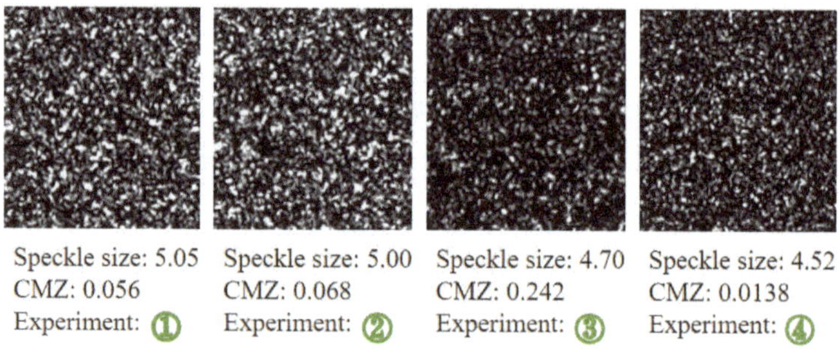

Figure 15. Four speckle patterns from the different experiments, their speckle sizes, and their CMZ values.

As shown in Figure 16, we used an object with a known size, such as a wire or wafer, to find the focal plane where the acquired image was clearest. The camera magnification factor was calculated through the real width of the wire or wafer and its corresponding pixel width. The rotation angle resolution, which was defined as the calculated displacement at the focal plane when rotating 1′, could be obtained through the magnification factor and the distance between the object and the focal plane. In the four experiments, we adjusted the rotation angle to acquire different speckle patterns, and then displacement fields containing 100 calculated points at the focal plane were calculated by normal DIC algorithm. Mean displacement of each speckle pattern was eventually converted to the measured angle using Equation (4). The range of the rotation angle was 4′–20′, and the step of rotation was set to be 2′.

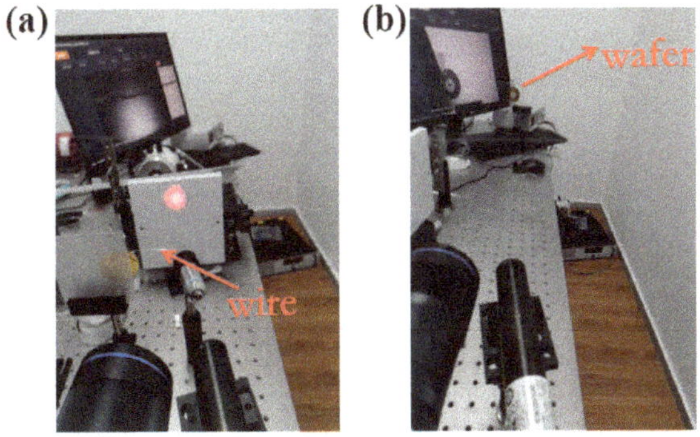

Figure 16. Physical diagram of focal plane determination in (**a**) the first two experiments and (**b**) the last two experiments.

The measurement parameters and calculated angle resolutions are shown in Table 1. Comparing experiments 1 and 2, we found that when the distance between the sensor plane and object was fixed and the object was placed between the focal and sensor planes, the larger Z_1 corresponded to a larger angle resolution. According to experiments 3 and 4, when the magnification was fixed, Z_1 was larger, and the angle resolution was higher. Thus, the results are consistent with the analysis based on Equation (1).

Table 1. Comparison of measurement parameters and calculated angle resolutions.

Experiment	Object at Focus Plane (Pixels/mm)	M	Z1 (mm)	Z2 (mm)	Angle Resolution (Pixels/Minutes)
1	36	0.4932	324	150	3.393
2	32	0.4384	313	161	2.913
3	9	0.1233	730	1200	1.800
4	9	0.1233	754	1200	1.971

Notice: (1 pixels/minute = 1 pixels/ $\frac{\pi}{180\times60}$ rad).

For all rotation angles, the real displacements at the focal plane are plotted as a line and the calculated displacements are plotted as scatter points in Figure 17, and results of the different experiments were all in good agreement. The angle resolution was considered to be the slope of the best fit line to the scatter points.

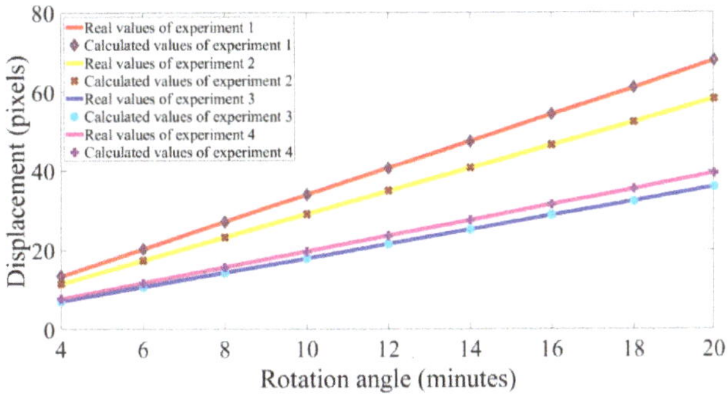

Figure 17. Comparison of displacement results at different rotation angles.

Table 2 gives the calculated slopes and errors. The calculated slopes were equal to the desired angle resolution. The mean errors of the displacements were less than 0.04 pixels, verifying the effectiveness of the laser speckle correlation method and further verifying the effectiveness of the CMZ speckle pattern quality criteria. Furthermore, the mean error and standard deviation of third experiment were the highest, which was consistent with the largest CMZ value of its speckle pattern. The mean errors of the rotation angle did not exceed 0.02 min, but the error included correlation calculation errors and calculated errors of the camera magnification.

Table 2. Comparison of error results from different experiments.

Experiment	Calculated Slope (Pixels/Minutes)	Displacement at Focus Plane (Pixels)		Rotation Angle (Minutes)	
		Mean Error	Standard Deviations	Mean Error	Standard Deviations
1	3.393	0.023	0.008	0.017	0.0064
2	2.913	0.030	0.012	0.017	0.0042
3	1.800	0.036	0.009	0.020	0.0053
4	1.971	0.021	0.011	0.006	0.0034

4.3. Vibration Experiment

Two real vibration experiments were conducted to further evaluate accuracy of the laser speckle correlation method. Using the same measurement system, a vibrating beam was used as the measured object. The vibration signal was passed through the signal generator, to the amplifier, and finally to the cantilever beam. To ensure that the speckle sizes were all around 5 pixels, the defocusing degree was determined using Equation (4), and then the exposure time was adjusted to ensure a low CMZ value to acquire a speckle pattern with a high quality. The size of the vibrating beam was 235 mm × 10 mm, as shown in Figure 18. Eight points (A–H) were measured in turn, where the distance from point A to the fixed end was 50 mm and the interval of the measurement points was 20 mm. The first experiment is illustrated in Figure 19a. At a distance of 30 mm from the free end, a simple harmonic excitation was applied to the cantilever beam with a frequency of 10 Hz, which was close to the first-order natural frequency of the beam. In the second experiment, which is shown in Figure 19c, the simple harmonic excitation was applied to point B with a frequency of 30 Hz. One speckle pattern coming from each experiment and the corresponding CMZ values are given in Figure 19b,d respectively. The laser power of the He-Ne laser was 20 mW, and the wavelength was 632.8 nm. However, the size of acquired speckle pattern was set to be 256 × 256 pixels to achieve a high sampling rate. The sampling rate was 2000 Hz, and the sampling time was 0.5 s. Meanwhile, laser doppler vibrometry (LDV) with a sampling rate of 40,000 Hz was used to collect the vibration signals. Finally, the calculated results of the speckle correlation method were compared with the signal processing results of the Doppler vibrator.

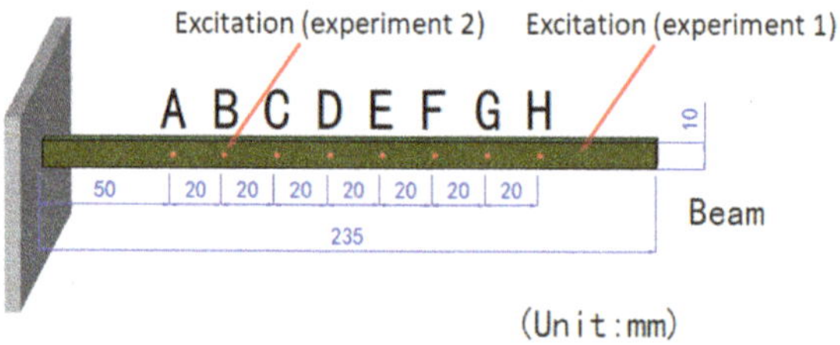

Figure 18. Dimension diagram of cantilever beam.

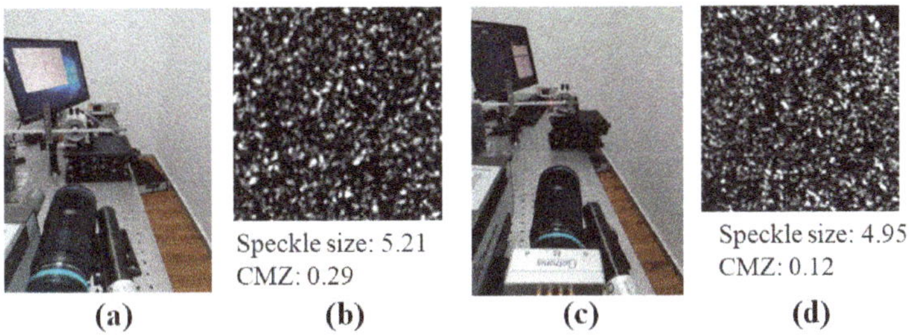

Figure 19. (**a**) Physical diagram of the first experiment. (**b**) One speckle pattern of the first experiment, its speckle size, and its CMZ value. (**c**) Physical diagram of the second experiment. (**d**) One speckle pattern of the second experiment, its speckle size, and its CMZ value.

Figures 20a and 21a show the frequency spectrum results from the two experiments obtained by the LDV, and Figures 20b and 21b show the frequency spectrum results from the two experiments calculated by the laser speckle correlation method. The focus of this study was the response frequency, so the pixel displacement was not translated into a physical displacement. The calculated response frequencies of the two methods were in good agreement, and response frequencies showed no obvious difference at different tested location. In the frequency spectrum of the first experiment, a primary energy peak at 10 Hz. Several secondary energy peaks were present at 20 Hz, 30 Hz, 40 Hz, and 50 Hz. The response frequency error of the laser speckle correlation method increased from 0.02 Hz to 0.1 Hz. Similarly, the second experiment showed primary peak at 30 Hz. Secondary peaks happened at 90 and 150 Hz, respectively. The response frequency error of the proposed method increased from 0.06 Hz to 0.3 Hz.

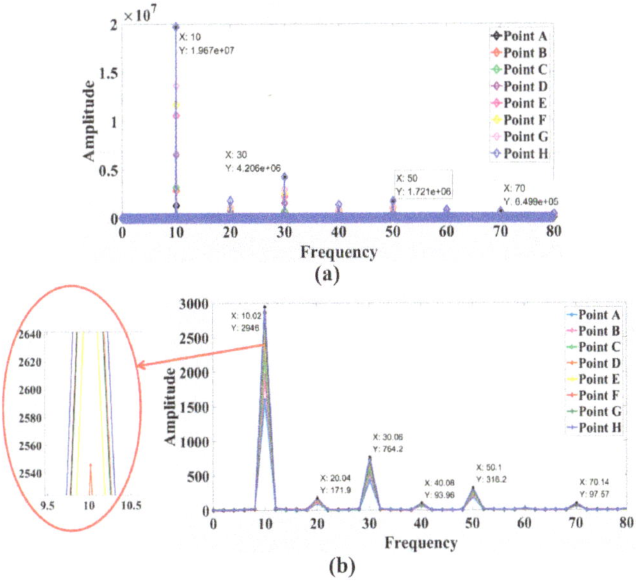

Figure 20. Comparison of the frequency spectrum obtained by (**a**) the LDV signals and (**b**) speckle patterns in the first experiment.

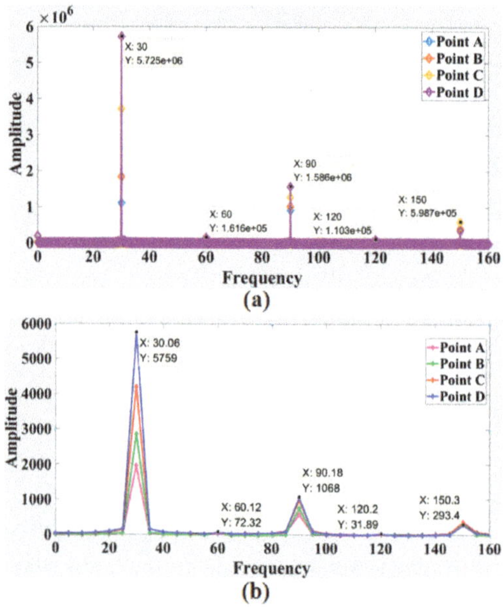

Figure 21. Comparison of the frequency spectrum obtained by (**a**) the LDV signals and (**b**) speckle patterns in the second experiment.

In order to further verify the quality assessment criterion, excitation signal frequency was kept 30Hz, and defocusing degree was changed to acquire four groups of vibration laser patterns with different CMZ value. Response frequencies at point G were calculated and compared. Results were shown in Table 3.

Table 3. Frequency results comparison of different pattern groups.

Patterns Group	CMZ Value	Frequency (Hz)		
		Primary Energy Peak	Secondary Energy Peak	Secondary Energy Peak
1	0.06	30.01	90.07	150.09
2	0.12	30.03	90.11	150.17
3	0.29	30.06	90.18	150.30
4	0.45	30.19	90.35	150.75

From the results we can see, the frequency error increases with the CMZ value rising, verifying the effectiveness of the proposed quality assessment criterion.

5. Conclusions

To balance the random error and the interpolation bias, a global assessment criterion CMZ was proposed, which combined the MIG and the ZSR. A CMZ closer to zero corresponded to a better quality of the speckle pattern and a smaller displacement error. Considering that the laser speckle has a more uniform distribution of speckle particles, the simulated speckle patterns with different speckle particle sizes and densities were used to illustrate the determination of the CMZ criterion.

During the application of laser speckle correlation method, the main influencing factors include defocusing degree, exposure time and measured location. Experimental speckle patterns at different defocusing degrees, exposure times and measured locations were analyzed, and the results validated the proposed assessment criterion using the CMZ. Defocusing degree affects speckle particle size. The analysis results also further showed that the optimal speckle particle size was 3–5 pixels, and the contrast of the speckle patterns was reduced due to the overexposure, causing a decline of the speckle pattern quality. Thus, high quality speckle patterns can be guaranteed based on the CMZ criterion.

The resolution and accuracy of the laser speckle correlation method were found to be related to the distances between the measurement planes. In the experimental setup stage, a suitable defocusing degree can be determined based on the resolution requirement and the optimal speckle size, and then other experimental parameters can be determined according to the CMZ criterion. In rotation experiments, the comparison of the angle resolution verified the relationship between the resolution and distance, and the accuracy of the calculated displacement was consistent with the results predicted by the CMZ value. The frequency spectrum results of the vibration experiments were in good agreement with the LDV results, which simultaneously verified the effectiveness of the laser speckle correlation method and the reliability of the experimental setup based on the proposed CMZ rule. Thus, adopting this experimental setup method can ensure the resolution and accuracy of the laser speckle correlation method and facilitate its widespread application.

Author Contributions: Conceptualization, W.H., Y.F. and H.M.; methodology, W.H.; software, W.H., Z.S. and K.Y.; validation, K.Y., Y.F. and H.M.; formal analysis, W.H., Z.S.; investigation, K.Y., Z.S.; writing—original draft preparation, W.H.; writing—review and editing, H.M.; visualization, K.Y.; supervision, Y.F.; funding acquisition, Y.F. All authors have read and agreed to the published version of the manuscript.

Funding: This research was funded by the National Natural Science Foundation of China, grant number 11972235, 11890683 and 11732009.

Institutional Review Board Statement: Not applicable.

Informed Consent Statement: Not applicable.

Data Availability Statement: Not applicable.

Conflicts of Interest: The authors declare no conflict of interest.

References

1. Huang, M.; He, Z.-N.; Lee, F.-Z. A novel methodology for enhancing the contrast of correlation fringes obtained by ESPI. *Measurement* **2004**, *36*, 93–100. [CrossRef]
2. Wang, S.; Lu, M.; Bilgeri, L.M.; Jakobi, M.; Bloise, F.S.; Koch, A.W. Temporal electronic speckle pattern interferometry for real-time in-plane rotation analysis. *Opt. Express* **2018**, *26*, 8744–8755. [CrossRef]
3. Barile, C.; Casavola, C.; Pappalettera, G.; Pappalettere, C. Analysis of the effects of process parameters in residual stress measurements on Titanium plates by HDM/ESPI. *Measurement* **2014**, *48*, 220–227. [CrossRef]
4. Pagliarulo, V.; Farroni, F.; Ferraro, P.; Lanzotti, A.; Martorelli, M.; Memmolo, P.; Speranza, D.; Timpone, F. Combining ESPI with laser scanning for 3D characterization of racing tyres sections. *Opt. Lasers Eng.* **2018**, *104*, 71–77. [CrossRef]
5. Falldorf, C.; Agour, M.; Bergmann, R.B. Digital holography and quantitative phase contrast imaging using computational shear interferometry. *Opt. Eng.* **2015**, *54*, 24110. [CrossRef]
6. Wang, X.; Gao, Z.; Yang, S.; Gao, C.; Sun, X.; Wen, X.; Feng, Z.; Wang, S.; Fan, Y.; Wang, X.; et al. Application of digital shearing speckle pattern interferometry for thermal stress. *Measurement* **2018**, *125*, 11–18. [CrossRef]
7. Frankovský, P.; Brodnianská, Z.; Bocko, J.; Trebuňová, M.; Kostka, J.; Kicko, M.; Čarák, P. Application of holographic interferometry in the analysis of stress states in a crack root area. *Appl. Opt.* **2020**, *59*, D170–D178. [CrossRef]
8. Abdelsalam, D. A comparison of digital holographic microscopy and on-axis phase-shifting interferometry for surface profiling. *Measurement* **2013**, *46*, 4121–4126. [CrossRef]
9. Hartlieb, S.; Tscherpel, M.; Guerra, F.; Haist, T.; Osten, W.; Ringkowski, M.; Sawodny, O. Highly accurate imaging based position measurement using holographic point replication. *Measurement* **2021**, *172*, 108852. [CrossRef]
10. Huang, T.; Li, X.; Fu, X.; Zhang, C.; Duan, F.; Jiang, J. Arbitrary phase shifting method for fiber-optic fringe projection profilometry based on temporal sinusoidal phase modulation. *Opt. Lasers Eng.* **2019**, *121*, 300–306. [CrossRef]
11. Hyun, J.-S.; Zhang, S. Enhanced two-frequency phase-shifting method. *Appl. Opt.* **2016**, *55*, 4395–4401. [CrossRef]
12. Gu, G.; Pan, Y.; Qiu, C.; Zhu, C. Real-time dual-channel speckle interferometry based on an improved dual-observation configuration with spatial phase-shifting. *Measurement* **2021**, *176*, 109188. [CrossRef]
13. Liu, C.-H.; Schill, A.; Raghunathan, R.; Wu, C.; Singh, M.; Han, Z.; Nair, A.; Larin, K.V. Ultra-fast line-field low coherence holographic elastography using spatial phase shifting. *Biomed. Opt. Express* **2017**, *8*, 993–1004. [CrossRef]

14. Kim, J.A.; Kim, J.W.; Kang, C.S.; Jin, J.; Eom, T.B. Interferometric profile scanning system for measuring large planar mirror surface based on single-interferogram analysis using Fourier transform method. *Measurement* **2018**, *118*, 113–119. [CrossRef]
15. Prabhakar, D.; Kumar, M.S.; Krishna, A.G. A Novel Hybrid Transform approach with integration of Fast Fourier, Discrete Wavelet and Discrete Shearlet Transforms for prediction of surface roughness on machined surfaces. *Measurement* **2020**, *164*, 108011. [CrossRef]
16. Fu, Y.; Guo, M.; Phua, P.B. Spatially encoded multibeam laser Doppler vibrometry using a single photodetector. *Opt. Lett.* **2010**, *35*, 1356–1358. [CrossRef] [PubMed]
17. Fu, Y.; Guo, M.; Phua, P.B. Multipoint laser Doppler vibrometry with single detector: Principles, implementations, and signal analyses. *Appl. Opt.* **2011**, *50*, 1280–1288. [CrossRef] [PubMed]
18. Halkon, B.J.; Rothberg, S.J. Establishing correction solutions for Scanning Laser Doppler Vibrometer measurements affected by sensor head vibration. *Mech. Syst. Signal Process.* **2020**, *150*, 107255. [CrossRef]
19. Sciuti, V.F.; Canto, R.B.; Veggers, J.; Hild, F. On the benefits of correcting brightness and contrast in global digital image correlation: Monitoring cracks during curing and drying of a refractory castable. *Opt. Lasers Eng.* **2021**, *136*, 106316. [CrossRef]
20. Wang, W.; Mottershead, J.E.; Siebert, T.; Pipino, A. Frequency response functions of shape features from full-field vibration measurements using digital image correlation. *Mech. Syst. Signal Process.* **2012**, *28*, 333–347. [CrossRef]
21. Rosakis, A.J.; Rubino, V.; Lapusta, N. Recent Milestones in Unraveling the Full-Field Structure of Dynamic Shear Cracks and Fault Ruptures in Real-Time: From Photoelasticity to Ultrahigh-Speed Digital Image Correlation. *J. Appl. Mech.* **2020**, *87*, 1–57. [CrossRef]
22. Hu, X.; Xie, Z.; Liu, F. Assessment of speckle pattern quality in digital image correlation from the perspective of mean bias error. *Measurement* **2021**, *173*, 108618. [CrossRef]
23. Chen, Z.; Quan, C.; Zhu, F.; He, X. A method to transfer speckle patterns for digital image correlation. *Meas. Sci. Technol.* **2015**, *26*, 095201. [CrossRef]
24. Song, J.; Yang, J.; Liu, F.; Lu, K. High temperature strain measurement method by combining digital image correlation of laser speckle and improved RANSAC smoothing algorithm. *Opt. Lasers Eng.* **2018**, *111*, 8–18. [CrossRef]
25. Zheng, Q.; Mashiwa, N.; Furushima, T. Evaluation of large plastic deformation for metals by a non-contacting technique using digital image correlation with laser speckles. *Mater. Des.* **2020**, *191*, 108626. [CrossRef]
26. Gregory, D. Basic physical principles of defocused speckle photography: A tilt topology inspection technique. *Opt. Laser Technol.* **1976**, *8*, 201–213. [CrossRef]
27. Hrabovsky, M.H.; Šmíd, P. Full theory of speckle displacement and decorrelation in the image field by wave and 27. Pgeometrical descriptions and its application in mechanics. *J. Mod. Opt.* **2004**, *51*, 725–742.
28. Jo, K.; Mohit, G.; Shree, K.N. Spedo: 6 dof ego-motion sensor using speckle defocus imaging. In Proceedings of the IEEE International Conference on Computer Vision, Santiago, Chile, 7–13 December 2015. [CrossRef]
29. Zalevsky, Z.; Beiderman, Y.; Margalit, I.; Gingold, S.; Teicher, M.; Mico, V.; Garcia, J. Simultaneous remote extraction of multiple speech sources and heart beats from secondary speckles pattern. *Opt. Express* **2009**, *17*, 21566–21580. [CrossRef] [PubMed]
30. Li, L.; Gubarev, F.A.; Klenovskii, M.S.; Bloshkina, A.I. Vibration measurement by means of digital speckle correlation. In Proceedings of the 2016 International Siberian Conference on Control and Communications (SIBCON), Moscow, Russia, 12–14 May 2016; pp. 1–5.
31. Beiderman, Y.; Horovitz, I.; Burshtein, N.; Teicher, M.; García, J.; Micó, V.; Zalevsky, Z. Remote estimation of blood pulse pressure via temporal tracking of reflected secondary speckles pattern. *J. Biomed. Opt.* **2010**, *15*, 061707. [CrossRef]
32. Wu, N.; Haruyama, S. Real-time audio detection and regeneration of moving sound source based on optical flow algorithm of laser speckle images. *Opt. Express* **2020**, *28*, 4475–4488. [CrossRef]
33. Sun, Y.F.; Pang, J.H.L. Study of optimal subset size in digital image correlation of speckle pattern images. *Opt. Lasers Eng.* **2007**, *45*, 967–974.
34. Pan, B.; Xie, H.; Wang, Z.; Qian, K.; Wang, Z. Study on subset size selection in digital image correlation for speckle patterns. *Opt. Express* **2008**, *16*, 7037–7048. [CrossRef]
35. Lecompte, D.; Smits, A.; Bossuyt, S.; Sol, H.; Vantomme, J.; Van Hemelrijck, D.; Habraken, A. Quality assessment of speckle patterns for digital image correlation. *Opt. Lasers Eng.* **2006**, *44*, 1132–1145. [CrossRef]
36. Crammond, G.; Boyd, S.; Dulieu-Barton, J. Speckle pattern quality assessment for digital image correlation. *Opt. Lasers Eng.* **2013**, *51*, 1368–1378. [CrossRef]
37. Pan, B.; Lu, Z.X.; Xie, H.M. Mean intensity gradient: An effective global parameter for quality assessment of the speckle patterns used in digital image correlation. *Opt. Lasers Eng.* **2010**, *48*, 469–477. [CrossRef]
38. Yu, H.; Guo, R.; Xia, H.; Yan, F.; Zhang, Y.; He, T. Application of the mean intensity of the second derivative in evaluating the speckle patterns in digital image correlation. *Opt. Lasers Eng.* **2014**, *60*, 32–37. [CrossRef]
39. Park, J.; Yoon, S.; Kwon, T.-H.; Park, K. Assessment of speckle-pattern quality in digital image correlation based on gray intensity and speckle morphology. *Opt. Lasers Eng.* **2017**, *91*, 62–72. [CrossRef]
40. Bossuyt, S. Optimized patterns for digital image correlation. In *Imaging Methods for Novel Materials and Challenging Applications*, 3rd ed.; Springer: Berlin/Heidelberg, Germany, 2013; pp. 239–248.
41. Stoilov, G.; Kavardzhikov, V.; Pashkouleva, D. A Comparative Study of Random Patterns for Digital Image Correlation. *J. Theor. Appl. Mech.* **2012**, *42*, 55–66. [CrossRef]

42. Bomarito, G.; Hochhalter, J.; Ruggles, T.; Cannon, A. Increasing accuracy and precision of digital image correlation through pattern optimization. *Opt. Lasers Eng.* **2017**, *91*, 73–85. [CrossRef]
43. Song, J.; Yang, J.; Liu, F.; Lu, K. Quality assessment of laser speckle patterns for digital image correlation by a Multi-Factor Fusion Index. *Opt. Lasers Eng.* **2020**, *124*, 105822. [CrossRef]
44. Su, Y.; Zhang, Q.; Gao, Z.; Xu, X.; Wu, X. Fourier-based interpolation bias prediction in digital image correlation. *Opt. Express* **2015**, *23*, 19242–19260. [CrossRef] [PubMed]
45. Charrett, T.; Tatam, R. Performance and Analysis of Feature Tracking Approaches in Laser Speckle Instrumentation. *Sensors* **2019**, *19*, 2389. [CrossRef] [PubMed]
46. Pan, B. An Evaluation of Convergence Criteria for Digital Image Correlation Using Inverse Compositional Gauss-Newton Algorithm. *Strain* **2013**, *50*, 48–56. [CrossRef]

sensors

MDPI

Article

A Fringe Phase Extraction Method Based on Neural Network

Wenxin Hu [1], Hong Miao [2], Keyu Yan [1] and Yu Fu [1,*]

[1] Shenzhen Key Laboratory of Intelligent Optical Measurement and Detection, College of Physics and Optoelectronic Engineering, Shenzhen University, 3688 Nanhai Avenue, Shenzhen 518060, China; huwenxin@szu.edu.cn (W.H.); yankeyu@szu.edu.cn (K.Y.)

[2] CAS Key Laboratory of Mechanical Behavior and Design of Materials, Department of Modern Mechanics, University of Science and Technology of China, Hefei 230027, China; miaohong@ustc.edu.cn

* Correspondence: fuyuoptics@gmail.com

Abstract: In optical metrology, the output is usually in the form of a fringe pattern, from which a phase map can be generated and phase information can be converted into the desired parameters. This paper proposes an end-to-end method of fringe phase extraction based on the neural network. This method uses the U-net neural network to directly learn the correspondence between the gray level of a fringe pattern and the wrapped phase map, which is simpler than the exist deep learning methods. The results of simulation and experimental fringe patterns verify the accuracy and the robustness of this method. While it yields the same accuracy, the proposed method features easier operation and a simpler principle than the traditional phase-shifting method and has a faster speed than wavelet transform method.

Keywords: phase extraction; U-net neural network; warped phase map; fringe pattern

Citation: Hu, W.; Miao, H.; Yan, K.; Fu, Y. A Fringe Phase Extraction Method Based on Neural Network. *Sensors* **2021**, *21*, 1664. https://doi.org/10.3390/s21051664

Academic Editor: Maria Lepore

Received: 26 January 2021
Accepted: 23 February 2021
Published: 28 February 2021

1. Introduction

Optical metrology has been widely used in various areas, such as 3D sensing, machine vision, intelligent robot control, industry monitoring, and dressmaking. In optical metrology, the output is usually in the form of a fringe pattern, from which a phase map can be determined. Once the phase map has been obtained, it can be converted into the desired parameters, such as the shape of the object, and in-plane or out-of-plane deformation. For instance, the fringe projection technique [1–4] is often used to measure the 3D-profile of objects [2]. When the fringe pattern is projected on a measured free surface, the phase of the fringe pattern is modulated by the height distribution of the object. A method to extract the phase map from the deformed fringe pattern is thus needed.

Thus far, many methods for phase calculation have been developed, including temporal phase-shifting [4,5], spatial phase-shifting [6,7], and Fourier transform [8]. The phase-shifting is a pointwise technique and it is sensitive to noise such as CCD random noise, environmental vibration, air disturbance, etc. The temporal phase-shifting method requires four images in one stage, which is unsuitable for real-time measurement, and the spatial phase-shifting method requires a complex optical path. The Fourier transform technique, on the contrary, is a global transform method that is hence more tolerant to noise. However, as the transform is global, an accurate frequency band containing effective information of the measured object needs to be determined to avoid large calculation errors. Some improvements have been proposed to overcome the shortcoming of the simple Fourier transform method. A windowed Fourier ridges algorithm [9–11] and a windowed Fourier filtering algorithm have been proposed to achieve a low standard deviation for local frequencies and phase distributions in fringe pattern analysis. Morlet wavelet transform has also been used for phase extraction on different types of fringe patterns [12–14].

In this research, we propose a fringe phase extraction method based on the neural network. As an important part of machine learning, neural networks have been widely used in various fields, such as object recognition [15–18], object segmentation [19], and speech recognition [20,21]. This method has also been introduced to optical measurement. Liu et al. employed the backpropagation (BP) [22,23] artificial neural network to directly build a nonlinear mapping relationship between the gray-gradient of speckle images before and after deformation, and sub-pixel displacement in the digital image correlation method. This method avoids the least squares analytical optimal solution of the correlation coefficient. Horisaki et al. have used support vector regression (SVR) to recover the image through the scattering layer [24]. This approach enables model-free sensing, where it is not necessary to know the sensing processes/models. Guan et al. have introduced a method of grating sub-division based on the radial-basis function (RBF) neural network. It converts displacement into a digital measure that is transmitted to the microprocessor of a neural network to obtain the sub-division value. This improves the accuracy of sub-division and the tracking speed of the displacement [22]. Rivenson et al. have proposed a holographic image reconstruction method based on the convolutional neural network (CNN) that can reconstruct the phase and amplitude of images of objects using only a hologram [25]. Pitkaaho et al. have employed the CNN to focus on automatic distance calculation in holographic image reconstruction [26]. Wang et al. have proposed a one-step end-to-end learning-based method for in-line holographic reconstruction that creates a network called eHoloNet to avoid phase shifting [27]. Deep-learning based temporal phase unwrapping (DL-TPU) is introduced by Wei Yin [28], which can substantially improve the unwrapping reliability compared with multi-frequency temporal phase unwrapping (MF-TPU). These results show that challenging problems in optical metrology can be overcome through machine learning, and provide new avenues for image analysis. Shijie Feng et al. has introduced a machine-learning-based fringe analysis method, which employs two convolutional neural networks (CNN1 and CNN2) to calculate phase information [29]. For CNN2, the inputs are the fringe pattern and the background image predicted by CNN1, and the outputs are the numerator and the denominator, which are then fed into the arctangent function to calculate phase. Some improvement and simplification have also been made by them. A micro deep learning profilometry using a single network is presented for high-speed 3D surface imaging [30]. However, three fringe patterns are needed for correct phase unwarpping. Haotian Yu et al. has introduced a novel phase retrieval method based on a deep neural network called FPPnet [31]. The FPPnet only requires one single image and one single network, and this network is used to achieve prediction of output fringes in the same period and different periods. Then, the phase calculation and the phase unwrapping can be achieved by these predicted fringes. However, these methods employ neural network to obtain intermediate calculation parameters such as numerator or denominator or related fringe pattern, not directly acquiring phase information. Furthermore, a deep-learning-based approach is proposed by Sam to extract height information from single deformed fringe patterns [32]. The fully CNN is trained on a large set of simulated height maps with corresponding deformed fringe patterns, so phase results rely too much on the complexity of these simulated height maps.

In this paper, we introduce a one-step deep-learning-based method to extract the wrapped phase map directly from a single fringe pattern. This method employs the U-net neural network to directly learn the correspondence between the gray level of a fringe pattern and the wrapped phase map. Once a stable network model has been obtained, the wrapped phase map of an arbitrary fringe pattern can be output directly, thus simplifying the phase extraction further. The mathematics problem is transformed into image processing problem, developing the advantage of neural network. Meanwhile, the network can be saved and shared. More and deeper training contributes to the network generalization ability, so as to solve more complex and different fringe patterns. Besides, experimental results verified effectiveness on different fringe pattern whether coming from fringe projection profilometry or interferometer. While it yields the same accuracy, the proposed method features easier operation and a simpler principle than the traditional phase-shifting method, and it owns faster computation speed and higher accuracy than wavelet transform method. Moreover, the results of simulated and experimental fringe patterns verify the efficiency and the robustness of the proposed method.

2. Method

2.1. Principle

The U-net is an end-to-end deep neural network that takes an image of any size (fringe patterns here) as input and outputs a specified image (corresponding wrapped phase maps here). The process of forward propagation, the training method and output principle used for the neural network are described in Sections 2.2–2.4, respectively. As our ultimate goal is to obtain a stable network model with effective parameters, a large amount of training data, including fringe patterns and corresponding wrapped phase maps, are needed in advance. Once this stable network model has been obtained, the wrapped phase map of an arbitrary fringe pattern can be obtained directly. The neural network method was programmed in Python based on the Tensorflow framework, and run on a desktop computer equipped with an Intel i5-4460 CPU and a GeForce GTX 1080 graphics card.

2.2. U-Net Neural Network

The size of the input fringe pattern is 512 × 512 pixels and the output maintain the same size. This network features a contracting path, a transition path and an expansive path. The contracting path is used to extract features of the fringe pattern, and the expansive path is applied for converting into corresponding warped phase map. With the deepening of the contracting layers, low-dimensional features including gray gradient of every pixel are changed into high-dimensional features including the location and the local gradient. More layers of each path mean more connection parameters, so as to fit more complex non-linear mapping relationship.

The principle of the contracting path is the same as that of the CNN [33]. The contracting path includes four repeated down-sampling process. Every down-sampling contains two convolution blocks and a pooling block. The feature channel doubles every two convolution blocks, and the image dimensions reduce the half after a pooling operation because the stride is two pixels.

The down-sampling operation is illustrated in Figure 1. As Figure 1 shows, the convolution kernel shifting stride is (1, 1) along two dimensions. This means that the convolution kernels shift 1 pixel along the x and y directions each time and multiple with the image. The convolution consists of a convolution layer and an activation function, and the principle [34] can be described by Formula (1):

$$v_{i,r_2}^{x,y} = f\left(\sum_{r_1=1}^{R}\sum_{p=0}^{P-1}\sum_{q=0}^{Q-1} w_{i,r_2,r_1}^{p,q} v_{i-1,r_1}^{x+p,y+q} + b_{i,r_2}\right), \tag{1}$$

where $v_{i,r_2}^{x,y}$ represents the value of the output at (x, y) for the r_2-th feature channel map of the i -th layer. $v_{i-1,r_1}^{x+p,y+q}$ represents the value of the result at $(x+p, y+q)$ for the r_1-th feature channel map of the $(i-1)$ -th layer, and R is the total number of feature channel in the $(i-1)$ -th convolution layer. b_{i,r_2} is a common basic term for the r_2-th feature channel map of the i -th layer. $w_{i,r_2,r_1}^{p,q}$ represents the weight of the convolution kernel at (p, q), and $P \times Q$ is the size of the convolution kernels in terms of pixels, which is 3 × 3 in all the convolution blocks of the contraction path. f represents the activation function, which uses rectified linear units (ReLUs) [33]. The principle of ReLU is described by Formula (2):

$$f(x) = RELU(x) = \max(0, x), \tag{2}$$

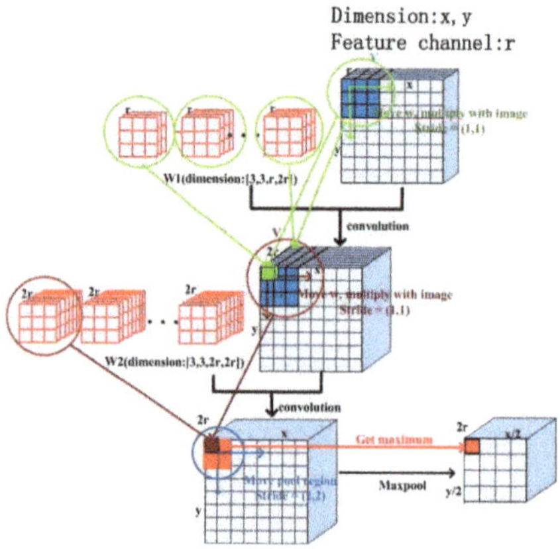

Figure 1. Process of down-sampling.

The pooling block employs the max-pooling method, and the region of every pooling operation is 2 × 2 pixels, which is intended to obtain the maximum pixel value in this small region. Additionally, the pooling stride is (2, 2) along the x and y directions so as to reduce the image size by a factor of 2.

The expansive path includes four repeated up-sampling process, and it aims to enlarge the image and decode the convolution process. The up-sampling operation including a transposed convolution block, a merge block and two convolution blocks is illustrated in Figure 2. This operation doubles the size of the image and reduces the number of feature channel by half.

The transposed convolution operation is identical to the convolution operation, but it enlarges the image from the previous block. Some zero-value pixels between neighboring image pixels are inserted, and a convolution operation on the up-sampled image is employed. The stride of the transported convolution layer is 2 × 2 pixels, which means that it inserts one zero pixel between neighboring image pixels, doubling the image size. The convolution kernel size in the transposed convolution layer is also 3 × 3 pixels.

The merge block is an image mosaic process. Once the result of the transposed convolution layer has been obtained, it is spliced into the feature image of the corresponding procedure in the contracting path. The principle of two convolution blocks is the same as the down-sampling operation, but it reduces the number of feature channel by half.

All the convolution kernel values are initialized with random numbers from a truncated Gaussian distribution and the values of biases are initialized as constant.

The whole process of network propagation is shown in Figure 3. This process features a contracting path (left), a transition path and an expansive path (right). The size of the input fringe pattern is 512 × 512 pixels. After once down-sampling operation, the size of the image is changed to 256 × 256 pixels and the number of feature channels to 64. By repeating this process four times, the size of the image is reduced to 32 × 32 pixels and the number of feature channel changes to 512.

The transition path is consisted of two convolution blocks. Additionally, the feature channel doubles after two convolution blocks. The size of the image maintains 32 × 32 pixels and the number of feature channel changes to 1024.

Then, the result is subjected to the expansive path including four repeated up-sampling operation. The size of image is 64 × 64 pixels, 128 × 128 pixels, 256 × 256, and 512 × 512 pixels after each up-sampling when the numbers of feature channels are 512, 256, 128, and 64, respectively. Finally, a convolution operation is applied, and the size of the convolution kernel is 1 × 1 pixels. The size of the image maintains 512 × 512 pixels, and the number of feature channel changes to 256.

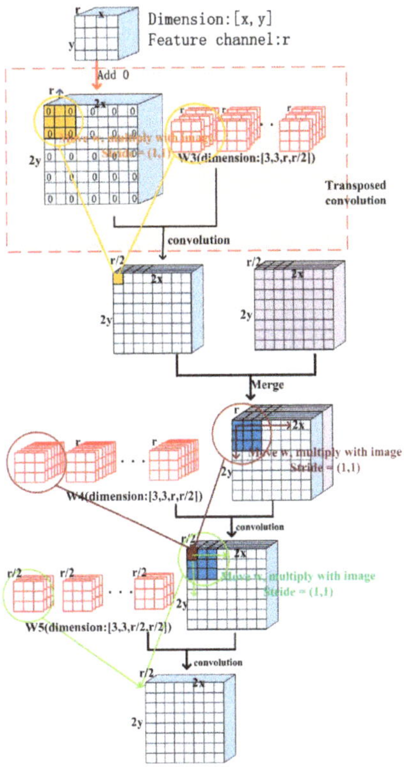

Figure 2. Process of up-sampling.

2.3. Network Training

Note that the warped phase results are periodic, when 1 output channel and MSE loss function are used, the result is easily restricted to the local optimal solution, where all output values tend to be 0. Thus, this problem is chosen to be converted into a classification problem. The result of the network is a three-dimensional matrix, and the size is 512, 512, 256 along x, y, and feature channel directions, respectively. For every pixel, the values along feature channel direction represent the possibility of being 0 to 255. A softmax function is used to reset the result to meet the requirement of probability distribution, so that cross-entropy [35] known as multi-class log loss can be used as loss function.

When the output of a pixel along the feature channel direction is $q_1, q_2, \cdots q_n$, the result of the softmax function can be described by

$$\mathrm{softmax}(q)_i = \frac{e^{q_i}}{\sum_{j=1}^{n} e^{q_j}}, \tag{3}$$

where n represents the number of feature channels, and there is 256.

The probability distribution of reset q meets the following condition:

$$\forall i;\ q_i \in [0,1]; \sum_{i=1}^{n} q_i = 1, \tag{4}$$

The ground truth g of each pixel supposed to be -pi to pi is scaled up to between 0 and 255. The probability distribution of training label p is given according to the formula:

$$\forall i;\ p_{i=g} = 1;\ p_{i \neq g} = 0, \tag{5}$$

Cross-entropy is defined by [35]

$$H(p,q) = -\sum_{x=1}^{w}\sum_{y=1}^{h}\sum_{i=1}^{n} p_i(x,y)\log q_i(x,y), \tag{6}$$

where p represents the training label and q represents the calculated result. The values of p and q are explained in the next section. n represents the number of feature channels, and w, h represents the width and height of the fringe pattern, respectively. The smaller the cross-entropy, the higher the probability that the actual and the calculated results are closer.

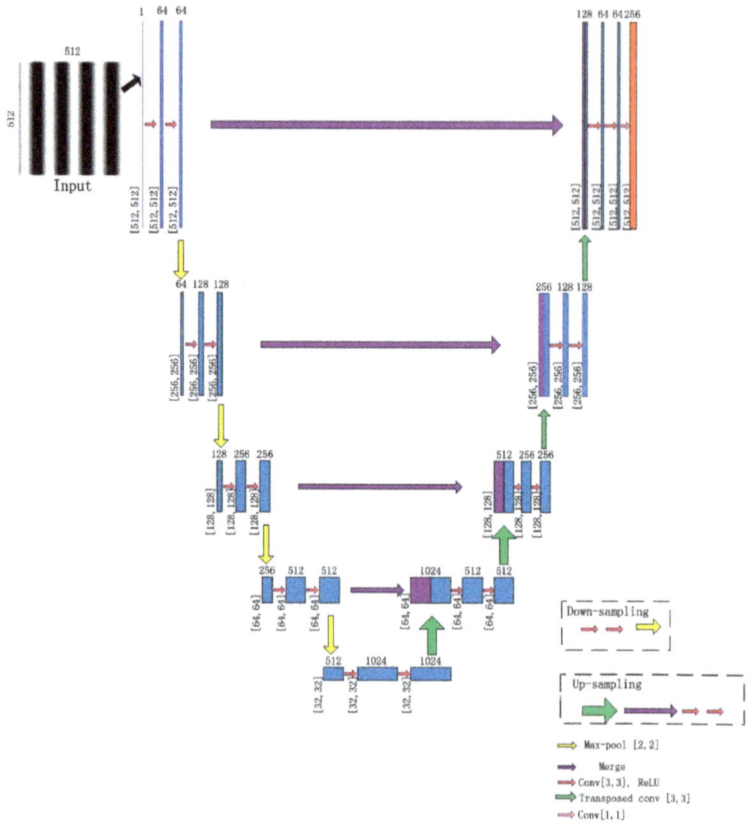

Figure 3. Process of network propagation.

The backpropagation algorithm [36] is used to back-propagate the error into the network, and adaptive moment estimation (Adam) [37]-based optimization is used to optimize the weights of convolution kernels(w) and common basic terms(b) of all layers. An input is first propagated through the network. Then, the difference between the calculated and the desired output is backpropagated from the output layer to the first layer of the network, thereby adjusting the network weights in the opposite direction of the derivative of the network error with respect to each individual network weight. By repeating this procedure multiple times for each data in a training set, the network can be taught to map the inputs on the correct outputs. The batch size was 10, and the epoch was 1000. The learning rate was 10^{-4}.

2.4. Output Principle

The values of each pixel along feature channel direction represent the possibility of being 0 to 255. As Figure 4 shows, the output is the position corresponding to maximum possibility, so the output value is between 0 and 255. Note that the number of feature channels in the last convolution operation can choose more than 256, corresponding to higher resolution and more calculated time. At last, the output is restored to between -pi and pi.

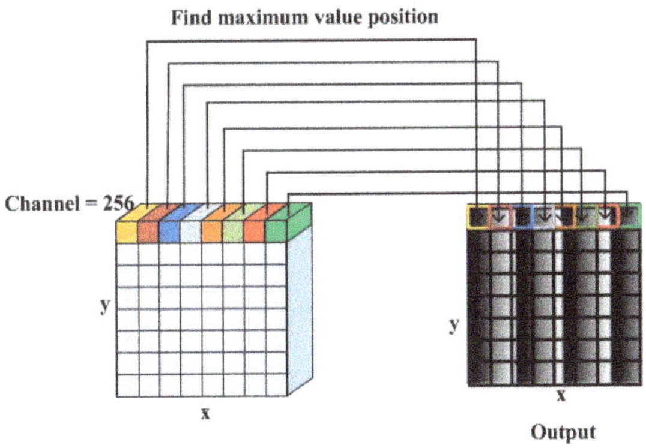

Figure 4. Output schematic.

3. Verification of Method

3.1. Simulation Image

Numerical simulations were carried out to test the performance of the proposed algorithm. From simple to complex, we used three equations to simulate fringe patterns, and the size of patterns was set to 512 × 512 pixels.

The gray level of the first kind of patterns was determined by Equation (7). A total of 1600 fringe patterns were obtained, in which the fringe number of a pattern was set between three and 44, and the fringe interval decreased gradually.

$$I(t,x,y) = 255 \times \cos(((44 \times pi/512 - 3 \times pi/512)/1600 \times (t-1) + 3 \times pi/512)) \times y), t = 1,2,\cdots 1600; y = 1,2,\cdots 512, \quad (7)$$

where t represents the series number of the pattern and y represents width in pixels.

The second and third kinds of fringe patterns were generated according to Equations (8) and (9). The 1600 fringe patterns with different fringe shapes were obtained through image cropping and rotation from a fringe pattern:

$$I(x,y) = 50 + 50 \times \cos(peaks(1000) + 20 \times pi/1000 \times y), y = 1,2\cdots 1000, \quad (8)$$

$$I(x,y) = 50 + 50 \times \cos(50 - \left((x-200)^2 - \frac{(y-200)^2}{10,000}\right)), x = 1,2,\cdots 1000; y = 1,2,\cdots 1000, \quad (9)$$

where x represents height in pixels and y represents width.

Figure 5 shows some simulation patterns. (a), (b), and (c) represent fringe patterns generated according to Equations (7)–(9), respectively. Of all simulation patterns, 100 fringe patterns were selected to evaluate the trained network and the rest were used to train the model. The gray level of the fringe patterns was set as the input to the network, and the wrapped phase data calculated by four-step phase-shifting method were set as the output.

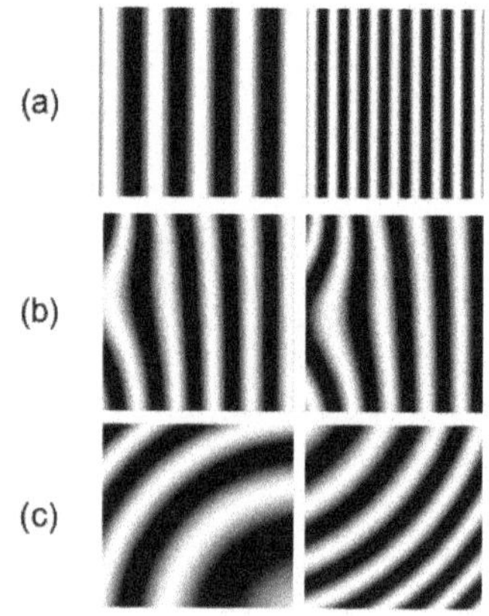

Figure 5. Simulated fringe patterns.

3.2. Experimental Image

The fringe projection technique is often used to determine the 3D-profile of an object. When the fringe pattern is projected on a measured free surface, its phase is modulated by the height distribution of the object. We used the fringe pattern captured from the fringe projection experiment to verify the ability of the neural network to extract the phase map. Figure 6a,b show schematic layout and physical diagram of the experiment, respectively.

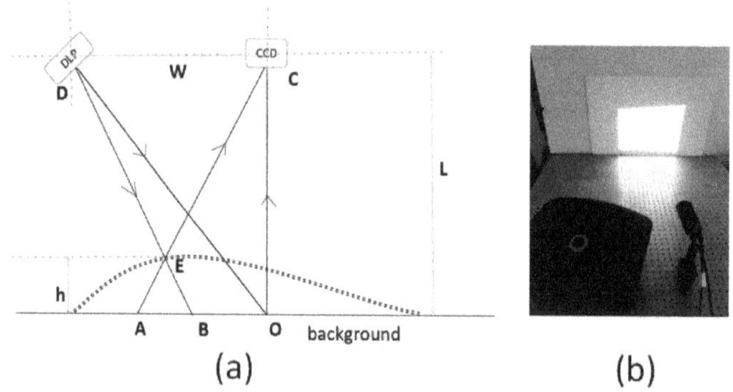

Figure 6. Fringe projection experiment: (**a**) Schematic layout; (**b**) physical diagram.

The optical path adopted oblique projection and vertical shooting. The digital fringe projector chosen was Vivitek-D5158HD at a resolution of 1920 × 1280 pixels. The camera used was the Basler ace 1600-20 g, with a resolution of 1600 × 1200 pixels. The optical axis of the projection and the receiving end intersected at point O. Moreover, the camera and the digital fringe projector were at the same height L. Due to modulation by the object's height, light that was supposed to obtain at point B was cast on point E, but the light point recorded by the camera was A. Finally, the height information of the object was recorded in the fringe pattern.

A 1-mm-thick disk was chosen as the measurement object. The position and angle of the disk were altered to obtain different fringe patterns. To obtain a sufficient number of images to meet the big data requirement of network training, such data extension as image cropping, translation, and rotation were used. One thousand fringe patterns were obtained. Figure 7 shows one of these and its corresponding phase-shifting fringe patterns.

Figure 7. A fringe pattern from the disk experiment and its corresponding phase-shifting fringe patterns.

Another complex object, a facial mask from opera, was measured, as shown in Figure 8. The same operations were performed as before to yield another 1000 experimental patterns.

Figure 8. A fringe pattern from the mask experiment and its corresponding phase-shifting fringe patterns.

Michelson interferometry is widely used to measure out-of-place displacement. An MI-based measurement system was setup to obtain different interferograms, and Figure 9 shows a schematic drawing of the measurement system. A light beam was emitted from the He–Ne laser generator, and expanded as parallel light beams after going through a spatial filter and convex lens. The parallel light beams were then divided into two identical parts by a beam splitter (BS), and one each was introduced to the reference arm and the objective arm. In the objective arm, the light beam propagated onto the surface of the object and was reflected. The reference arm had a reflector coupled with a PZT used for phase-shifting. Finally, two reflected light beams were returned to the BS and interference onto the surface of a CCD. The phase of the captured interferogram recorded the out-of-place displacement information of the object. By changing the fringe interval or the position of the reflector, 100 interferograms were obtained directly, and the other 900 interferograms were obtained through data extension. Figure 10a,b show physical diagram and some interferograms, respectively.

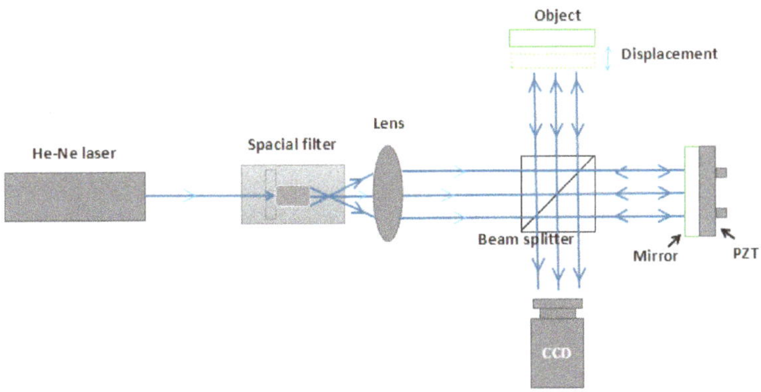

Figure 9. The schematic drawing of the measurement system.

The 100 fringe patterns from the fringe projection experiment and 100 interferograms were selected to evaluate the trained network, and the remainder was used to training the model. During the training, the wrapped phase dataset as the output of the neural network was calculated by the four-step phase-shifting method.

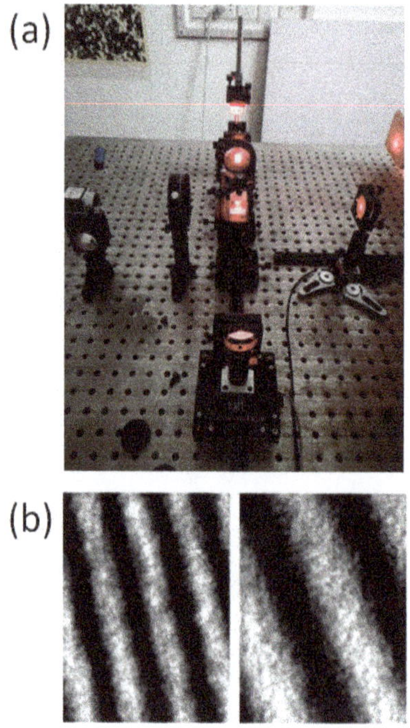

Figure 10. (**a**) Physical diagram. (**b**) Interferograms.

4. Results and Discussion

4.1. Computation Accuracy

4.1.1. Results on Simulation Image

The 100 simulation fringe patterns were used to test the accuracy and robustness of the method. Figure 11 shows some simulated fringe patterns and corresponding calculated wrapped phase maps. Figure 11b was calculated by the four-step phase-shifting method and Figure 11c was obtained through the trained neural network. The results show that the two types of measurement aligned well, and patterns with different fringe intervals and types yielded the correct values. In order to further illustrate the accuracy of our proposed method, wavelet transform method was used to make comparison. The two-dimension wavelet transform was implemented according to approach proposed by Wang [38], and the window-modifying parameter chose to be 2. The error maps coming from different methods are shown in Figure 11d,e, respectively. The whole error level verified the high accuracy.

To evaluate the accuracy of this method, we defined two types of error: systematic error (E) and standard deviation error (S) [31].

E is defined as

$$E = \frac{1}{N}\frac{1}{M}\sum_{j=1}^{M}\sum_{i=1}^{N}\left|d_{j_{i_{cal}}} - d_{j_{i_{real}}}\right|, \tag{10}$$

where $d_{j_{i_{real}}}$ represents the wrapped phase data of the i-th pixel in the j-th calculated image, and $d_{j_{i_{cal}}}$ represents the calculated phase data of the i-th pixel in the j-th calculated image, which was also warped. M represents the number of calculated images, and N represents the number of pixels in an image. S reflects the average error, and its best possible score was zero.

S is defined as

$$S = \sqrt{\left(\frac{1}{N}\frac{1}{M}\sum_{j=1}^{M}\sum_{i=1}^{N}(d_{ji_{cal}} - d_{ji_{real}} - E)^2\right)}, \quad (11)$$

where S reflects the deviation in the measured displacement corresponding to the mean value, and has is related to random error.

For the phase data of the simulation fringe patterns, the value of E was 0.03 rad and that of S was 0.07 rad.

Figure 11. (**a**) Simulated fringe patterns, wrapped phase maps calculated from (**b**) four-step phase shifting method, and (**c**) neural network, and corresponding error maps from (**d**) wavelet transform method, and (**e**) neural network.

4.1.2. Results on Experimental Image

The 100 fringe patterns obtained from the experiment on fringe projection were used to test the accuracy and the robustness of the method. Figure 12 shows some patterns and their results of wrapped phase maps. The patterns in Figure 12a were chosen from different projection experiments that used an empty background, a disk, and different parts of a mask as measured object. One hundred interferograms were also used to test the accuracy and the robustness of the method. Some interferograms and their wrapped phase map results are shown in Figure 13. The results shown in Figures 12 and 13b were calculated by the four-step phase-shifting method, and those shown in Figures 12 and 13c were obtained through the trained neural network. Through the comparison, we see that the results of the two methods were consistent. They also show that both experimental fringe projection patterns and interferograms yielded the correct results, and verified the robustness of the trained neural network and the feasibility of the machine learning method. Additionally, error maps are presented in Figure 12d,e and Figure 13d,e. Through contrast, errors of our proposed method were reduced obviously, demonstrating its improved performance in measuring complex objects under environmental noise. Thus, the proposed method owns higher noise resistance ability compared with wavelet transform method.

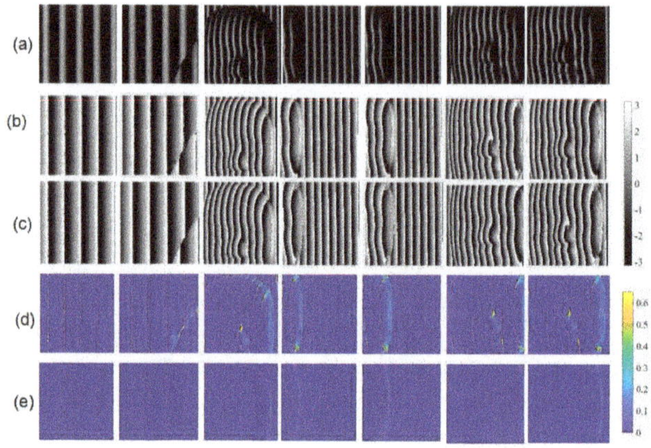

Figure 12. (**a**) Experimental fringe patterns, wrapped phase maps calculated from (**b**) four-step phase-shifting method, and (**c**) neural network, and corresponding error maps from (**d**) wavelet transform method, and (**e**) neural network.

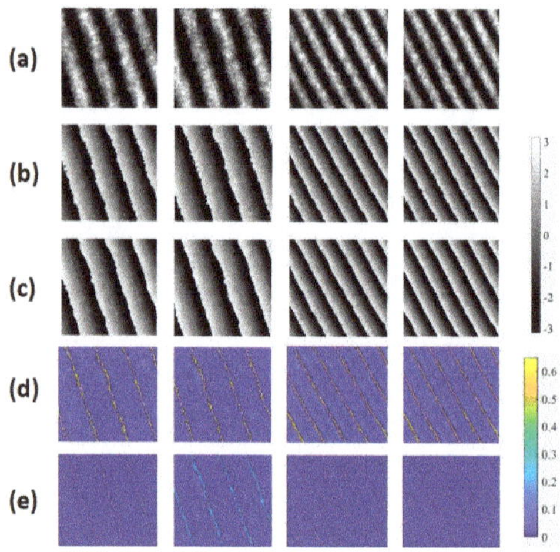

Figure 13. (**a**) Interferograms, wrapped maps calculated from (**b**) four-step phase-shifting method, and (**c**) neural network, and corresponding error maps from (**d**) wavelet transform method, and (**e**) neural network.

By referring again to the results of calculation of the four-step phase-shifting method, we used E and S to evaluate the accuracy of the experimentally obtained fringe pattern. For the wrapped phase data of the experimental fringe patterns, the value of E was 0.10 rad and S was 0.08 rad. For the wrapped phase data of interferograms, the value of E was 0.22 rad and S was 0.24 rad. The results verified the precision of this trained neural network and the accuracy of the learning-based method, as well as the system's ability to resist environmental noise. We also see that with decreasing quality of the fringe pattern, error increased.

4.2. Computation Efficiency

In order to illustrate the computation efficiency of the proposed method, the wavelet transform method was used to make comparison, which also only needed a single fringe pattern. The wavelet transform method was also programming using python language. 10

simulation patterns, 10 fringe projection patterns and 10 interferograms were calculated, respectively, and the values of E and S were shown in Table 1. The average calculated time of one pattern were recorded in Table 2.

Table 1. E and S of the proposed method and wavelet transform method.

Value	E(rad)			S(rad)		
Patterns	Simulation	Fringe Projection	Interferograms	Simulation	Fringe Projection	Interferograms
The proposed method	0.03	0.10	0.22	0.07	0.08	0.24
Wavelet transform method	0.05	0.15	0.24	0.08	0.14	0.29

Table 2. Calculated time of the proposed method and wavelet transform method.

Time (s)	Simulation Patterns	Fringe Projection Patterns	Interferograms
The proposed method	0.069	0.066	0.071
Wavelet transform method	1.154	3.152	2.850

From Table 2, we can see, once the neural network was determined, the calculated time of the neural network method have no matter with the fringe pattern quality, only depending on the network structure and the input size. Under the same accuracy, the calculated speed of the proposed method is 20 times faster than the wavelet transform method.

4.3. Discussion

From Figure 12c, we find that different measured objects can yield correct results regardless of the change in the position or the angles of objects. Figure 13c shows that this trained neural network can be applied to interferograms with lower pattern quality. To verify the network's ability to handle more complex fringe patterns with different shapes and intervals of fringes, some fringe obtained from such data extensions as image extension and rotation were calculated, and consistent results were obtained as shown in Figure 14. The results of Figure 14 further verify the robustness of this trained neural network.

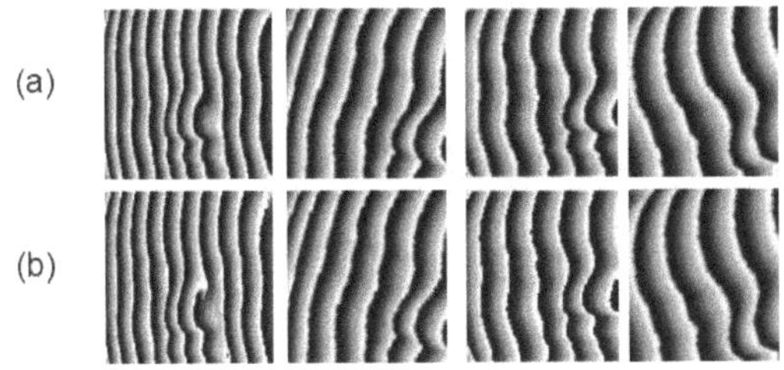

Figure 14. Calculated wrapped phase maps from (**a**) four-step phase-shifting method, and (**b**) neural network.

This learning-based method is an improving process. The more different fringe patterns are learned, the more complex model can be calculated. A sufficient amount of big data can support this neural network to adapt to all kinds of fringe patterns. Different fringe patterns verified its precision with the four-step phase-shifting method, however, this end-to-end neural network only needs one original fringe pattern to obtain the corresponding wrapped phase map.

This method requires a considerable amount of training data, which is time-consuming in data preparation. This training can be a continuous process. Once a batch of fringe patterns has been trained, the obtained neural network can be saved. The next batch of images can then be trained on the saved neural network, which can significantly reduce the time needed for training.

5. Conclusions

In this paper, we proposed a fringe wrapped-phase map extraction method based on the U-net neural network that can obtain the wrapped phase map directly from a fringe pattern. The results of simulated and experimental fringe patterns verified the efficiency and the robustness of this method. At the same accuracy, the proposed method boasts easy operation and a simple principle compared with the traditional phase-shifting method and owns faster speed than wavelet transform method.

Author Contributions: Conceptualization, W.H., Y.F. and H.M.; methodology, W.H.; software, W.H. and K.Y.; validation, K.Y., Y.F. and H.M.; formal analysis, W.H.; investigation, K.Y.; writing—original draft preparation, W.H.; writing—review and editing, H.M.; visualization, K.Y.; supervision, Y.F.; funding acquisition, Y.F. All authors have read and agreed to the published version of the manuscript.

Funding: This research was funded by the National Natural Science Foundation of China, grant number 11972235, 11890683 and 11732009. This research was supported by Shenzhen Key Laboratory of Intelligent Optical Measurement and Detection (ZDSYS20200107103001793).

Institutional Review Board Statement: Not applicable.

Informed Consent Statement: Not applicable.

Data Availability Statement: Not applicable.

Conflicts of Interest: The authors declare no conflict of interest.

References

1. Huang, P.S.; Hu, Q.; Jin, F.; Chiang, F.P. Color-encoded digital fringe projection technique for high-speed 3-D surface contouring. *Opt. Eng.* **1999**, *38*, 1065–1071. [CrossRef]
2. Silva, A.; Flores, J.L.; Muñoz, A.; Ayubi, G.A.; Ferrari, J.A. Three-dimensional shape profiling by out-of-focus projection of colored pulse width modulation fringe patterns. *Appl. Opt.* **2017**, *56*, 5198–5203. [CrossRef]
3. Bhaduri, B.; Mohan, N.K.; Kothiyal, M.P. Simultaneous measurement of out-of-plane displacement and slope using a multiaperture DSPI system and fast Fourier transform. *Appl. Opt.* **2007**, *46*, 5680–5686. [CrossRef] [PubMed]
4. Schreiber, W.; Theory, N.G. Arrangements of self-calibrating whole-body 3-D-measurement systems using fringe projection technique. *Opt. Eng.* **2000**, *39*, 159–169. [CrossRef]
5. Wang, Y.; Basu, S.; Li, B. Binarized dual phase-shifting method for high-quality 3D shape measurement. *Appl. Opt.* **2018**, *57*, 6632–6639. [CrossRef]
6. Wang, Y.; Zhang, S. Superfast multifrequency phase-shifting technique with optimal pulse width modulation. *Opt. Express* **2011**, *19*, 5149–5155. [CrossRef] [PubMed]
7. Suja Helen, S.; Kothiyal, M.P.; Sirohi, R.S. Analysis of spectrally resolved white light interferograms: Use of phase shifting technique. *Opt. Eng.* **2011**, *40*, 1329–1336. [CrossRef]
8. Li, J.; Su, X.; Guo, L. Improved Fourier transform profilometry for the automatic measurement of three-dimensional object shapes. *Opt. Eng.* **1990**, *29*, 1439–1445.
9. Kemao, Q.; Wang, H.; Gao, W. Windowed Fourier transform for fringe pattern analysis: Theoretical analyses. *Appl. Opt.* **2008**, *47*, 5408–5419. [CrossRef] [PubMed]
10. Huang, L.; Kemao, Q.; Pan, B.; Asundi, A.K. Comparison of Fourier transform, windowed Fourier transform, and wavelet transform methods for phase extraction from a single fringe pattern in fringe projection profilometry. *Opt. Lasers Eng.* **2010**, *48*, 141–148. [CrossRef]
11. Agarwal, N.; Kemao, Q. Windowed Fourier ridges as a spatial carrier phase-shifting algorithm. *Opt. Eng.* **2017**, *56*, 080501. [CrossRef]
12. Fu, Y.; Tay, C.J.; Quan, C.; Chen, L.J. Temporal wavelet analysis for deformation and velocity measurement in speckle interferometry. *Opt. Eng.* **2004**, *43*, 2780–2788. [CrossRef]
13. Tay, C.J.; Quan, C.; Fu, Y.; Huang, Y. Instantaneous velocity displacement and contour measurement by use of shadow moiré and temporal wavelet analysis. *Appl. Opt.* **2004**, *43*, 4164–4171. [CrossRef] [PubMed]
14. Braga, R.A.; González, R.J. Accuracy in dynamic laser speckle: Optimum size of speckles for temporal and frequency analyses. *Opt. Eng.* **2016**, *55*, 121702. [CrossRef]

15. Ghosh, R.; Mishra, A.; Orchard, G.; Thakor, N.V. Real-time object recognition and orientation estimation using an event-based camera and CNN. In Proceedings of the 2014 IEEE Biomedical Circuits and Systems Conference (BioCAS), Lausanne, Switzerland, 22–24 October 2014; pp. 544–547.
16. Bevilacqua, V.; Daleno, D.; Cariello, L.; Mastronardi, G. Pseudo 2D Hidden Markov Models for Face Recognition Using Neural Network Coefficients. In Proceedings of the 2007 IEEE Workshop on Automatic Identification Advanced Technologies, Alghero, Italy, 26 January 2017; pp. 107–111.
17. Herrmann, C.; Willersinn, D.; Beyerer, J. Low-resolution Convolutional Neural Networks for video face recognition. In Proceedings of the 2016 13th IEEE International Conference on Advanced Video and Signal Based Surveillance (AVSS), Colorado Springs, CO, USA, 23–26 August 2016; pp. 221–227.
18. Mondal, M.; Mondal, P.; Saha, N.; Paramita Chattopadhyay. Automatic number plate recognition using CNN based self-synthesized feature learning. In Proceedings of the 2017 IEEE Calcutta Conference (CALCON), Kolkata, India, 2–3 December 2017; pp. 378–381.
19. Acharya, U.R.; Fujita, H.; LihOh, S.; Hagiwara, Y.; Tan, J.H. Muhammad Adam Application of deep convolutional neural network for automated detection of myocardial infarction using ECG signals. *Inf. Sci.* **2017**, *190*, 415–416.
20. Heeht Nielsen, R. Theory of the Back ProPagation Neural Network. *Int. Jt. Conf. Neural Netw.* **1989**, *1*, 593–606.
21. Hu, X.; Saiko, M.; Hori, C. Incorporating tone features to convolutional neural network to improve Mandarin/Thai speech recognition. In Proceedings of the 2014 Asia-Pacific Signal and Information Processing Association Annual Summit and Conference (APSIPA), Chiang Mai, Thailand, 9–12 December 2014; pp. 1–5.
22. Yang, B.B.; Ju, J.B.; Zhao, Y.G. Technical Research of Anti-submarine Patrol Aircraft Magnetic Compensation Based on BP Neural Network. *Comput. Meas. Control* **2016**, *24*, 146–152.
23. Xiao, H.; Wang, G.L.; Wang, B.D. Application of BP Neural Network Based on Factor Analysis to Prediction of Rock Mass Deformation Modulus. *J. Eng. Geol.* **2016**, *1*, 87–95.
24. Horisaki, R.; Takagi, R.; Tanida, J. Learning-based imaging through scattering media. *Opt. Express* **2016**, *24*, 13738–13743. [CrossRef]
25. Rivenson, Y.; Zhang, Y.; Günaydin, H.; Teng, D.; Ozcan, A. Phase recovery and holographic image reconstruction using deep learning in neural networks. *Light. Sci. Appl.* **2018**, *7*, 17141. [CrossRef]
26. Pitkäaho, T.; Manninen, A.; Naughton, T.J. Performance of autofocus capability of deep convolutional neural networks in digital holographic microscopy. In *Digital Holography and Three-Dimensional Imaging*; OSA Technical Digest; Optical Society of America: Washington, DC, USA, 2017; p. W2A-5.
27. Wang, H.; Lyu, M.; Situ, G. eHoloNet: A learning-based end-to-end approach for in-line digital holographic reconstruction. *Opt. Express* **2018**, *26*, 22603–22614. [CrossRef]
28. Yin, W.; Chen, Q.; Feng, S.; Tao, T.; Huang, L.; Trusiak, M.; Zuo, C. Temporal phase unwrapping using deep learning. *Sci. Rep.* **2019**, *9*, 1–12. [CrossRef] [PubMed]
29. Feng, S.; Chen, Q.; Gu, G.; Tao, T.; Zhang, L.; Hu, Y.; Zuo, C. Fringe pattern analysis using deep learning. *Adv. Photonics* **2019**, *1*, 025001. [CrossRef]
30. Feng, S.; Zuo, C.; Yin, W.; Gu, G.; Chen, Q. Micro deep learning profilometry for high-speed 3D surface imaging. *Opt. Lasers Eng.* **2019**, *121*, 416–427. [CrossRef]
31. Yu, H.; Zhang, Z.; Chen, X.; Zheng, D.; Han, J.; Zhang, Y. Novel phase retrieval based on deep learning for fringe projection profilometry by only using one single fringe. *arXiv* **2019**, arXiv:1906.05652.
32. Van der Jeught, S.; Dirckx, J.J. Deep neural networks for single shot structured light profilometry. *Opt. Express* **2019**, *27*, 17091–17101. [CrossRef] [PubMed]
33. Schmidhuber, J. Deep learning in neural networks: An overview. *Neural Netw.* **2015**, *61*, 85–117. [CrossRef]
34. LeCun, Y.; Bengio, Y.; Hinton, G. Deep learning. *Nature* **2015**, *521*, 436–444. [CrossRef]
35. Pan, B.; Xie, H.; Xu, B.-Q.; Dai, F.-L. Performance of sub-pixel registration algorithms in digital image correlation. *Meas. Sci. Technol.* **2006**, *17*, 1615.
36. Rumelhart, D.E.; Hinton, G.E.; Williams, R.J. Learning representations by back-propagating errors. *Nature* **1986**, *323*, 533–536. [CrossRef]
37. Diederik, K.; Ba, J. Adam: A method for stochastic optimization. *arXiv* **2014**, arXiv:1412.6980.
38. Wang, Z.; Ma, H. Advanced continuous wavelet transform algorithm for digital interferogram analysis and processing. *Opt. Eng.* **2006**, *45*, 045601.

sensors

Review

Continuous In-Line Chromium Coating Thickness Measurement Methodologies: An Investigation of Current and Potential Technology

Adam Jones [1,*], Leshan Uggalla [1], Kang Li [1], Yuanlong Fan [1], Ashley Willow [2], Christopher A. Mills [2] and Nigel Copner [1]

1 Wireless and Optoelectronic Research and Innovation Centre, University of South Wales, Treforest CF37 1DL, UK; leshan.uggalla@southwales.ac.uk (L.U.); kang.li@southwales.ac.uk (K.L.); yuanlong.fan@southwales.ac.uk (Y.F.); nigel.copner@southwales.ac.uk (N.C.)
2 Tata Steel, Research and Development, Swansea Technology Centre, Swansea SA2 8PP, UK; Ashley.Brew@tatasteeleurope.com (A.W.); Christopher.Mills@tatasteeleurope.com (C.A.M.)
* Correspondence: adam.jones@southwales.ac.uk

Abstract: Coatings or films are applied to a substrate for several applications, such as waterproofing, corrosion resistance, adhesion performance, cosmetic effects, and optical coatings. When applying a coating to a substrate, it is vital to monitor the coating thickness during the coating process to achieve a product to the desired specification via real time production control. There are several different coating thickness measurement methods that can be used, either in-line or off-line, which can determine the coating thickness relative to the material of the coating and the substrate. In-line coating thickness measurement methods are often very difficult to design and implement due to the nature of the harsh environmental conditions of typical production processes and the speed at which the process is run. This paper addresses the current and novel coating thickness methodologies for application to chromium coatings on a ferro-magnetic steel substrate with their advantages and limitations regarding in-line measurement. The most common in-line coating thickness measurement method utilized within the steel packaging industry is the X-ray Fluorescence (XRF) method, but these systems can become costly when implemented for a wide packaging product and pose health and safety concerns due to its ionizing radiation. As technology advances, nanometer-scale coatings are becoming more common, and here three methods are highlighted, which have been used extensively in other industries (with several variants in their design) which can potentially measure coatings of nanometer thickness in a production line, precisely, safely, and do so in a non-contact and non-destructive manner. These methods are optical reflectometry, ellipsometry and interferometry.

Keywords: coating thickness measurement; chromium; steel substrate; optical metrology

Citation: Jones, A.; Uggalla, L.; Li, K.; Fan, Y.; Willow, A.; Mills, C.A.; Copner, N. Continuous In-Line Chromium Coating Thickness Measurement Methodologies: An Investigation of Current and Potential Technology. *Sensors* **2021**, *21*, 3340. https://doi.org/10.3390/s21103340

Academic Editor: Maria Lepore

Received: 12 April 2021
Accepted: 8 May 2021
Published: 11 May 2021

Publisher's Note: MDPI stays neutral with regard to jurisdictional claims in published maps and institutional affiliations.

1. Introduction

1.1. Background of Functional Coatings, Applications, and Coating Thickness Methodologies

When applying a coating to a substrate, it is important to consider whether the measurement technique for coating thickness is sensitive in the thickness range required by the coating specification, as the thickness of a coating is one of the key variables in determining the effectiveness for its given function [1]. Coating weight, and hence thickness specifications, is chosen for optimum performance with regards to a particular application and it is common that under-coating and over-coating can both lead to negative effects [2]. Coatings are typically important for waterproofing applications [3], corrosion resistance [4,5], adhesion performance [6,7], aesthetic effects [8], painting applications, and several more needs. There are many different methodologies for monitoring coating thickness and each method is usually specific for an application dependent on the substrate and coating materials. There are several reported methodologies for thin film characterization for application in the photonics industry [9], metal finishing industry [10,11], medical industry [12–14], and semi-conductor industry [15,16]. The successful monitoring of coating thickness can produce results such as creating a product to the intended specification for the customer, cost savings resultant from the prevention of material wastage, and a greater yield output from the implementation of in-line control.

Coating thickness test methods are split into two main categories, destructive and non-destructive measurement [10]. There are also other factors to consider when applying a proposed method to a fast-paced production setting with a harsh environment. There are contact and non-contact coating thickness measurement methods which can be used for specific coating and substrate applications. A contact method for this application would be unfeasible, as the production line is continuous in nature, and these methods may cause damage to the product or even the production process. The need for high-speed measurement rules out several other methods such as X-ray Photoelectron Spectroscopy (XPS) and Transmission Electron Microscopy (TEM) [17]. Most accurate methods, such as XPS and TEM, require sample preparation [17], making them unfeasible for in-line measurement. Another requirement for this application is that the measurement method must have the ability to measure coating thicknesses under 50 nanometers in thickness. Although there are other important factors to consider for the measurement methods, these factors have been considered a priority for this application.

1.2. Background of Electrolytic Chromium Coated Steel (ECCS)/Tin Free Steel (TFS)

This review emerges from a requirement to measure chromium coatings on a steel substrate passing through a production line running at full speed, which can be more than 6 m s^{-1}. For this application, metallic chromium (Cr) is electroplated onto both the top and bottom surfaces of a stainless-steel substrate in a series of chromium baths, this layer then passivates to form a chromium oxide (Cr_2O_3) layer. This product is widely known as ECCS or TFS, and its primary function is for the steel packaging industry. ECCS is used in many applications such as food and beverage containers, personal hygiene products, industrial packaging products and paint containers, as well as containers for gift products [18]. Steel packaging is highly recyclable and offers a long shelf-life which makes it a highly sustainable packaging solution, both for the present and the future [19]. Figure 1 illustrates the typical coating structure of ECCS.

Figure 1. On-line ECCS layer structure for packaging steel. ECCS comprises a steel substrate, a metallic chromium layer, a chromium oxide layer and a Dioctyl Sebacate (DOS) oil layer. ECCS was developed in the 1980s for the packaging market, providing cost savings in comparison to tinplated packaging products as ECCS requires a much thinner coating to provide similar corrosion protection [20]. The steel substrate for this application is typically less than 0.5 mm in thickness and varies with customer specifications. The total chromium layer for this application (summation of both the metallic and oxide layers) has a maximum thickness of 50 nm and again varies with customer specifications, but the thickness of these coatings must also comply to European standards, EN 10202:2001, to ensure product quality requirements [21].

1.3. Developments and Challenges in ECCS/TFS

The traditional ECCS manufacturing process utilizes Cr (VI) compounds in the passivation process of the steel. These Cr (VI) compounds are highly oxidizing substances which are easily reduced to Cr (III) compounds [18]. Unfortunately, with Cr (VI) compounds, there have been several studies that indicate their high toxicity compared to Cr (III) compounds in humans and animals, causing allergenicity and carcinogenicity through ingestion, dermal contact, and inhalation [22]. High concentrations of Cr (VI) have been found in US tap water, which may be resultant from the discharge of steel, pulp, metal-plating, and leather-tanning facilities, as well as through erosion of soil and rock [23–25]. REACH (Registration, Evaluation, Authorization and Restriction of Chemicals) legislation require the withdrawal of the use of chromium (VI) compounds due to safety concerns [18], which requires the steel packaging industry to develop new substrate variants [20]. One of the most recent developments in this area is Trivalent Chromium Coating Technology (TCCT), which is based on a chromium (III) process [6,26].

1.4. Continuous In-Line Chromium Coating Thickness Measurement

In addition to the ECCS application, thin chromium and chromium oxide films have been studied in several other applications, such as electrode materials for solar energy conversion and electrochromic windows [27], masking for lithography [28], and improvement of adhesion to transparent substrates [29]. The overall aim of this research is to investigate current and potentially novel in-line coating thickness measurement methods, applicable to chromium coatings on a steel substrate within a steel production setting, which could potentially be used for other applications highlighted above, or even other thin film coatings, with the following requirements:

- Non-destructive;
- Non-contact;
- Ability to measure nanometer coating thickness;
- High speed measurement for a continuous production setting;
- Feasibility for measuring multilayer coating structures;
- Feasibility of the methodologies in terms of the substrate material and coating material.

This research has been driven by a need for a new in-line chromium coating thickness measurement method for ECCS and TCCT coatings for packaging steels. The most common in-line coating thickness test methods for this application are based on X-ray Fluorescence (XRF) methods [30] however, these systems can become extremely costly, especially when implemented for a wide ECCS product, and impose health considerations, as they involve ionizing radiation. In terms of chromium thickness measurement, there is very little literature for an in-line measurement technique, except from similar application cases implementing several variants of ellipsometry [31–36]. Here, we examine the potential methods for in-line coating thickness measurement methods for ECCS and TCCT [33], and includes not only the measurement of the thin metallic chromium layer, but also the chromium oxide and DOS oil layers [31,32]. Therefore, this article aims to review potential methods for the in-line measurement of nanometrically thin chromium layers on a steel substrate for packaging applications, which could potentially be implemented into other industries. This review will briefly cover the operating principles of traditional, offline, and potentially novel in-line coating thickness measurement methods and summarize the advantages and limitations of these methodologies. Development of an accurate in-line coating measurement system would determine the precise coating thickness, removing the need for over-plating (excess electrical current), and consequently providing both cost and environmental benefits.

2. Traditional Coating Thickness Test Methods

Traditional coating thickness test methods are well known techniques that have been used previously and currently to measure coating thickness in several applications.

2.1. Coulometry

Coulometry is a destructive method in which the coating weight is measured by stripping the coating off the substrate and a mass per unit area relative to the coating thickness is calculated [10,37,38]. This method uses the reverse method of the electroplating process used to deposit coatings and the measurement of the coating is calculated using an electrolysis cell which contains electrolyte specifically designed for stripping the intended coating [38]. Constant current runs through the test cell and strips the coating (as the coating surface acts as the anode). As the current and surface area remains constant, the coating thickness can be calculated relevant to the time taken to strip the coating off the substrate. This method is very effective for electrically conductive coatings on a conductive substrate [10]. Schneider et al. [37], conducted a study to compare coulometry and optical reflectometry for thickness determination on titanium oxide films, which lead to the potential further study of an in situ spectroelectrochemical cell for this particular application This method has high measurement accuracy, is a low-cost system, can measure multilayer coatings and can be useful for a wide range of offline coating thickness applications. As this is a destructive process, this method is not feasible for in-line measurement. Figure 2 illustrates an electrochemical cell used in coulometry.

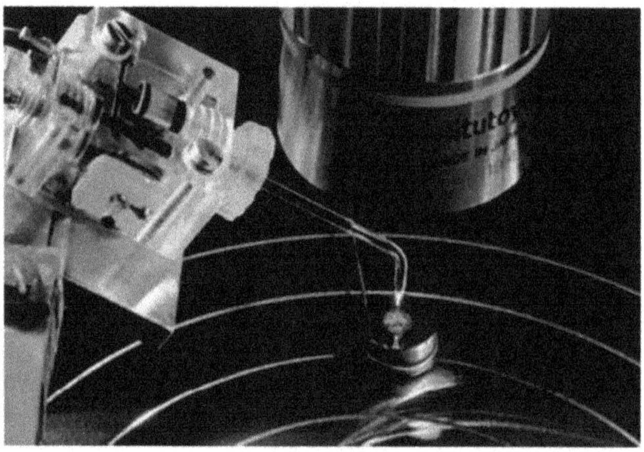

Figure 2. Photograph of an electrochemical cell used in coulometry. The tip of the cell has direct action with the coating structure, with a crater formed on dissolution on the sample. (Reprinted from [37], Copyright © 2011 John Wiley & Sons, Ltd.).

2.2. Beta Particle-Backscattering

For the Beta particle backscattering method, the sample under test is exposed to beta particles from a beta emitting isotope. Particles are directed though an aperture onto the sample with the coating to be measured. A percentage of these particles are backscattered back through an aperture into a Geiger–Muller tube [10]. Upon interaction of the backscattered particles with the gas within this tube, an ionization event occurs which is detected by the tube electrodes, held at a high potential difference, forming an electrical pulse which is then electronically counted. Materials of low atomic number backscatter at a lower rate than materials with a high atomic number. The change in the rate of electrons backscattered is a measure of the coating thickness. However, for this method to work effectively, the atomic number difference between the coating and the substrate must differ by at least 5 atomic units. Chromium's elemental atomic number is 24, whereas steel, predominantly iron, has an atomic number of 26, so the atomic number difference between these elements is only 2, rendering this method not feasible for this measurement application. Figure 3 illustrates the operating principle of the beta-backscatter method.

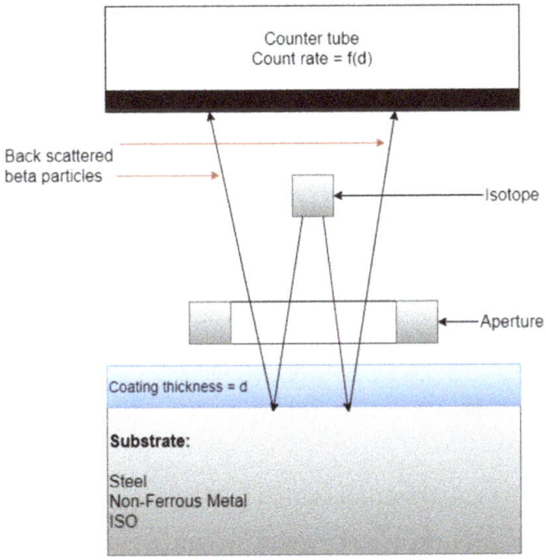

Figure 3. Operating principle of Beta-backscatter method. (Adapted from [10], Copyright © 2001 Published by Elsevier Inc.).

2.3. Eddy Current

The eddy current method uses a probe which contains a current carrying coil which is driven by a high frequency oscillator to generate an AC high frequency magnetic field. When the probe comes into proximity with the sample, eddy currents are generated at the sample, which alter in amplitude and phase to the original magnetic field. This difference generates an impedance signal between the excitation coil and the sample which is related to the coating thickness. This impedance change is dependent on the distance between the probe and the conductive substrate material. There are two main alternatives to this method, measuring the phase or the amplitude of the impedance changes [10]. Eddy current techniques are widely available commercially and offer micron coating thickness resolution. There is literature available for this technique, including the development of Pulsed Eddy Current (PEC) techniques [39] and research into varying probe constructions [40]. Eddy current measurements can be made on nonconductive coatings on nonferrous conductive substrates, nonferrous conductive coatings on nonconductive substrates and nonferrous metal coatings on nonferrous metals, rendering this method not feasible for this application, as the steel substrate is ferromagnetic. Figure 4a illustrates the operating principle of the eddy current method for coating thickness evaluation.

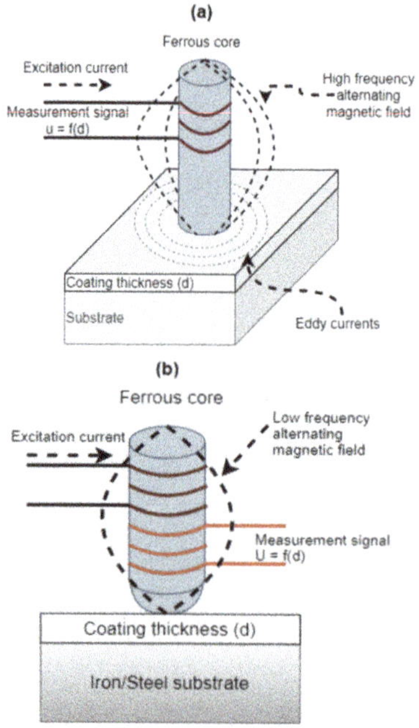

Figure 4. (**a**) Eddy current operating principle. (Adapted from [10], Copyright © 2001 Published by Elsevier Inc.). (**b**) Magnetic induction operating principle. (Adapted from [10], Copyright © 2001 Published by Elsevier Inc.).

2.4. Magnetic Induction

This method consists of a measurement probe containing a low frequency AC coil which is placed on the sample surface. By bringing a ferromagnetic substrate into the magnetic field, the core magnetic flux density is changed and is captured by a secondary sensing coil. The difference between the contact point at the sample and the distance to the substrate would equal the coating thickness. The typical time for a measurement ranges between 50 and 100 ms [10], which has potential for in-line measurement; however, this method generally requires contact with the sample, and has only micron level thickness determination. The magnetic induction method is used for measuring thickness of nonferrous coatings on ferro-magnetic substrates, thickness of magnetic coatings on non-magnetic substrates and thickness of magnetic coatings on magnetic substrates if the permeabilities of the coating and substrate are different and constant. Both Petrilli [10] and Hinken et al. [41] express the application of the eddy current and magnetic induction methods unfeasible for this application (ferromagnetic stainless-steel substrate) due to the limitations of the substrate/coating material combinations along with the thickness limitations, rendering this method not feasible. Figure 4b illustrates the operating principle of the magnetic induction method for coating thickness evaluation.

2.5. X-ray Fluorescence (XRF)

XRF is a measurement method which can be used to determine the compositional analysis of solids, liquids, powders, and can also be used to determine the thickness of coatings. This method can measure elements from Beryllium (Be) to Uranium (U) at sub-ppm levels. Elements with higher atomic numbers have better detection than lighter elements. This method is fast, accurate and non-destructive and can measure in a continuous production process for in-line measurement. The measurement time of a sample can range between seconds and minutes [30] and is based on the number of elements to be examined within a sample and the required accuracy. As the sample is irradiated with X-ray excitation, the sample will emit fluorescent radiation with discrete energies that are related to the elements of the sample. These energies differ by color, and this determines the different elements of the sample which provides qualitative analysis of the sample. Measuring the intensity of this fluorescent radiation can determine the quantity of each element present within the sample, which provides quantitative analysis. Fluorescent radiation from a sample occurs when the incident source expels an electron from an atom with a higher energy level. This produces a vacancy in the inner electron orbit, causing instability within the atom. The atom naturally wants to restore stability by transferring an electron from the outer to the inner orbit. This process emits fluorescent radiation in the form of a photon by the energy required for this orbit transfer. XRF methods are split into two main groups, Energy dispersive (ED-XRF) and Wavelength dispersive (WD-XRF). ED-XRF can measure from Sodium (Na) to Uranium (U). This method is comprised of detectors that measure the different energies of the characteristic radiation from the sample. WD-XRF can measure from Beryllium (Be) to Uranium (U). This method uses an analysis crystal that disperse the differing energies from the irradiated sample. This radiation is diffracted off the crystal in differing directions like a visible spectrum prism. ED-XRF systems feature fast measurement times, which is commonly used for in-line coating thickness measurement. Both ED-XRF systems and WD-XRF systems are currently being used in this research application to measure chromium coating thickness in-line and offline; however, there have been issues regarding the in-line system accuracy. These systems are very costly, and to implement an ED-XRF system for a wide product in a continuously fast-moving production environment, can become extremely complex. There are many applications for this measurement method [10,11,30,41,42], although this research originates from a need to investigate another potential in-line method, rendering this method not feasible for this application.

2.6. X-ray Reflectometry (XRR)

XRR operates on a similar theory as reflectance spectroscopy [43]. This reflectance measurement comprises of an incident ray, which reflects off the different interfaces at the sample, which form interference patterns through a range of incident angles. The incident ray in XRR consists of a beam of X-ray photons. The incident angle of this ray can be varied to give variations in the amplitude of the reflected beam, which can provide information on the coating and substrate. The incident ray will reflect or transmit through the sample at differing optical path lengths and amplitudes. Varying the incident angle can supply information on the sample, and to vary the incident angle, a Goniometer [10] can be used with reference to the source and detector, this slow process can take between 30 and 60 min. There are advancements within focused beam optics that can decrease this measurement time from 1 to 10 s. Typical additions to the system to increase performance are the monochromator, entrance aperture and collimators. This method is a very costly solution but does feature all the requirements needed for this application. Research has already been conducted with this method, investigating nanometer thick chromium coating thickness on a silicon substrate [44], which provided positive results when comparing with SEM methodologies. This method looks highly feasible for this application, however, to implement this system for a full product width would not provide the detection times needed for a continuous production line, and therefore it can be considered not feasible for this application.

2.7. Ultrasonic Detection

The ultrasonic method is a non-destructive technology (NDT) for the coating thickness measurement of primarily wood, concrete, and plastic applications. This method comprises of a single element transducer and numerical techniques to filter sound wave echoes from a sample. Current instruments on the market typically need to apply contact to the sample. The principle of operation is that a probe is placed upon the sample surface, a sound wave is propagated through the sample which results in differing vibrations at each layer interval. These vibrations are received at the probe source and numerical calculation is conducted to investigate the appropriate thickness of each layer. Novel advances to this method include high frequency scanning acoustic microscopy to measure sprayed coating thickness, non-destructively [45]. This method has been reported not ideal for thin metallic coatings or metallic substrates and conventional gauges require contact with the sample, which renders this method not feasible for continuous in-line implementation [46].

2.8. Overview of Traditional Coating Thickness Methodologies

Table 1 summarizes the feasibility for in-line implementation for this research application. These methods are the most common and traditional coating thickness measurement methods, surveyed within the current literature. It can be noted that there are clearly several key variables to consider for these technologies, and the only traditional test method outlined that could be used is the XRF method (in this case, ED-XRF for the fast measurement response). This test method is currently implemented in-line for this application and has several disadvantages, such as being a costly system to implement for a full product width, the measurement of total chromium only (unable to differentiate between oxide and metallic components) and has reported inaccuracies when comparing with offline test methods. Therefore, novel methods were needed to be explored to address these disadvantages, to provide a potential complimentary system which could be low-cost and not utilizing ionization radiation for safer working practices. Seven traditional coating thickness measurement methodologies are reviewed for this steel packaging application within Table 1. These are Coulometry (A1), Beta-backscatter (A2), Eddy current (A3), Magnetic induction (A4), XRF (A5), XRR (A6), and Ultrasonic detection (A7). All the traditional methodologies highlighted are currently available commercially, however, the typical cost and pricing of most of the systems is provided on a case-by-case quotation. The eddy current, magnetic induction and ultrasonic methodologies are widely available on the current market for several coating thickness applications, and the price is dependent on the complexity of the application requirements. The thickness ranges, sampling rate, spatial resolution and measurement accuracy of each system are also dependent on their individual system design.

Table 1. Overview of traditional coating thickness methodologies.

Methodology	A1	A2	A3	A4	A5	A6	A7
Thickness ranges	1 nm–50 µm	1–800 µm	1–10,000 µm	1–10,000 µm	0.5 nm–10 µm	1 nm–1 µm	10–7500 µm
Measurement accuracy (%)	0.1–0.5	0–5	0.1–0.7	1–3	0.2–0.5	0.33–0.65	0.53–0.7
Multi-layer	✓	✓	X	X	✓	✓	✓
Sampling rate	1–500 m	0.5–15 s	0.6–1 s	0.6–1 s	0.5–100 s	1 s–60 m	0.5–2.5 s
Detection area (ø)	1.5–3.2 mm	0.63–20 mm	5–8 mm	5–8 mm	0.1–15 mm	2 mm	5 mm
Commercial availability	[47]	[48,49]	[50]	[50]	[51]	[52]	[53]
Typical cost (GBP)	>3000	Unspecified	>1000	>1000	>35,000	>30,000	>1000
Non-contact	X	✓	✓	X	✓	✓	X
Non-destructive	X	✓	✓	✓	✓	✓	✓
Materials	✓	X	X	✓	✓	✓	X

3. Offline Coating Thickness Test Methods

Offline measurement methodologies have been characterized as methods which do not possess detection times fast enough for continuous production settings, are destructive in nature, require contact with the sample under test, require sample preparation or also require vacuum sealed environments.

3.1. X-ray Photoelectron Spectroscopy (XPS)

XPS is a measurement method which can be used to quantitatively measure coating thickness by identifying elemental composition of a sample via the photoelectric effect.

A sample under test is irradiated by a beam of X-ray radiation (photons) which excites an electron spectra resultant from the kinetic energy of the transmitted electrons. The typical measurement time of an XPS system differs dependent on the type of analysis performed, but the minimum analysis time is roughly 1 min, which renders this technique an offline measurement solution [17]. This method, however, is very accurate and can directly quantify coatings in nanometer scale thickness; however, this system is a very costly solution. Although this method requires sample preparation and requires a vacuum sealed environment, the measurement process is typically non-contact and non-destructive [17]. XPS studies have been conducted for monitor similar application areas, such as studying the surface chemistry of Zn-Al alloy coatings on steel [54] and studying the effects of passive films on stainless steel [55]. XPS is a well-known technique for investigating several different coatings, such as oxide thickness determination [56] and is also a method used currently for this research application to investigate the thicknesses of chrome oxide and metallic chromium layers, offline. Due to the cost and requirement for a vacuum environment, this technique is not feasible for in-line measurement.

3.2. Scanning Electron Microscopy (SEM)

Electron microscopes permit the observation of material composition at the nm and µm level. Electron microscopes are instruments which use a beam of electrons to examine an object on a magnified scale. This examination can provide information of the topography, morphology, composition, and crystallographic properties of a sample. A source of electrons is beamed over the surface of a sample. When the electrons penetrate the surface, several interactions occur that result in the emission of electrons or photons through the surface. These electrons can be detected, and an output is derived by modulating the brightness of a cathode ray tube [9] in terms of voltage. Many kinds of samples can be analyzed such as metals, ceramics, glass, hair, bones, and plastics. The main constraints of SEM are that the sample must be conductive (non-conductive materials must be carbon coated) and materials with a smaller atomic number than carbon cannot be detected. The vertical resolution of current SEM instruments ranges from between 0.6 and 1.5 nm [9], depending on the primary voltage. Giurlani et al., provide a comprehensive review on SEM technology, the advancement of SEM over time and the different detection methodologies used to create an SEM image, such as using secondary electrons, backscattered electrons or through a microanalysis map [26]. This method is considered not feasible for in-line measurement as it requires extensive sample preparation and highly controlled environmental conditions that are not suitable for in-line inspection. Figure 5 illustrates the operating principle of SEM.

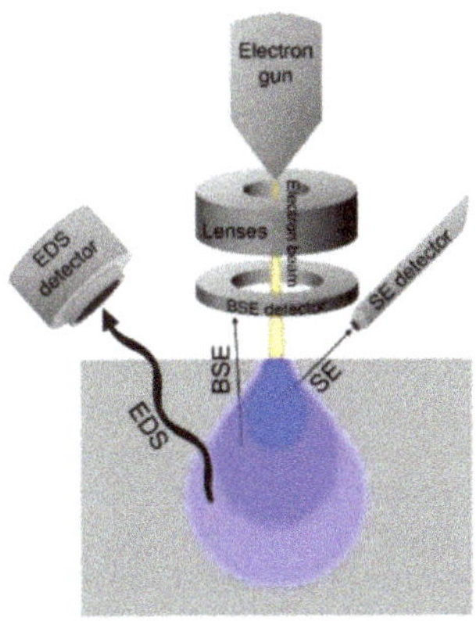

Figure 5. Operating scheme of SEM with SE, BSE, and EDS detection. (Reprinted from [26], CC BY 4.0 license).

3.3. Atomic Force Microscopy (AFM)

The field of Profilometry has a wide range of methods for investigating surface topographical analysis with sub-nanometer resolution. AFM measures the surface profile and film thickness of a sample through direct mechanical action. This method functions by directly measuring the deflection and oscillation of a flexible, microscopic cantilever tip caused by attractive and repulsive atomic forces when the tip is moved vertically in respect to the film surface. The AFM system uses a cantilever, tip, laser source, and a scanner. The laser source focuses a beam of light onto the rear face of the cantilever tip and the photodetector measures the intensity of the beam off the tip. The system may also use a voltmeter or a waveform generator to measure the oscillation of the cantilever or drive the oscillation with a waveform. An AFM system may be used in two modes, a contact mode in where the tip remains in contact with the sample during testing and a tapping mode in where a spring force mechanism is implemented on the cantilever tip. This measurement method is highly accurate and allows for calculation of surface roughness, however, this method requires contact with the sample under test and the resolution depth is limited by its magnification range [9]. There are advancements within AFM which requires no contact with the sample under test [57], and there is an abundant amount of literature covering the different operating methodologies This method is considered **not** feasible for in-line measurement due to the need for sample preparation and the detection speed to determine coating thickness. AFM has been used in several thickness measurement applications, such applications include the thickness measurement of graphene onto oxidized Si wafers [58], thin metal films on silicon substrates and polymer films on silicon substrates [59].

3.4. Glow Discharge Optical Emission Spectroscopy (GDOES)

This technique comprises of an optical emission spectrometer (OES) coupled with glow discharge (GD), and allows for the monitoring of both the surface and depth profiling of the elemental composition of solid materials, with high sensitivity. This technique is destructive in nature and involves sputtering an area of a sample with GD plasma, whilst observing optical emission. This technique has vertical resolution at the nm level and can identify which elements are present within a sample, quantify the concentration of elements in a sample, and in this research application case, measure coating thickness within a sample. GDOES has been used to measure coating thickness and elemental composition in several applications [60–62]; however, as this technique is destructive in nature, this method is assessed as not feasible for this application.

3.5. Overview of Offline Coating Thickness Methodologies

Table 2 provides a summary of offline coating thickness measurement methodologies. These methods have been considered not feasible for in-line measurement due to their limitations outlined and have been briefly researched to demonstrate and convey why they are considered unfeasible. All the offline measurement systems outlined have high accuracy and can determine nanometer scale thickness, but all methods require sample preparation. In terms of the detection speeds listed above, these are relevant for this application (coatings under 100 nm in thickness). From the literature, each of these methods and method variants have advanced greatly over the past two decades, with results in faster detection speeds. Four offline coating thickness measurement methodologies are reviewed for this steel packaging application within Table 2. These are XPS (B1), SEM (B2), AFM (B3), and GDOES (B4). As noted from the review table for the offline coating methodologies, the thickness ranges and measurement accuracy have an increased vertical resolution over the traditional methodologies, at the expense of the increase in system cost. There are many commercial systems available for these techniques which have been surveyed and highlighted, and XPS technology is by far the most expensive (where systems can be in the cost excess of GBP 500,000). It must also be noted that all offline methodologies can determine the coating weight of multi-layer structures through direct measurement, at the expense of slow measurement speeds (typically within the range of minutes to hours).

Table 2. Overview of offline coating thickness methodologies.

Methodology	B1	B2	B3	B4
Thickness ranges	0.5–20 nm	0.1 nm–2 μm	0.1–100 nm	1 nm–50 μm
Measurement accuracy (%)	0.05	0.2	2–5	<5
Multi-layer	✓	✓	✓	✓
Sampling rate	1 m–4 h	1–5 m	5 m–1 h	3 s–12 m
Detection area (ø)	10 μm–5 mm	50 nm–1 cm	10–100 μm	Unspecified
Commercial availability	[63]	[64]	[65]	[66]
Typical cost (GBP)	>200,000	>50,000	>20,000	Unspecified
Non-contact	✓	✓	X	X
Non-destructive	✓	✓	✓	X
Materials	✓	✓	✓	✓

4. Potential In-Line Coating Thickness Test Methods

In-line measurement methodologies have been characterized as methods which have the potential to measure real time, non-destructively and require no contact with the sample under test.

4.1. Thermoelectric Method with Magnetic Readout

The thermoelectric effect with magnetic readout takes in principle the Seebeck effect of bi-metallic materials coupled with a non-contact magnetic readout [41]. The Seebeck effect is mostly known for its application in thermocouples, an instrument that provides a voltage output approximating the applied heat at the junction of bi-metallic materials [67]. This method has been studied for Non-Destructive Testing (NDT) like the magnetic induction method, without contact with the sample [41,68] and is operated in the following steps:

- At the junction of dissimilar electrical conductors, a thermoelectric voltage is present when the junctions are at differing temperature.
- As this thermoelectric voltage would create a closed circuit regarding the coating and substrate, a current will flow from the hot junction to the cold junction.
- In turn, this electrical current will generate a magnetic field with a flux density that would extend to the outside of the material and into the air interface.

This method is largely dependent on the magnetic field strength that occurs at the air interface and the sensitivity of the magnetic sensor. The strength of the magnetic field can be influenced by the temperature gradient applied to the sample. A reverse process of this effect is known as the Peltier effect [41]. Applying the sample to an external magnetic field will induce currents and therefore voltages. This difference in voltage at the interface junctions will give rise to a change in temperature which could potentially be monitored with a thermoreflectance approach [69], but this technique will be discussed separately later. This method would provide non-destructive, non-contact analysis, and a high potential for in-line measurement. For nanometer scale measurement, it would be expected that the process heating phase would require a large differential to excite a magnetic field strong enough to detect with present magnetometers, based on the results from Hinken et al., experiments at the micrometer thickness ranges [41].

4.2. Terahertz Time Domain Spectroscopy (THz-TDS)

This method involves the use of high frequency radio waves (typically pulsed laser radiation) to reflect off the sample to provide information on the coating thickness [70]. A typical example of this method utilizes pulsed-echo terahertz thickness measurement by sending terahertz energy via a transceiver and to reflect off a conductive material substrate back to the transceiver. The terahertz transceiver is separated from the sample by an air path which makes this method non-contact and non-destructive in nature [71]. There is little research regarding this method in terms of nanometer scale resolution; however, this technique looks like a promising method for continuous in-line implementation [72,73]. The operating principle of this method involves the time domain analysis of the delay between the received echoes from the top surface of the coating and the top surface of the substrate. This method has a very similar operating principle to the optical reflectometry method; however, the detection and analysis of this method is not a function of optical intensity as in reflectometry, but as a function of the difference in time. The time delay between the coating surface echo (FS) and the substrate surface echo (BS) is directly affected by the thickness variation of the coating [73].

4.3. Optical Reflectometry

Optical reflectometry is a method which analyzes the reflectance spectra of a sample to measure characteristics such as optical constants and thin film thickness and is a well-known and powerful technique for measuring film thickness quickly in several industries [12,74–77]. There are extensive variants of this method, which are widely available for review in the literature, but the key operating principle of this methodology is the illumination of a sample with an optical light source, detection of the reflected intensity and fitting this intensity as a function of coating thickness [76]. Most technologies utilizing this method measure a range of wavelengths to calculate the film thickness, and one of the key limitations to this measurement method in the past is that the technique is only able to measure one point at a time [76]. Variants of this method have included volumetric detection using Charge Coupled Devices (CCD) to monitor an area of a sample, simultaneously, with the disadvantage of decreased detection speeds, but there is literature available into techniques which can reduce these detection times, such as direct phase extraction techniques [78]. Reflectometry methods usually operate at normal incident angles for measurement; however, varying incident angles can be used which again increases the system complexity in terms of monitoring both parallel and perpendicular polarizations, based on the Fresnel equations [79]. This method also consists of a form of self-interferometry, in where there are several internal reflections between the coating and substrate interfaces. These coating and substrate interfaces are often modelled using the Fresnel reflection coefficients to determine the reflected intensity present at the optical detector, however, for several layers these calculations become complex, and researchers tend to model these interfaces using Transfer Matrix Methodologies (TMM) [80]. The main operating principle for this method relies on reflection intensity, which is calculated by the ratio of the reflected intensity at the detector to the incident intensity. In terms of measurement limitations for this application, to measure metallic coatings, there is a requirement that the incident radiation can transmit through the coating to the substrate, and for chromium coatings, the calculated limitation through Beer's law is roughly below 70 nm coating thickness. As this application intends to monitor thicknesses of less than 50 nm, this method has high feasibility [29]. This method has high speed capability, has the potential to measure the thickness ranges of the application area, is a non-contact and non-destructive approach which renders this method high feasibility for future research. There are several factors that must be considered taking this method forward, such as the surface roughness of the coating and the substrate, the coating uniformity and the production process parameters that could hinder measurement, most importantly, the vibration of the electroplated product running continuously through the production line. Figure 6 illustrates the operating principle of optical reflectometry for coating thickness evaluation.

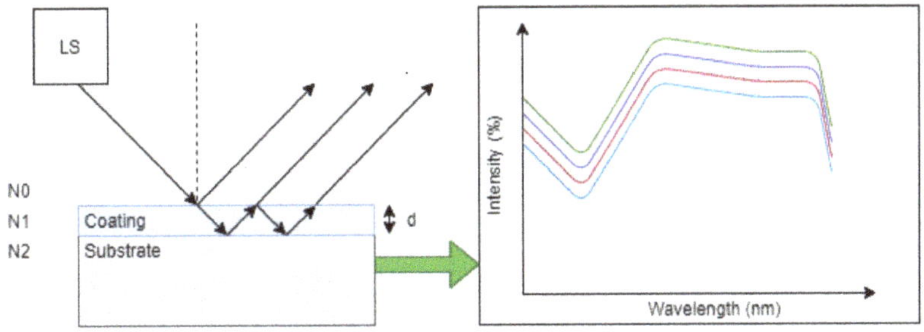

Figure 6. Principle of optical reflectometry. A light source (LS) illuminates the sample of interest and as a result, there are several reflections returned to the detector due to layer interference. As the layer thickness is increased or decreased, dependent on the optical properties of the layer structure, the resultant reflection spectrum can determine coating thickness as a function of reflection intensity.

4.4. Optical Interferometry

Interferometry is a very similar optical technique to the reflectometry method in that they both use the phenomena of optical reflection or transmission to gain an understanding of a sample under measurement. Where reflectometry measures the intensity of a reflected or transmitted spectrum and takes into consideration the self-interference of the layers of a sample, the interferometry method takes into consideration the optical path difference in relation to a single light source between a sample under test and a reference mirror via a beam splitter [81]. As the optical path difference will change relative to the reference mirror and the sample as a function of coating thickness, differing phases based on the optical path difference will provide constructive or destructive interference for the output signal. There are many variants of optical interferometry, such as using differing light sources such as white light or singular wavelength [16,82], differing optical path layouts [83,84], and differing detection elements [85,86]. There are many interferometric solutions for the measurement of thin film thickness which operate on the principle of interference but utilize different components within the system for different applications. An interferometric and reflectometric method which may be suitable for this project's application was reported to be non-destructive, non-contact and has high-speed data acquisition, has published results within the nanometer thickness range in a microelectronics production setting [81]. Typically, interferometry is limited by a factor of the incident wavelength of light, however, the proposed system [81] utilizes Fast Fourier Transforms for the surface topography coupled with a reflectometric approach which has highly promising continuous in-line implementation. The setup proposed is basically a Michelson interferometer with two arms, one to generate the reference wave from a flat mirror and one to generate the reflection from the sample of interest. Two similar objective lenses are added into the system to adjust the lateral measuring magnification. The thickness ranges for this method have been reported down to a single nanometer [87], but like reflectometry, this setup is limited to metallic layers within the optical transparency region, which renders interferometry highly promising for this application. Figure 7 illustrates the operating principle of optical interferometry for coating thickness evaluation.

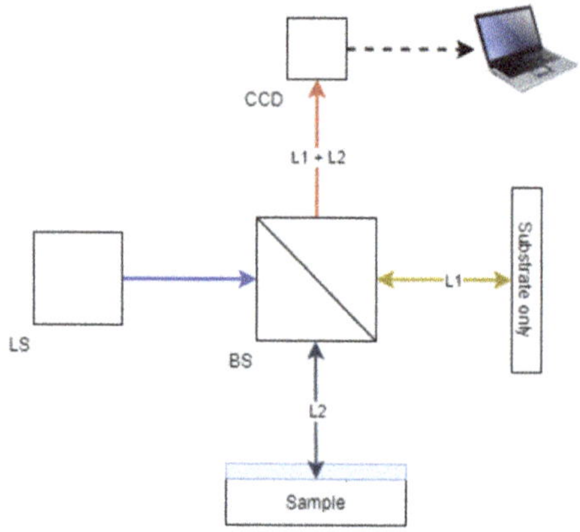

Figure 7. Principle of optical interferometry (Michelson setup) for coating thickness applications. A non-polarizing Beamsplitter (BS) separates a light source into two separate elements determined by the beam splitting ratio (typically 50:50). The transmitted light reflects from a reference sample (substrate only) at L1. The reflected light from the Beamsplitter irradiates the sample and undergoes varying self-interference dependent on the thickness of the coating outlined within the reflectometry method, coupled with L2. Both elements combine at the Beamsplitter and are then transmitted to a Charge Coupled Device (CCD). Signal processing is the conducted on the interference signal to evaluate coating thickness.

4.5. Optical Ellipsometry

Ellipsometry is an optical measurement method using light reflection or transmission of a sample. This method is very similar to the reflectometry method; however, ellipsometry also measures the change in polarization regarding reflection or transmission on a sample. This method can be high speed in operation with the ability to measure nanometer scale coating thickness, giving rise to real time measurement [32]. This method measures two variables, Δ and Ψ, where Δ is the phase difference between parallel (p) and perpendicular planes (s) and Ψ is the amplitude ratio of parallel (p) and perpendicular (s) planes [88]. To put the method simply, a light source generates a beam of light that is transmitted through a linear polarizer and compensator to control the incident polarization of the light at the sample surface. Reflections occur at each interface within the sample at a given amplitude and phase. These reflections are analyzed and detected and can determine specific properties of each layer of the sample, whether it is to evaluate coating thickness, coating uniformity, surface roughness or even to determine the optical constants for the coating structure [89,90]. There are several variants of ellipsometry and its optical setup, such as using spectroscopic or singular wavelength light sources [91,92], using numerous passive optical components setups [93,94] and using imaging techniques [1]. This method is covered extensively in the literature for application for in-line coating thickness measurement in several industries, such as the measurement of thin film thickness of photovoltaic development on production lines [34], the monitoring of optical constants and layer thickness of organic photovoltaics in a roll-to-roll (r2r) production setting [35] and the evaluation of optical constants, layer thickness with nanometer precision and uniformity of organic electronic (OE) devices [36]. Extending in-line ellipsometry further for this steel packaging application, there are two publications in which researchers applied ellipsometry to extremely similar steel packaging applications to this research [31,33]. Firstly, Izumidate et al. [31] researched and developed an in-line ellipsometry system to measure the ultra-thin oil layer that is applied to the steel packaging product after the coating process, and to measure the hydrated chromium oxide layer that occurs through passivation after the metallic chromium layer has been electroplated onto the steel substrate. This research was reported to have been successfully implemented to measure these layers, continuously in-line with nanometer thickness ranges, which gives high potential for this application. However, the difference for this application is that the metallic chromium layer also needs to be measured in-line, which may have been a challenge as the metallic chromium thickness on their Tin Free Steel (TFS) product had reported millimeter thickness ranges, which would result in no optical transmission through the coating layer to the substrate interface. Secondly, Rischmueller et al. [33] conducted research utilizing the ellipsometry method for the inspection of future REACH (Registration, Evaluation, Authorization, and Restriction of Chemicals) compliant Trivalent Chromium Conversion Coatings (TCCC) on an aluminum substrate. This research article provided great depth on measuring this coating thickness within the optical transparency region of chromium (<70 nm), researching the most effective wavelength for their product with angular dependency and mitigation controls regarding the surface roughness resultant from the rolling process. It was reported [33] that this method could be implemented in-line, with a non-contact approach which is like this research application. Given the requirements of this application for nanometer chromium thickness measurement on steel with a measurable surface roughness, further research would need to be conducted into the feasibility of this method, however judging from the literature available, this method looks highly promising. Figure 8 illustrates the operating principle of optical ellipsometry for coating thickness evaluation.

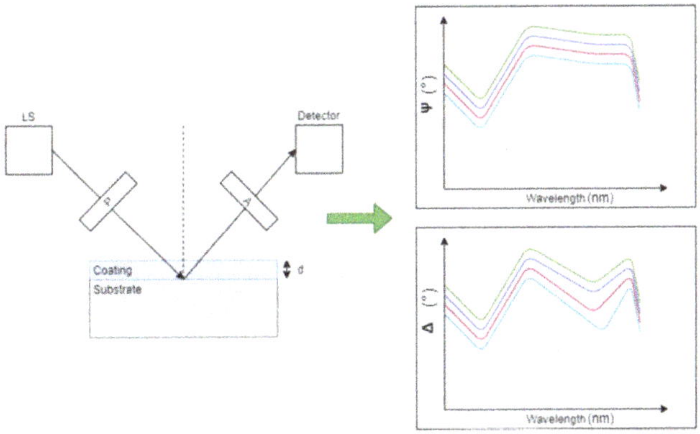

Figure 8. Principle of optical reflection ellipsometry for coating thickness applications. A polarizer (P) linearly polarizes a light source into a known polarization state at incidence of a sample. As light interacts with the sample, a change of polarization for both parallel and perpendicular polarizations will occur, dependent on the sample properties and coating thickness. An analyzer (A) is used to detect the polarization state of both polarizations, which is then fed to a detection element for further analysis. A ratio of both the parallel and perpendicular polarization determines the amplitude and phase difference generated from the sample of interest, which can then be compared to pre-determined optical model to measure several parameters, such as coating thickness, coating uniformity, and optical constants [95].

4.6. Stimulated Brillouin Scattering (SBS)

Stimulated Brillouin scattering is non-linear scattering which involves the artificial generation of acoustic phonons in a transparent medium. This type of scattering comes from light interaction with acoustic waves in a material and are generated by thermodynamic fluctuations [96]. This scattering method has been used in a spectroscopic nature to non-invasively provide general imaging solutions for applications in biology and materials science [97]. The SBS method utilizes two incident light waves at two differing frequencies, and the waves generated in the sample are the density variations (acoustic waves) that are resultant from the incident light waves which develop a beat frequency, which can then be monitored to provide information on the coating thickness. For this method to work for this application, the thin film layers and the substrate will need to be transparent [97], rendering this method unfeasible; however, this method is non-contact and non-destructive, which could potentially be applied to other coating/substrate material combinations for in-line measurement.

4.7. *Self-Mixing Interferometry (SMI)*

Self-mixing interferometry is a non-contact optical method which has been employed for the measurement of refractive indices and thickness measurement of optical components [98]. This method operates on similar principles to standard interferometry, using the measurement of optical path length; however, this method utilizes the phase difference of a back scattered beam without a beam splitter and reference measurement [99]. The self-mixing interferometer emits a collimated beam from a laser and is passed through a transparent sample. Upon reflection at a second photodiode [100], some of the reflected light is re-entered into the laser cavity where the weak signal interacts with the incident beam. This results in modulation of amplitude and frequency of the laser field where the driver is the optical path length. This method provides non-contact, non-destructive measurement and has proven research at the micron thickness level [98–100]. For this application, this method is unfeasible as SMI requires optical transmission through both the coating and substrate but could potentially be applied to other transparent coating/substrate material combination applications for in-line measurement.

4.8. *Chromatic Confocal Microscopy (CCM)*

The CCM method is a non-destructive and non-contact optical technique that has been researched for film thickness measurement for transparent and non-transparent films [101,102]. An incident white light source is focused into its spectrum of wavelengths for varying distances to the sample, and in terms of film thickness measurement, two peaks relational to the wavelengths and distance to the sample are calculated, the difference between these peaks would result in the optical thickness of the measured material. When the optical properties of the material are known, the physical thickness can be calculated via the dispersion properties as a function of wavelength in terms of the sample optical properties [102]. This method has been reported to have micron thickness resolution, and feasible for in-line implementation [102]. There are several researched variants of the CCM technique, one such variant is Chromatic Confocal Spectral Interferometry (CCSI), which combines the operating principles of CCM and optical interferometry for increased resolution [103]. Although current literature reports only micron thickness resolution, it is assessed that this method may be developed further, rendering this method potentially feasible for this application.

4.9. *Infrared Thermography*

Infrared imaging has been previously used to measure the coating thickness of paints on steel substrates with a coating thickness at the micron level [104]. The principle of operation is to apply external heating of a sample and measure the thermal radiation via a high-resolution thermal camera within the IR spectrum [104]. The benefits of this measurement method are that it can image a large spatial area concurrently, allowing for the monitoring of the coating uniformity real time. The underlying principle of this method is the correlation between the heat transfer of each coating layer with the layer thickness [104]. As most elements have differing heat transfer properties, this method makes it suitable to measure many combinations of substrate and coating materials. Some of the key requirements for this method is the introduction of an external heating source, a high-resolution thermal IR detector capable of nanometer resolution and the heat transfer properties between coating layers and substrate are diverse. This method is non-contact, non-destructive and can potentially be used for a vast number of coatings and substrates with high-speed detection [104]. In terms for this application, for coating thickness at the nanometer scale, the detection elements to measure the full width of the electroplated product would require great expense, not to mention, the heating element may also cause complications to the production process dependent on the amount of heating required to trigger accurate detection. One variant of this technique discussed briefly previously was a thermoreflectance imaging approach [69]. This technique operates on the reverse process of the Seebeck effect, known as the Peltier effect. By exciting the sample under test with an alternating current supply, this in turn would generate an oscillating temperature field in the sample, which could then be detected using a CCD. This technique has been used in the semi-conductor industry [69], providing a topographical image of the heat propagation at sub-micron thickness resolution.

4.10. Overview of Potential In-Line Coating Thickness Methodologies

Table 3 summarizes the potential and novel in-line coating thickness test methods that have high potential for this research application. Nine potential in-line coating thickness measurement methodologies are reviewed for this steel packaging application, these are thermoelectric magnetic method (C1), THz-TDS (C2), Optical reflectometry (C3), Optical interferometry (C4), Optical ellipsometry (C5), SBS (C6), SMI (C7), CCM (C8), and infrared thermography (C9). From a commercial product survey, the thermoelectric magnetic, SBS, and SMI methodologies were not found to have a specific product for a coating thickness application. It must also be noted that at this current time, only three methods have been found through research for the capability to measure coatings of nanometer in thickness, these are optical reflectometry, ellipsometry, and interferometry. There are several variants of these three optical methods as highlighted previously in this review paper, and there are also commercial systems available for these methodologies; however, none of these systems specify their unique ability to be applied to this steel packaging application, specifically the metallic chromium layer. It was interesting to find through the reflectometry commercial product analysis [105] that it states that reflectometry and ellipsometry have the capability to measure metallic coatings of under 50 nm in thickness, which directly relates to the outlined literature, which reinforces the high potential of these methodologies in particular for further research and development or this steel packaging application, as we are intending investigate metallic chromium coatings under these thickness limits. In terms of the detection area of these methodologies, these are highly dependent on the system construction, which can be clearly noted from the commercial product survey and the appropriate citations. The interferometry method highlighted from the commercial product survey is based on Optical Coherence Tomography (OCT), where the method provides similar output results to AFM, non-destructively, and without contact with the sample under test.

Table 3. Overview of potential in-line coating thickness methodologies.

Methodology	C1	C2	C3	C4	C5	C6	C7	C8	C9
Thickness ranges	1–200 μm	300–1400 μm	0.5 nm–3 mm	0.1 nm–10 μm	0.5 nm–1 mm	>1 μm	>1 μm	>1 μm	>1 μm
Measurement accuracy (%)	10	0.43	0.1–0.2	<1	0.1	Unspecified	Unspecified	<3%	Unspecified
Multi-layer	Unspecified	✓	✓	✓	✓	✓	✓	✓	✓
Sampling frequency	0.1–4 s	>50 ms	0.1–5 s	<3 s	0.1–300 s	Unspecified	Unspecified	<1 s	<1 s
Detection area (ø)	Unspecified	Unspecified	50 μm–1 mm	Unspecified	50 μm–1 mm	Unspecified	Unspecified	Unspecified	Unspecified
Commercial availability	Unspecified	[106]	[105]	[107]	[108]	Unspecified	Unspecified	[109]	[110]
Typical cost (GBP)	Unspecified	Unspecified	>13,000	>45,000	>40,000	Unspecified	Unspecified	>10,000	>10,000
Non-contact	✓	✓	✓	✓	✓	✓	✓	✓	✓
Non-destructive	✓	✓	✓	✓	✓	✓	✓	✓	✓
Materials	✓	✓	✓	✓	✓	X	X	✓	✓

From this review table, there are three clear methodologies to pursue for future research into this steel packaging application, these are optical reflectometry, optical ellipsometry and optical interferometry. Taking into consideration the system complexity, cost and potential, reflectometry and its variants are highlighted as the focus of future research, followed by interferometry and ellipsometry.

5. Conclusions

This review has attempted to identify several potential in-line coating thickness measurement methods that could be applied to measure nanometer chrome coatings on a steel substrate. Each method outlined within this paper has been categorized into their potential for: continuous, non-contact, non-destructive, in-line measurement; at speed of sub 50 nm coating thickness measurement of applicable substrate/coating material combinations. From this interim investigation there are three main optical methods which have been assessed as highly feasible for this research application, optical reflectometry, ellipsometry, and interferometry. There are also other novel methodologies which may satisfy this application requirements such as a thermoelectric method with a magnetic readout approach, THz-TDS, CCM, and infrared thermography; however, there is very little research regarding these methods for nano-scale precision. The X-ray methods outlined such as XRR and XRF are also highly feasible for this application; however, this research is based upon finding a novel approach from the current coating thickness test methods implemented at the industrial partner. There is an abundance of research in the literature suitable to establish a novel method to measure metallic coating thickness for steel packaging applications. All three optical methodologies identified have also been reviewed for potential in-line monitoring for other applications [111], and it can clearly be seen through the literature that optical techniques are being highlighted as the successors for potential in-line monitoring of nanometric coatings. For the optical methods outlined, there is the potential to: create a low-cost system; implement this low-cost system at multiple points at within a production line; multiplex measurements across the width of a production line to ensure full width coverage; and to provide redundancy to currently used XRF-based systems that are unable to differentiate between metallic and oxide chromium layers. In terms of the applicable optical methods, there are several variants that have been researched, developed, and implemented in other industries. Consequently, this research aims to develop these techniques, or variants thereof, for steel packaging applications with metallic sub-50 nm coating thicknesses. In addition to this review paper, further research will be conducted into the three optical methods highlighted via simulations and experimental research for this application based on the requirements set out within this paper, with emphasis placed on other significant factors that may hinder the feasibility of these methods such as the effect of illumination wavelength, effect of incident angle, effect of chrome oxide layer within the sample, effect of differing substrate chemistry, effect of substrate surface roughness resultant from the steel rolling process, and the effect of vibrational parameters on a continuous production line.

Author Contributions: Conceptualization, A.J., A.W. and N.C.; Methodology, A.J., K.L. and L.U.; Validation, C.A.M., A.W. and L.U.; Investigation, A.J.; Resources, K.L.; Writing—Original Draft Preparation, A.J.; Writing—Review & Editing, L.U., Y.F., A.W. and C.A.M.; Supervision, L.U., K.L. and Y.F.; Project Administration, L.U.; Funding Acquisition, A.W. and C.A.M. All authors have read and agreed to the published version of the manuscript.

Funding: Knowledge Economy Skills Scholarships (KESS) is a pan-Wales higher-level skills initiative led by Bangor University on behalf of the HE sectors in Wales. It is part funded by the Welsh Government's European Social Fund (ESF) program for East Wales.

Acknowledgments: The authors of this paper would like to extend extreme gratitude to the technical and production teams at Tata Steel, Trostre Works, for their continued commitment to aid and deliver this research, your efforts are greatly appreciated.

Conflicts of Interest: The authors declare no conflict of interest.

References

1. Bammer, F.; Huemer, F. Inline Thickness Measurement with Imaging Ellipsometry. Available online: https://www.spiedigitallibrary.org/conference-proceedings-of-spie/11144/111440H/Inline-thickness-measurement-with-imaging-ellipsometry/10.1117/12.2531940.full?SSO=1 (accessed on 2 April 2021).
2. Tator, K.; Tator, K. Coating deterioration. *ASM Handb.* **2015**, *5*, 462–473.
3. Li, L.; Bai, Y.; Li, L.; Wang, S.; Zhang, T. A Superhydrophobic Smart Coating for Flexible and Wearable Sensing Electronics. *Adv Mater* **2017**, *29*, 1702517. [CrossRef] [PubMed]
4. Hosking, N.; Ström, M.; Shipway, P.; Rudd, C. Corrosion resistance of zinc–magnesium coated steel. *Corros. Sci.* **2007**, *49*, 3669–3695. [CrossRef]
5. Jehn, H.A. Improvement of the corrosion resistance of PVD hard coating–substrate systems. *Surf. Coat. Technol.* **2000**, *125*, 212–217. [CrossRef]
6. Allman, A.; Jewell, E.; de Vooys, A.; Hayes, R. Inter-layer adhesion performance of steel packaging materials for food cans under retort conditions. *J. Packag. Technol.* **2018**, *2*, 115–124. [CrossRef]
7. Lee, J.-W.; Tien, S.-K.; Kuo, Y.-C.; Chen, C.-M. The mechanical properties evaluation of the CrN coatings deposited by the pulsed DC reactive magnetron sputtering. *Surf. Coat. Technol.* **2006**, *200*, 3330–3335. [CrossRef]
8. Scrinzi, E.; Rossi, S. The aesthetic and functional properties of enamel coatings on steel. *Mater. Des.* **2010**, *31*, 4138–4146. [CrossRef]
9. Whiteside, P.J.; Chininis, J.A.; Hunt, H.K. Techniques and challenges for characterizing metal thin films with applications in photonics. *Coatings* **2016**, *6*, 35. [CrossRef]
10. Petrilli, C. The basics of coating thickness measurement. *Met. Finish.* **2001**, *99*, 8–13. [CrossRef]
11. Sokolov, A.; Hasikova, J.; Pecerskis, A.; Gostilo, V.; Lee, K.; Jung, H.; Lim, J. Application of Industrial XRF Coating Thickness Analyzer for Phosphate Coating Thickness on Steel. *Coatings* **2018**, *8*, 126. [CrossRef]
12. Lu, H.; Wang, M.R.; Wang, J.; Shen, M. Tear film measurement by optical reflectometry technique. *J. Biomed. Opt.* **2014**, *19*, 027001. [CrossRef]
13. Sacher, S.; Wahl, P.; Weißensteiner, M.; Wolfgang, M.; Pokhilchuk, Y.; Looser, B.; Thies, J.; Raffa, A.; Khinast, J.G. Shedding light on coatings: Real-time monitoring of coating quality at industrial scale. *Int. J. Pharm.* **2019**, *566*, 57–66. [CrossRef]
14. Wolfgang, M.; Peter, A.; Wahl, P.; Markl, D.; Zeitler, J.A.; Khinast, J.G. At-line validation of optical coherence tomography as in-line/at-line coating thickness measurement method. *Int. J. Pharm.* **2019**, *572*, 118766. [CrossRef] [PubMed]
15. Ghim, Y.-S.; Suratkar, A.; Davies, A. Reflectometry-based wavelength scanning interferometry for thickness measurements of very thin wafers. *Opt. Express* **2010**, *18*, 6522–6529. [CrossRef] [PubMed]
16. Hlubina, P.; Ciprian, D.; Luňáček, J.; Lesňák, M. Thickness of SiO 2 thin film on silicon wafer measured by dispersive white-light spectral interferometry. *Appl. Phys. B* **2006**, *84*, 511–516. [CrossRef]
17. Shallenberger, J.; Cole, D.; Novak, S.; Moore, R.; Edgell, M.; Smith, S.; Hitzman, C.; Kirchhoff, J.; Principe, E.; Biswas, S. Oxide thickness determination by XPS, AES, SIMS, RBS and TEM. Available online: https://ieeexplore.ieee.org/document/812056 (accessed on 2 April 2021).
18. Słowik, M.; Cępa, P.; Czapla, K.; Żabiński, P. Steel Packaging Production Process and a Review of New Trends. *Arch. Metall. Mater.* **2021**, 135–143.
19. Brimacombe, L.; Coleman, N.; Honess, C. Recycling, reuse and the sustainability of steel. *Millenium Steel* **2005**, *446*, 3–7.
20. Melvin, C.; Jewell, E.; de Vooys, A.; Lammers, K.; Mc Murray, N. Surface and adhesion characteristics of current and next generation steel packaging materials. *J. Packag. Technol. Res.* **2018**, 2, 93–103. [CrossRef]
21. EN 10202:2001 Cold Reduced Tinmill Products—Electrolytic Tinplate and Electrolytic Chromium/Chromium Oxide Coated Steel. Available online: https://standards.iteh.ai/catalog/standards/cen/eb353920-e059-4b31-a83c-2b0deb0069d0/en-10202-2001 (accessed on 2 April 2021).
22. Shekhawat, K.; Chatterjee, S.; Joshi, B. Chromium toxicity and its health hazards. *Int. J. Adv. Res.* **2015**, *3*, 167–172.
23. Costa, M.; Klein, C.B. Toxicity and carcinogenicity of chromium compounds in humans. *Crit. Rev. Toxicol.* **2006**, *36*, 155–163. [CrossRef]
24. Sutton, R. *Chromium-6 in US Tap Water*; Environmental Working Group: Washington, DC, USA, 2010.
25. Zhang, J.D.; Li, X.L. Chromium pollution of soil and water in Jinzhou. *Zhonghua Yu Fang Yi Xue Za Zhi* **1987**, *21*, 262–264.
26. Giurlani, W.; Berretti, E.; Innocenti, M.; Lavacchi, A. Measuring the Thickness of Metal Coatings: A Review of the Methods. *Coatings* **2020**, *10*, 1211. [CrossRef]
27. Cheng, C.S.; Gomi, H.; Sakata, H. Electrical and optical properties of Cr_2O_3 films prepared by chemical vapour deposition. *Phys. Status Solidi* **1996**, *155*, 417–425. [CrossRef]
28. Chiu, P.-K.; Liao, Y.-T.; Tsai, H.-Y.; Chiang, D. Effect of electron-beam deposition process variables on the film characteristics of the CrOx films. *AIP Adv.* **2018**, *8*, 025016. [CrossRef]
29. Lozanova, V.; Lalova, A.; Soserov, L.; Todorov, R. Optical and Electrical Properties of Very Thin Chromium Films for Optoelectronic Devices. Available online: https://iopscience.iop.org/article/10.1088/1742-6596/514/1/012003 (accessed on 2 April 2021).
30. Brouwer, P. *Theory of XRF*; PANalytical BV: Almelo, The Netherlands. Available online: https://www.iotcco.com/uploads/VirtualTeaching/Articles/PANanalytical/PANanalytical%20XRF%20theory.pdf (accessed on 2 April 2021).

31. Izumidate, A.; Yamamoto, H.; Shiki, S.; Nomura, Y. On–Line System for Measuring Thickness of Ultra-Thin Non-Metallic Layer on Strip Surface. In *Information-Control Problems in Manufacturing Technology*; Elsevier: Amsterdam, The Netherlands, 1978; pp. 25–32.
32. Miyazaki, T.; Yamada, Y.; Komine, I. High-Speed 3-Channel Ellipsometer for Industrial Uses. *Trans. Soc. Instrum. Control. Eng.* **1988**, *24*, 670–677. [CrossRef]
33. Rischmueller, J.; Toschke, Y.; Imlau, M.; Schlag, M.; Brüning, H.; Brune, K. Inspection of Trivalent Chromium Conversion Coatings Using Laser Light: The Unexpected Role of Interference on Cold-Rolled Aluminium. *Sensors* **2020**, *20*, 2164. [CrossRef]
34. Fried, M. On-line monitoring of solar cell module production by ellipsometry technique. *Thin Solid Film.* **2014**, *571*, 345–355. [CrossRef]
35. Logothetidis, S.; Georgiou, D.; Laskarakis, A.; Koidis, C.; Kalfagiannis, N. In-line spectroscopic ellipsometry for the monitoring of the optical properties and quality of roll-to-roll printed nanolayers for organic photovoltaics. *Sol. Energy Mater. Sol. Cells* **2013**, *112*, 144–156. [CrossRef]
36. Laskarakis, A.; Logothetidis, S. In-Line Quality Control of Organic Thin Film Fabrication on Rigid and Flexible Substrates. In *Ellipsometry of Functional Organic Surfaces and Films*; Springer: Berlin/Heidelberg, Germany, 2018; pp. 437–458.
37. Schneider, M.; Langklotz, U.; Michaelis, A. Thickness determination of thin anodic titanium oxide films-a comparison between coulometry and reflectometry. *Surf. Interface Anal.* **2011**, *43*, 1471–1479. [CrossRef]
38. Szklarska, M.; Dercz, G.; Smołka, A.; Popczyk, M.; Łosiewicz, B. A Coulometric Method By Local Anodic Dissolution for Measuring the Thickness of Ni/Cu Multi-Layer Electrocoatings. Available online: https://www.scientific.net/SSP.228.319 (accessed on 2 April 2021).
39. Sophian, A.; Tian, G.; Fan, M. Pulsed eddy current non-destructive testing and evaluation: A review. *Chin. J. Mech. Eng.* **2017**, *30*, 500–514. [CrossRef]
40. AbdAlla, A.N.; Faraj, M.A.; Samsuri, F.; Rifai, D.; Ali, K.; Al-Douri, Y. Challenges in improving the performance of eddy current testing: Review. *Meas. Control.* **2018**, *52*, 46–64. [CrossRef]
41. Hinken, J.H.; Barenthin, B.; Halfpaap, J.; Moebes, C.; Wrobel, H.; Ziep, C.; Hekli, M. Thickness Measurement of Chromium Layers on Stainless Steel Using the Thermoelectric Method with Magnetic Readout(TEM). *J. Nondestruct. Test.* **2005**, *10*, 1.
42. Vrielink, J.A.M.; Tiggelaar, R.M.; Gardeniers, J.G.E.; Lefferts, L. Applicability of X-ray fluorescence spectroscopy as method to determine thickness and composition of stacks of metal thin films: A comparison with imaging and profilometry. *Thin Solid Film.* **2012**, *520*, 1740–1744. [CrossRef]
43. Clark, R.N.; Roush, T.L. Reflectance spectroscopy: Quantitative analysis techniques for remote sensing applications. *J. Geophys. Res. Solid Earth* **1984**, *89*, 6329–6340. [CrossRef]
44. Matyi, R.; Hatzistergos, M.; Lifshin, E. X-ray reflectometry analyses of chromium thin films. *Thin Solid Film.* **2006**, *515*, 1286–1293. [CrossRef]
45. Xu, C.; He, L.; Xiao, D.; Ma, P.; Wang, Q. A Novel High-Frequency Ultrasonic Approach for Evaluation of Homogeneity and Measurement of Sprayed Coating Thickness. *Coatings* **2020**, *10*, 676. [CrossRef]
46. Beamish, D. Using ultrasonic coating thickness gauges. *Mater. Perform.* **2004**, *43*, 30–33.
47. Multi-Layers Coating Thicknesses Measurement | Electrolysis. 2021. Available online: https://www.helmut-fischer.com/products/benchtop-measurement-instruments (accessed on 6 May 2021).
48. IMS. ß-ray Coating Weight—IMS. 2021. Available online: https://www.ims-gmbh.de/product-catalog/coating-thickness-measurement/cwg-rray/?lang=en (accessed on 6 May 2021).
49. Fischerindia.net. 2021. Available online: https://www.fischerindia.net/pdf/betascope.pdf (accessed on 6 May 2021).
50. Defelsko.com. Coating Thickness Gages—PosiTector 6000 | DeFelsko. 2021. Available online: https://www.defelsko.com/positector-6000#resources (accessed on 6 May 2021).
51. XRF Instruments XDL & XDLM | For Chrome Coatings. 2021. Available online: https://www.helmut-fischer.com/products/fischerscope-x-ray-xdl-and-xdlm (accessed on 6 May 2021).
52. Rigaku.com. X-ray Reflectometry (XRR) | Rigaku Global Website. 2021. Available online: https://www.rigaku.com/techniques/x-ray-reflectometry-xrr (accessed on 6 May 2021).
53. QuintSonic—Ultrasonic Coating Thickness Gauge. [online]. Checkline.eu. Available online: https://www.checkline.eu/prod/coating-thickness-gauges-paint-thickness-gauges/quintsonic#buy-now (accessed on 6 May 2021).
54. Feliu, S., Jr.; Barranco, V. XPS study of the surface chemistry of conventional hot-dip galvanised pure Zn, galvanneal and Zn–Al alloy coatings on steel. *Acta Mater.* **2003**, *51*, 5413–5424. [CrossRef]
55. Natarajan, R.; Palaniswamy, N.; Natesan, M.; Muralidharan, V. XPS analysis of passive film on stainless steel. *Open Corros. J.* **2009**, *2*, 114–124. [CrossRef]
56. Hantsche, H. Comparison of basic principles of the surface-specific analytical methods: AES/SAM, ESCA (XPS), SIMS, and ISS with X-ray microanalysis, and some applications in research and industry. *Scanning* **1989**, *11*, 257–280. [CrossRef]
57. Lauritsen, J.V.; Reichling, M. Atomic resolution non-contact atomic force microscopy of clean metal oxide surfaces. *J. Phys. Condens. Matter* **2010**, *22*, 263001. [CrossRef]
58. Nemes-Incze, P.; Osváth, Z.; Kamarás, K.; Biró, L. Anomalies in thickness measurements of graphene and few layer graphite crystals by tapping mode atomic force microscopy. *Carbon* **2008**, *46*, 1435–1442. [CrossRef]

59. Crozier, K.; Yaralioglu, G.; Degertekin, F.; Adams, J.; Minne, S.; Quate, C. Thin film characterization by atomic force microscopy at ultrasonic frequencies. *Appl. Phys. Lett.* **2000**, *76*, 1950–1952. [CrossRef]
60. Daviðsdóttir, S.; Shabadi, R.; Galca, A.C.; Andersen, I.H.; Dirscherl, K.; Ambat, R. Investigation of DC magnetron-sputtered TiO_2 coatings: Effect of coating thickness, structure, and morphology on photocatalytic activity. *Appl. Surf. Sci.* **2014**, *313*, 677–686. [CrossRef]
61. Marin, E.; Guzman, L.; Lanzutti, A.; Ensinger, W.; Fedrizzi, L. Multilayer Al_2O_3/TiO_2 Atomic Layer Deposition coatings for the corrosion protection of stainless steel. *Thin Solid Film.* **2012**, *522*, 283–288. [CrossRef]
62. Mouche, P.A.; Ang, C.; Koyanagi, T.; Doyle, P.; Katoh, Y. Characterization of PVD Cr, CrN, and TiN coatings on SiC. *J. Nucl. Mater.* **2019**, *527*, 151781. [CrossRef]
63. Kratos.com. Coatings & Thin Film | Kratos. 2021. Available online: https://www.kratos.com/applications/coatings-thin-film (accessed on 6 May 2021).
64. Microscopy, E.; Microscopes, E.; Microscopes, D.; ProX, P. Desktop SEM | Phenom ProX | Thermo Fisher Scientific—UK. [online]. Thermofisher.com, 2021. Available online: https://www.thermofisher.com/uk/en/home/electron-microscopy/products/desktop-scanning-electron-microscopes/phenom-prox.html (accessed on 6 May 2021).
65. Parksystems.com. Park NX10—Specifications | Park Atomic Force Microscope. 2021. Available online: https://www.parksystems.com/index.php/products/small-sample-afm/park-nx10/specifications (accessed on 6 May 2021).
66. Horiba.com. Detail. 2021. Available online: https://www.horiba.com/en_en/products/detail/action/show/Product/gd-profiler-2tm-1401/ (accessed on 6 May 2021).
67. Edwards, J.E.; Otterson, D.W. Tech Talk: (5) Temperature Measurement Basics. *Meas. Control.* **2014**, *47*, 276–282. [CrossRef]
68. Hinken, J.H.; Tavrin, Y. Basics of the Thermoelectric Effect with Magnetic Readout. Available online: http://www.bookbootusers.co.uk/N6e.pdf (accessed on 2 April 2021).
69. Bahk, J.-H.; Favaloro, T.; Shakouri, A. Thin film thermoelectric characterization techniques. *Annu. Rev. Heat Transf.* **2013**, *16*, 51–99. [CrossRef]
70. Roth, D.J.; Cosgriff, L.M.; Harder, B.; Zhu, D.; Martin, R.E. Absolute Thickness Measurements on coatIngs without Prior Knowledge of Material Properties Using Terahertz Energy. Available online: https://searchworks.stanford.edu/view/10747265 (accessed on 2 April 2021).
71. Taschin, A.; Bartolini, P.; Tasseva, J.; Torre, R. THz time-domain spectroscopic investigations of thin films. *Meas. Control.* **2018**, *118*, 282–288. [CrossRef]
72. Bayati, E.; Winebrenner, D.P.; Arbab, M.H. Measuring the Thickness of Ultra-Thin Film Layers Using Terahertz Time-Domain Polarimetry (THz-TDP). Available online: https://ieeexplore.ieee.org/document/8067109 (accessed on 2 April 2021).
73. Tu, W.; Zhong, S.; Shen, Y.; Incecik, A. Nondestructive testing of marine protective coatings using terahertz waves with stationary wavelet transform. *Ocean. Eng.* **2016**, *111*, 582–592. [CrossRef]
74. Döring, S.; Hertlein, F.; Bayer, A.; Mann, K. EUV reflectometry for thickness and density determination of thin film coatings. *Appl. Phys. A* **2012**, *107*, 795–800. [CrossRef]
75. Hirth, F.; Rossner, M.; Jakobi, M.; Koch, A.W. Impact of Angle Ranges on Thickness Resolution in Thin Film Reflectometry. Available online: https://ieeexplore.ieee.org/document/5326107 (accessed on 2 April 2021).
76. Kim, K.; Kim, S.; Kwon, S.; Pahk, H.J. Volumetric thin film thickness measurement using spectroscopic imaging reflectometer and compensation of reflectance modeling error. *Int. J. Precis. Eng. Manuf.* **2014**, *15*, 1817–1822. [CrossRef]
77. Urbanek, M.; Spousta, J.; Navratil, K.; Szotkowski, R.; Chmelík, R.; Buček, M.; Šikola, T. Instrument for thin film diagnostics by UV spectroscopic reflectometry. *Surf. Interface Anal. Int. J. Devoted Dev. Appl. Tech. Anal. Surf. Interfaces Thin Film.* **2004**, *36*, 1102–1105. [CrossRef]
78. Kim, K.; Kwon, S.; Pahk, H.J. Fast Analysis of Film Thickness in Spectroscopic Reflectometry using Direct Phase Extraction. *Curr. Opt. Photonics* **2017**, *1*, 29–33. [CrossRef]
79. Wu, F.; Zheng, C. Illumination Model for Two-Layer Thin Film Structures. Available online: https://dl.acm.org/doi/10.5220/0005261401990206 (accessed on 2 April 2021).
80. Mohammed, Z.H. The Fresnel Coefficient of Thin Film Multilayer Using Transfer Matrix Method TMM. Available online: https://iopscience.iop.org/article/10.1088/1757-899X/518/3/032026 (accessed on 2 April 2021).
81. Ghim, Y.-S.; Kim, S.-W. Fast, precise, tomographic measurements of thin films. *Appl. Phys. Lett.* **2007**, *91*, 091903. [CrossRef]
82. Hernández, M.; Juarez, A.; Hernandez, R. Interferometric thickness determination of thin metallic films. *Superf. Y Vacío* **1999**, *9*, 283–285.
83. De Groot, P.J.; de Lega, X.C.; Fay, M.F. Transparent Film Profiling and Analysis by Interference Microscopy. Available online: https://www.spiedigitallibrary.org/conference-proceedings-of-spie/7064/70640I/Transparent-film-profiling-and-analysis-by-interference-microscopy/10.1117/12.794936.short (accessed on 2 April 2021).
84. Gao, F.; Muhamedsalih, H.; Jiang, X. Surface and thickness measurement of a transparent film using wavelength scanning interferometry. *Opt. Express* **2012**, *20*, 21450–21456. [CrossRef]
85. Jo, T.; Kim, K.; Kim, S.; Pahk, H. Thickness and surface measurement of transparent thin-film layers using white light scanning interferometry combined with reflectometry. *J. Opt. Soc. Korea* **2014**, *18*, 236–243. [CrossRef]

86. Manallah, A.; Bouafia, M.; Meguellati, S. Optical Coherence Tomography as Film Thickness Measurement Technique. Available online: https://www.spiedigitallibrary.org/conference-proceedings-of-spie/9450/945006/Optical-coherence-tomography-as-film-thickness-measurement-technique/10.1117/12.2061387.short (accessed on 2 April 2021).
87. Jeong, H.; Park, B.; Kim, Y.; Kim, H.; Ghim, Y.-S.; You, J.; Kim, S.-W. Oxide Thickness Profile Measurement by Dispersive White-Light Interferometry in CMP Process. Available online: https://ieeexplore.ieee.org/document/5760426 (accessed on 2 April 2021).
88. Tompkins, H.G.; Tasic, S.; Baker, J.; Convey, D. Spectroscopic ellipsometry measurements of thin metal films. *Surf. Interface Anal.* **2000**, *29*, 179–187. [CrossRef]
89. Maulana, L.Z.; Megasari, K.; Suharyadi, E.; Anugraha, R.; Abraha, K.; Santoso, I. Inexpensive Home-Made Single Wavelength Ellipsometer (λ = 633 nm) for Measuring the Optical Constant of Nanostructured Materials. *IOP Conf. Ser. Mater. Sci. Eng.* **2017**, *202*, 012031. [CrossRef]
90. Pascu, R.; Dinescu, M. Spectroscopic Ellipsometry. *Rom. Rep. Phys.* **2012**, *64*, 135–142.
91. Cain, J.P.; Robie, S.; Zhang, Q.; Singh, B.; Emami, I. Combined Use of X-ray Reflectometry and Spectroscopic Ellipsometry for Characterization of Thin Film Optical Properties. Available online: https://www.spiedigitallibrary.org/conference-proceedings-of-spie/6155/61550P/Combined-use-of-x-ray-reflectometry-and-spectroscopic-ellipsometry-for/10.1117/12.660088.short (accessed on 2 April 2021).
92. Hauge, P.; Dill, F. Design and operation of ETA, an automated ellipsometer. *IBM J. Res. Dev.* **1973**, *17*, 472–489. [CrossRef]
93. Garcia-Caurel, E.; De Martino, A.; Gaston, J.P.; Yan, L. Application of spectroscopic ellipsometry and Mueller ellipsometry to optical characterization. *Appl. Spectrosc.* **2013**, *67*, 1–21. [CrossRef]
94. Taya, S.A.; El-Agez, T.M.; Alkanoo, A.A. Ellipsometric configurations using a phase retarder and a rotating polarizer and analyzer at any speed ratio. *Chin. Phys. B* **2012**, *21*, 110701. [CrossRef]
95. Tompkins, H.G.; Hilfiker, J. Spectroscopic Ellipsometry. Available online: https://www.momentumpress.net/books/spectroscopic-ellipsometry-practical-application-thin-film-characterization (accessed on 2 April 2021).
96. Pang, W.; Every, A.; Comins, J.; Stoddart, P.; Zhang, X.; Crowhurst, J.; Pietersen, D. Brillouin Scattering as a Tool for Characterizing Surfaces, Interfaces and Thin Films. In *Review of Progress in Quantitative Nondestructive Evaluation*; Springer: Berlin/Heidelberg, Germany, 1998; pp. 1315–1322.
97. Ballmann, C.W.; Thompson, J.V.; Traverso, A.J.; Meng, Z.; Scully, M.O.; Yakovlev, V.V. Stimulated Brillouin Scattering Microscopic Imaging. *Sci. Rep.* **2015**, *5*, 18139. [CrossRef]
98. Xu, L.; Zhang, S.; Tan, Y.; Sun, L. Simultaneous measurement of refractive-index and thickness for optical materials by laser feedback interferometry. *Rev. Sci. Instrum.* **2014**, *85*, 083111. [CrossRef]
99. Fathi, M.T.; Donati, S. Thickness measurement of transparent plates by a self-mixing interferometer. *Opt. Lett.* **2010**, *35*, 1844–1846. [CrossRef]
100. Chen, W.; Zhang, S.; Long, X. Thickness and refractive-index measurement of birefringent material by laser feedback technique. *Opt. Lett.* **2013**, *38*, 998–1000. [CrossRef] [PubMed]
101. Boettcher, T.; Gronle, M.; Osten, W. Multi-layer topography measurement using a new hybrid single-shot technique: Chromatic Confocal Coherence Tomography (CCCT). *Meas. Sci.* **2017**, *25*, 10204–10213. [CrossRef]
102. Niese, S.; Quodbach, J. Application of a chromatic confocal measurement system as new approach for in-line wet film thickness determination in continuous oral film manufacturing processes. *Int. J. Pharm.* **2018**, *551*, 203–211. [CrossRef]
103. Lyda, W.; Gronle, M.; Fleischle, D.; Mauch, F.; Osten, W. Advantages of chromatic-confocal spectral interferometry in comparison to chromatic confocal microscopy. *Meas. Sci. Technol.* **2012**, *23*, 054009. [CrossRef]
104. Decker, C.A.; Mackin, T.J. Measuring film thickness using infrared imaging. *Thin Solid Film.* **2005**, *473*, 196–200. [CrossRef]
105. Filmetrics, I. In-Line Thickness Measurement System—Filmetrics F32. [online]. Filmetrics.com, 2021. Available online: https://www.filmetrics.com/thickness-measurement/f32 (accessed on 7 May 2021).
106. Thorlabs.com. Fiber-Coupled Terahertz Spectrometer. 2021. Available online: https://www.thorlabs.com/newgrouppage9.cfm?objectgroup_id=4713 (accessed on 7 May 2021).
107. Thorlabs.com. Ganymede™ Series SD-OCT Systems. 2021. Available online: https://www.thorlabs.com/newgrouppage9.cfm?objectgroup_id=8214 (accessed on 7 May 2021).
108. Horiba.com. Detail. 2021. Available online: https://www.horiba.com/gbr/products/detail/action/show/Product/uvisel-plus-in-situ-1362/ (accessed on 7 May 2021).
109. NANOVEA. IN-LINE PROFILER—QC ROUGHNESS—NANOVEA. 2021. Available online: https://nanovea.com/instruments/in-line-profiler/ (accessed on 7 May 2021).
110. Stressphotonics.com. 2021. Available online: https://www.stressphotonics.com/Product_Pages/pdf_files/DT_1000.pdf (accessed on 7 May 2021).
111. Bugnicourt, E.; Kehoe, T.; Latorre, M.; Serrano, C.; Philippe, S.; Schmid, M. Recent Prospects in the Inline Monitoring of Nanocomposites and Nanocoatings by Optical Technologies. *Nanomaterials* **2016**, *6*, 150. [CrossRef] [PubMed]

sensors

Review

Advanced Optical Sensing of Phenolic Compounds for Environmental Applications

Ines Delfino [1,*], Nadia Diano [2] and Maria Lepore [2]

[1] Dipartimento di Scienze Ecologiche e Biologiche, Università della Tuscia, I-01100 Viterbo, Italy
[2] Dipartimento di Medicina Sperimentale, Università della Campania "Luigi Vanvitelli", I-80138 Napoli, Italy; nadia.diano@unicampania.it (N.D.); maria.lepore@unicampania.it (M.L.)
* Correspondence: delfino@unitus.it

Abstract: Phenolic compounds are particularly dangerous due to their ability to remain in the environment for a long period of time and their toxic effects. They enter in the environment in different ways, such as waste from paper manufacturing, agriculture (pesticides, insecticides, herbicides), pharmaceuticals, the petrochemical industry, and coal processing. Conventional methods for phenolic compounds detection present some disadvantages, such as cumbersome sample preparation, complex and time-consuming procedures, and need of expensive equipment. Therefore, there is a very large interest in developing sensors and new sensing schemes for fast and easy-to-use methods for detecting and monitoring the phenolic compound concentration in the environment, with special attention to water. Good analytical properties, reliability, and adaptability are required for the developed sensors. The present paper aims at revising the most generally used optical methods for designing and fabricating biosensors and sensors for phenolic compounds. Some selected examples of the most interesting applications of these techniques are also proposed.

Keywords: phenols; optical methods; biosensors; sensors

Citation: Delfino, I.; Diano, N.; Lepore, M. Advanced Optical Sensing of Phenolic Compounds for Environmental Applications. *Sensors* **2021**, *21*, 7563. https://doi.org/10.3390/s21227563

Academic Editor: Shyqyri Haxha

Received: 12 October 2021
Accepted: 9 November 2021
Published: 14 November 2021

1. Introduction

Over the last several decades, water pollution has threatened both quality of life and public health worldwide. In particular, phenol and related phenolic compounds (chloro, bromo, nitro, and alkyl phenol) that are discharged into the wastewater and can contaminate surface, ground, and sometimes drinking water have recently drawn attention due their potential impact on human and environmental health. They derive from urban, agro-industrial, and livestock-related human activities [1,2], and they can cause adverse effects in all the food chain rings, even at low concentration (µg/L–ng/L). By way of example, Table S1 shows the permissible concentration limit allowed for some phenolic compounds.

The US Environmental Protection Agency (EPA) and European Commission have already placed them on the Priority Pollutants List that must be monitored by local governments in the next years [3,4]. However, it must be considered that the types and mixtures of pollutants can change according to the urban and industrial activities specific for each territory and that synergic effects can strengthen toxic effects. For these reasons, it is crucial to monitor their levels in samples deriving from local wastewaters, to test their biological activity and to develop new sustainable and green strategies to remove them.

Due to their persistence in the environment and their toxicity, the phenolic compounds can induce acute and chronic hazardous health effects [5–8]. Long-term exposure to phenols can cause irregular breathing, muscle weakness, and respiratory arrest at lethal doses in humans. Chronic exposure to phenols leads to disorders of the gastrointestinal and central nervous systems and the liver, as well as growth retardation and abnormal development and reproduction in animals. It is known that some of phenol compounds affect the endocrine system, altering the hormones balance within the human body [9]. The alteration of correct levels of endogenous hormones or the introduction of chemicals that can mimic their effect in the cell has been related not only to estrogen-dependent tumors but also to an increased risk for a tremendous number of pathologies, such as Alzheimer's disease, metabolic syndrome, and cardiovascular disorders [10]. The xenoestrogen compound bisphenol A (BPA) is one of the most widely used chemicals, commonly used in the production of epoxy resins, polycarbonates, dental fillers, food storage containers, baby milk containers, and mineral water containers. Due to the extensive manufacturing of these products, human exposure to BPA through several routes, such as food and the environment, is ubiquitous [11,12]. Biomonitoring studies around the world have shown that BPA exposure is common among the general population, with a detectable concentration in more than 80% of the considered cases [13–15].

The most largely used techniques for phenolic compounds detection are gas chromatography, high-performance liquid chromatography, and capillary electrophoresis [16]. These well-established methods present some disadvantages, such as cumbersome sample preparation complex and time-consuming procedures. and the need for expensive equipment; consequently. they cannot be used for routine analysis.

For all these reasons, there is a very large interest in developing sensors and new sensing schemes for monitoring the phenolic compound concentration in environments, with special attention to water. Good analytical properties (sensitivity, accuracy, reproducibility, rapidness, and signal-to-noise ratio), reliability (long life, resistance to the environment, and operational safety), and adaptability (small size, light weight, simple structure, and low cost) are required to the developed sensors [17].

In this framework, the optical sensing scheme can play a pivotal role since sensors can be small, light, chemically inert, non-toxic, and immune to electromagnetic interferences. In fact, optical sensors are among the most versatile sensing devices; they can detect a large class of physical and chemical parameters, such as temperature, pressure, force, electric and magnetic field, pH, strain, chemical concentration, displacement, humidity, and many others.

In 2013, Rodionov et al. [17] revised the development and applications of optical sensors for the determination of phenolic compounds in the period of 1993–2013. They presented an overview of the different approaches for optical sensing with particular attention to spectrophotometric and fluorescent sensors using optical fibers for the delivery and collection of light signals. In their Tables 1 and 2, the authors reported the spectrophotometric and fluorescence sensors for different phenolic compounds with the information about linearity range, limit-of-detection (LOD), and response time for the most representative results reported in the literature in the examined period (see Ref. [17] and cited references). In Table 3 of the cited paper, many optical sensors used in the determination of phenolic compounds in real samples, such as residential and industrial wastewater and river water, were summarized.

The aim of the present paper is to revise the most generally used optical methods for designing and fabricating biosensors and sensors for phenolic compounds with particular attention to the methodologies that we personally used in our previous investigations [18–22]. We also discuss some representative examples for each of them. As far as the cited references, the list is far from being exhaustive, but it is indicative of the large amount of literature available in this field.

2. Optical Detection Techniques

A large number of optical techniques can be used for developing fast, accurate, and sensitive sensors for phenol detection. In this section, we briefly summarize some aspects of the most-used ones.

The simplest method is based on the visual inspection of the investigated samples after their interaction with a proper probe. For example, in Figure 1 it is possible to note the color change occurring in water samples with different concentrations of phenols after the interaction with extracellular gold nanoparticles on which *Streptomyces tuirus* DBZ39 has been synthetized (see Ref. [23] for further details).

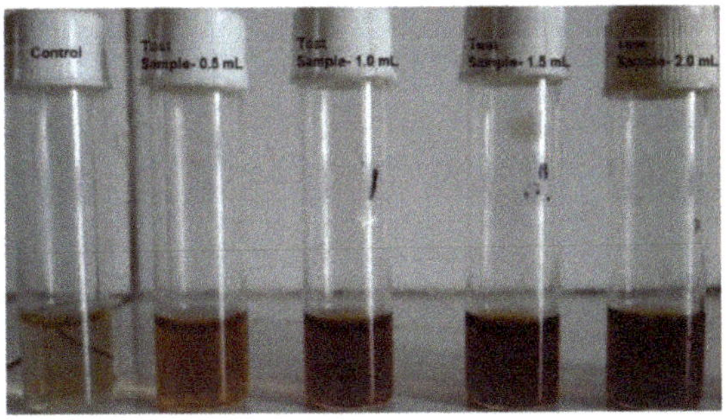

Figure 1. An example of the use of *Streptomyces*-mediated gold nanoparticles for the detection of phenols from industrial wastewater. (Reprinted from [23] under an Open Access condition).

The qualitatively appreciable color change upon visual inspection can be quantified thanks to absorption measurements. Using a spectrophotometer, it is possible to quantitatively determine absorbance changes linearly related to the presence of different concentrations of phenolic compounds. Bayram et al. [24] show the results obtained by a colorimetric assay based on the formation of quinone-type complexes in an alkaline medium for different concentrations of BPA (see Figure 3 in Ref. [24]). As is evident, different absorbances are obtained in correspondence of color variations, and considering the absorbance values at a particular wavelength, it is possible to obtain a calibration curve that can be used for estimating analyte concentration in other samples of interest.

Diffuse reflectance measurements often coupled with the use of optical fibers are also widely used for developing phenolic compounds sensors [25,26]. A typical set-up, where a tungsten halogen broadband is used as a light source and a fiber-optic probe is in contact with the sample used for this approach, is shown in Figure 2.

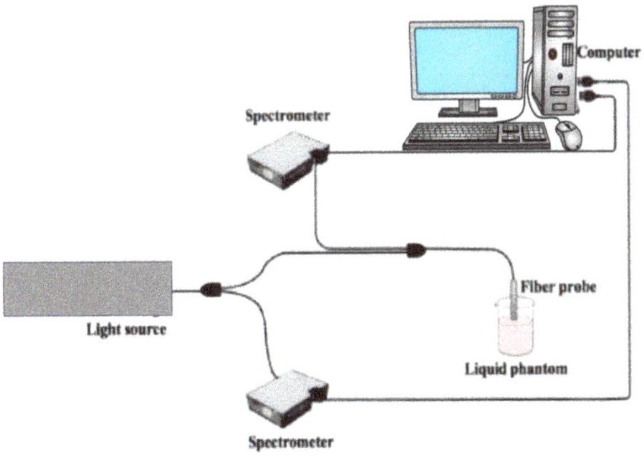

Figure 2. Schematic layout of an experimental set-up for diffuse reflectance measurements. A broadband light source is used, and a spectrometer allows wavelength selection of the incident beam. The second spectrometer is employed for monitoring the reflected signal.

One of the spectrometers is used for recording the lamp spectrum and its background, while the second spectrometer detects the signal reflected the sample. CCD array detectors and a computer allowed the detection and processing of optical signals. In Figure 3a, typical results obtained using this reflectance approach are reported. In these cases, the authors [25] have developed a fiber-optic sensor for p-aminophenol (PAP) based on the use of 25,26,27,28-tetrahydroxycalix[4]arene (CAL4) immobilized onto Amberlite XAD-16 and reflectance spectrometry. The sensor is based on the reaction of PAP with CAL4 in the presence of an oxidant to produce an indophenol dye. The measurements were carried out at 620 nm since it gives the largest differences in reflectance spectra before and after reaction with the analyte. Figure 3b shows the changes in the reflectance spectra of immobilized CAL4 before and after reaction with different concentrations of PAP. In Figure 3c,d, the calibration curves for the wide range of PAP concentrations are reported.

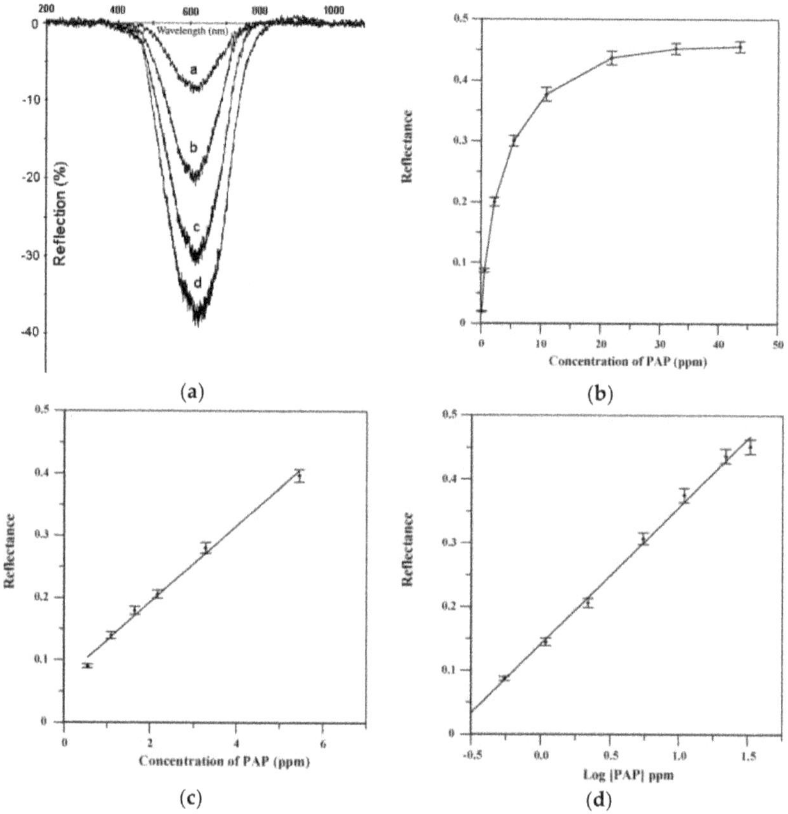

Figure 3. (**a**) Reflectance spectra of immobilized CAL4 before (Rf) and after (a–d) reaction with different concentrations of PAP, being 0.545 mg/L (curve a), 2.18 mg/L (curve b), 5.45 mg/L (curve c), and 10.9 mg/L (curve d). (**b**) Response curve to a wide range of PAP concentrations. The panels (**c**,**d**) show the linear and non-linear portions of the response curve. (Reprinted-adapted-with permission from [25]).

Fluorescence and luminescence are certainly the most-used methods to develop sensor devices for phenolic substance and to conceive new sensing schemes. See, for example, Table 2 in the review of Rodianov et al. [17] to realize the variety of the available schemes. The fluorescence approach makes available a large cohort of parameters to be used; among them, the most largely used are intensity, decay time, anisotropy, quenching efficiency, and luminescence energy transfer. The components of a basic fluorescence experimental set-up are the light source, two wavelength selectors, and the detector. Usually, the light sources are tungsten-halogen and xenon lamps. As far as the wavelength selection, the simplest apparatus uses fixed filters to isolate both the excitation and emission wavelengths, but it is better to use monochromators to select both the excitation and emission wavelengths. Most modern instruments of this type employ diffraction grating monochromators for this purpose. Using monochromators, both excitation and emission spectra can be recorded making full use of the analytical potential of the fluorescence approach. Regarding detectors, the most largely used are photomultipliers exploiting the wide variety of types available nowadays. Usually, the different components can be chosen and assembled in accordance with the user's needs, or they are combined in a commercial spectrofluorometer, such as the one reported in Figure 4, in which two monochromators are used as wavelength selectors.

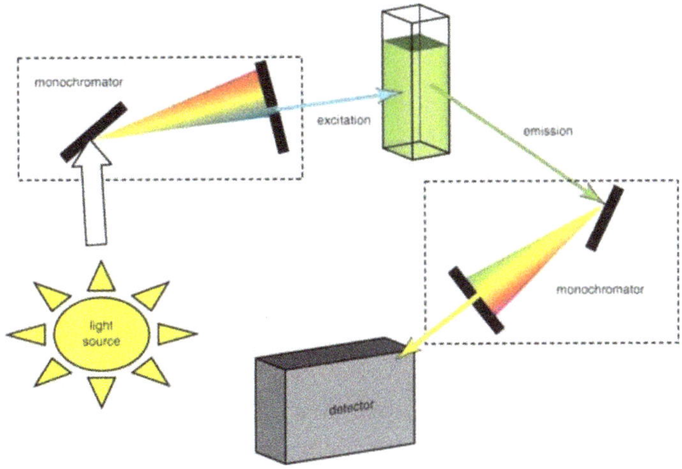

Figure 4. Schematic diagrams of a spectrofluorometer (image taken from https://chem.libretexts.org/Bookshelves/Analytical_Chemistry/Physical_Methods_in_Chemistry_and_Nano_Science_(Barron) (accessed on 8 November 2021).

As is evident from the literature, optical fibers and waveguide-based optical sensors have gained a large interest since they show enormous potential for applications in various fields [27–31]. In fact, fiber-optic sensors can have small size; are sensitive to multiple environmental parameters; allow remote sensing also into normally inaccessible areas; do not strictly require contact; are independent from radio frequency and electromagnetic interference; avoid contamination of their surrounding area; are characterized by high sensitivity, resolution, and dynamic range; and can be connected with data communication systems. A particularly interesting use of optical fibers is related to the development of fiber-optic chemical sensors (FOCS) [32,33]. These devices usually include three main components: an active sensing element that recognizes the analyte and generates an optical signal, a detector that measures one of the characteristics of the optical signal (intensity, frequency, phase) that can be employed for evaluating the concentrations of the analyte of interest, and a computer and a software for data acquisition and processing. In Figure 5, basic sensing schemes for FOCS are shown [34]. In detail, in Figure 5a, one end of the fiber is made sensitive to the chemical substance to be detected. The interaction between the tip of the fiber and the substance causes fluorescence or other signals that can be revealed in the coming back light. In Figure 5b, another configuration for FOCS is reported. In this case, the interaction occurs between the evanescent portion of the light propagating in the fiber and the chemical substances that are near the fiber surface. In Figure 5c, a transmission geometry is adopted for studying the investigated interaction process.

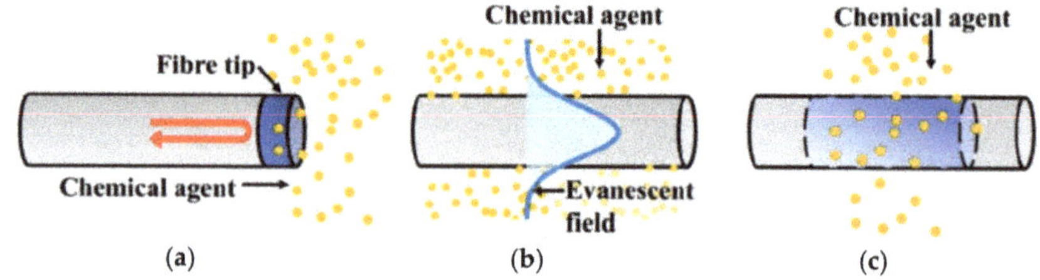

Figure 5. Basic sensing schemes for typical fiber-optic chemical sensors based on (**a**) fiber tip, (**b**) evanescent field sensing, and (**c**) simple transmission setup. (Reprinted from [34] under Open Access conditions).

Another largely employed approach that uses optical fiber is represented by fiber Bragg gratings (FBGs) that nowadays are used for many applications in different fields by adopting one of the most common configurations, shown in Figure 6 [35,36].

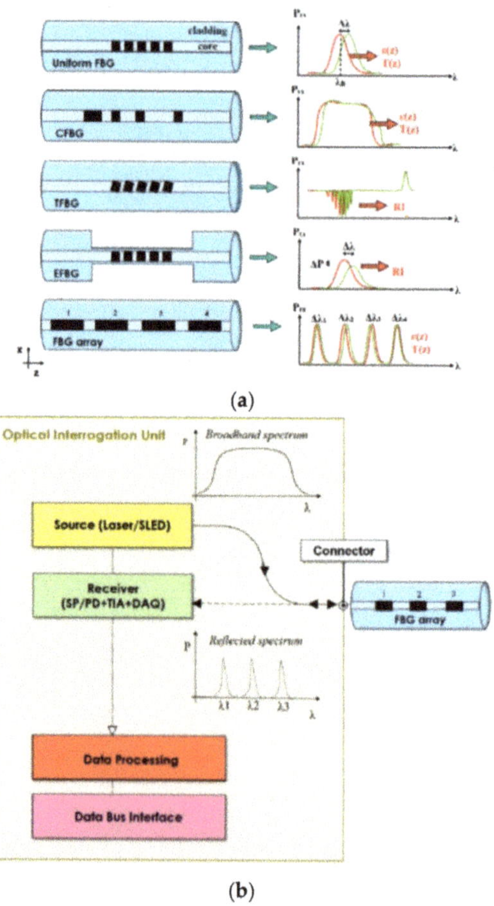

Figure 6. In the (**a**) panel, the main configurations of FBGs are shown. The spectral responses of the measured parameters (strain (ε), temperature (T), refractive index (RI)) are also reported. In the (**b**) panel, a schematic representation of the developed device with all the components used with FBGs is presented. (Reprinted-rearranged-with permission from [36] under Open Access conditions).

Basically, an FBG-based sensor consists of an optical resonator located inside the fiber-optic core. By using an external laser source and a proper optical element (an interferometer or a diffraction grating), it is possible to periodically perturb the refractive index of a single-mode optical fiber, inducing an interference between waves propagating in the opposite direction in the fiber. The refractive index and the periodicity of the interference pattern can be modified by the external environment, and these changes can be used for mechanical, biomechanical, surgical, physiological, and chemical applications [35–38].

Surface plasmon resonance (SPR) optical sensors offer large advantages in terms of detection limit, sensitivity, and selectivity when compared with other sensing schemes, and for this reason, many researchers have focused their attention on these sensing devices, starting in the 1990s [39]. The SPR effect was observed by Wood in 1902, when he sent a monochromatic polarized light beam on a diffraction grating and he noticed a pattern of white and dark bands. The physical explanation of these effects was reported by Otto and Kretschmann [40]. The SPR configuration proposed by Daniyal et al. [39] is the most-used (see Figure 7) one, and it is based on the "angular interrogation" approach, since the wavelength of the incident light is kept constant, and the angle of incidence of the light is varied. In fact, in this configuration, a monochromatic and p-polarized light is used for exciting a surface plasmon that propagates along a metal surface. At a certain angle, named the resonance angle, the intensity of the reflected light decreases due to resonance, which occurs when the momentum of the surface plasmon wave is equivalent to that of the incident light. This sensing method is based on the incident angle interrogation, since the wavelength of the light is kept constant, and the angle of incidence of the light is varied (angular interrogation) while in the other method, the wavelength of light is varied, and the angle of incidence is kept constant and greater than the critical angle (wavelength interrogation). An SPR optical sensor is based on the measurement of the refractive index near the metal surface. Any changes in refractive index will also change the resonance angle.

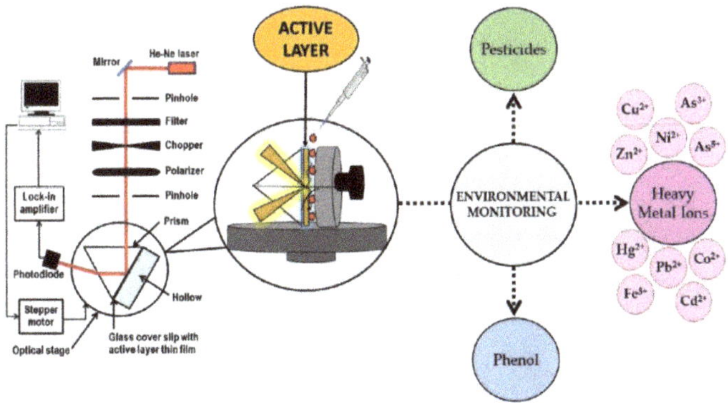

Figure 7. Schematic representation of an SPR optical sensor for phenol and pesticide detection in environmental applications. (Reprinted from [39] under Open Access conditions).

To further enhance their sensitivity and selectivity towards a specific target pollutant, various active layers to place on the top of a metal surface have been investigated in the last several years [40].

In the last several decades, there has also been a very large use of vibrational spectroscopies techniques for developing very sensitive optical sensing schemes for phenolic substances. These spectroscopies represent a class of analytical techniques that give information on vibrational energy levels associated with the functionality of the examined sample. Fourier Trasform Infrared (FT-IR) spectroscopy measures the light absorption using a broadband light source in the 4000–500 cm^{-1} wavenumber region (or in the 2500–20,000 nm wavelength range), while Raman spectroscopy (RS) is related to the inelastic scattering process occurring when light interacts with matter [41]. When FT-IR and the Raman spectrometer are equipped with a microscope, they allow the biochemical characterization of samples also at microscopic levels [42,43].

In Figure 8, a typical commercial instrument for FT-IR spectroscopy is reported. This system adopts a long-life source with proprietary hot-spot stabilization as a source, and it is equipped with two different detectors for the acquisition at macroscopic and microscopic levels. It allows the acquisition of infrared spectra in transmission and attenuated total reflection (ATR) mode for macroscopic samples. At the microscopic level, spectra can be acquired in transmission, reflectance, transflection, and micro-ATR collection geometry. The availability of all these different approaches makes FT-IR spectroscopy very versatile and suitable for all types of samples [41].

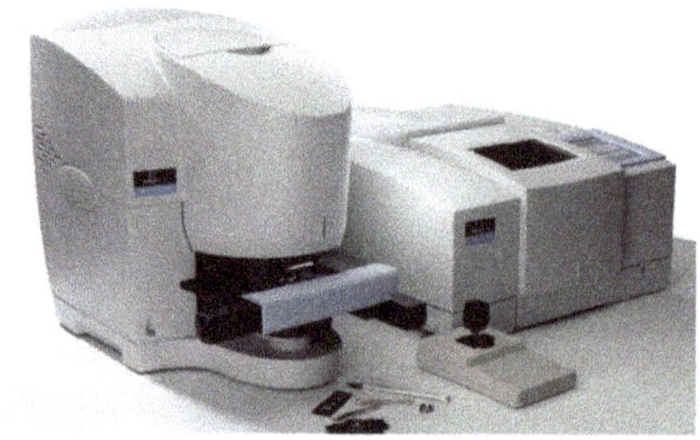

Figure 8. Photo of a commercial Fourier transform infrared (FT-IR) spectrometer equipped with different options for acquisition of spectra in different geometries at micro and macroscopic levels.

In Figure 9, a typical micro-Raman set-up is shown. Typical equipment for micro-Raman spectroscopy requires a laser source, a microscope objective to view the sample, focusing the laser beam on the sample and collecting the weak light from it in back-scattering geometry, a filtering system for rejecting the exciting light (Notch filter), a dispersive apparatus (monochromator), a high-sensitivity detector for the detection of the output signal (typically a nitrogen-cooled CCD), an electronic counting chain for the acquisition of the Raman signal, and a software for recording and processing Raman spectra. The adoption of confocal microscope stages allows the acquisition of appreciable Raman signals in experimental conditions of interest for phenolic compound sensing. See Ref. [43] and references therein for additional information.

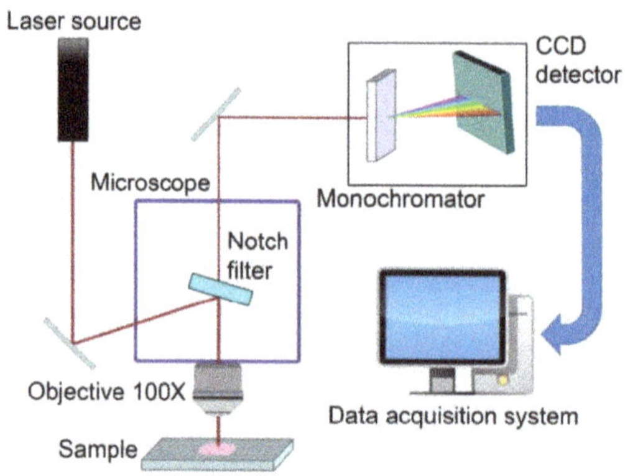

Figure 9. Experimental set-up for micro-Raman spectroscopy.

RS and FT-IR can give complementary information since the related signals are due to nonpolar and polar functional groups, respectively. In addition, the design and fabrication of nanostructured substrates has allowed the development of companion techniques, such as Surface-Enhanced Infrared Radiation Absorption (SEIRA) [22,44,45] and Surface-Enhanced Raman Spectroscopy (SERS) [46], that are currently used for the preparation of a new generation of sensors.

3. Biosensors

In discussing optical biosensors, we followed the classification made by Borisov and Wolfbeis in ref. [47] that divide biosensors in two classes: catalytic and affinity biosensors. Catalytic biosensors use biocomponents capable of recognizing biochemical species and causing their transformation in a product by means of chemical reaction. The most relevant examples of this class are enzymatic biosensors. Affinity biosensors use analyte able to bind to a biorecognition element. This class is divided into immunosensors (which exploit the specific interactions between an antibody and an antigen), biosensors based on interactions between an analyte and a bioreceptor, nucleic acid biosensors (which use the affinity between complementary oligonucleotides), and whole-cell biosensors that behave as recognition elements that respond to substances by expressing a specific gene. Some examples of these two different classes of biosensors that have been specifically designed for phenolic species detection in the environment will be reviewed in the following sections.

3.1. Enzymatic Biosensors

The enzymes generally used for phenolic compounds biosensing are laccase, tyrosinase, and, in a few cases, horseradish peroxidase. Tyrosinase is characterized by low stability and a relevant inhibition caused by reaction products. Horseradish peroxidase requires hydrogen peroxidase for carrying out its catalytic action. Laccase can catalyze electron-transfer reactions without the presence of cofactors; it is stable and can oxidize phenols and o,m,p-benzenediol compounds when molecular oxygen is present. These characteristics are particularly useful for the development of high-quality biosensors.

3.1.1. Laccase-Based Biosensors

Laccases (benzenediol oxygen oxidoreductase; EC 1.10.3.2) are cuproproteins and are also called polyphenol oxidase or blue multicopper proteins and are widely distributed in higher plants, fungi, and bacteria [48]. Most laccases are characterized by four copper atoms per functional unit, which are crucial for catalytic activity. Copper atoms are allocated in different binding sites and have different spectroscopic, functional, and paramagnetic features that enable their classification into three groups. The copper of type 1 (T1) shows a maximum in the absorption spectrum around 600 nm, and it causes the typical blue color of these cuproproteins. It is the primary site of oxidation. The copper of type 2 (T2) shows only a weak absorption in the visible, and it is electron paramagnetic resonance (EPR) active. The two copper atoms of type 3 (T3) are characterized by an absorption band around 330 nm and are EPR silent [49,50]. In a typical laccase reaction, a phenolic substrate is subjected to one-electron oxidation. The obtained species can be converted to a quinone in the second step of the oxidation process. Low-substrate specificity is typical of laccases, and they show a large variability in their catalytic properties in dependence of the source. Laccase can catalyze the oxidation of many compounds, such as hydroquinone, catechol, guaiacol, 2,6-dimethoxyphenol, polyphenols, aromatic amines, benzenethiols, and a series of other compounds [51].

In 2015, M.M. Rodriguez-Delgado et al. [52] revised the laccase-based biosensors for the detection of phenolic compounds. The authors briefly revised the different immobilization procedures available for the laccase enzyme (see Table 1 of Ref. [52]) and presented also the different optical biosensors reported in the literature.

For phenols detection in environmental analysis, Abdullah et al. [53] developed an optical biosensor exploiting the ability of laccase to oxidize methoxy phenols in the presence of 3-methyl-2 benzothiazolinonehydrazone (MBTH) to produce azo-dye compounds. Stacked films of MBTH in hybrid Nafion/sol-gel silicate and laccase in chitosan were used for fabricating the proposed biosensor that is more selective to catechol, as compared with another analyte, such as guaiacol, o-cresol, and m-cresol, that were also investigated. This characteristic is due to the immobilization of laccase in a hybrid material. The linear range obtained for this biosensor was 0.5–8 mM, and synthetic samples were also used for mimicking real samples.

Sanz et al. [54] developed a laccase-polyacrilamide sensor exploiting absorption and fluorescence and characterized by a linear range of 0.109–2.5 mM and a limit of detection (LOD) of 100 μM. This biosensor was also tested on wastewater samples.

Andreu-Navarro et al. [55] proposed another interesting optical sensing approach for the determination of polyphenols, such as catechol, resorcinol, hydroquinone, pyrogallol, hydroxyhydroquinone, phloroglucinol, and gallic acid. This biosensor is based on the inhibition of the green indocyanine fluorescence in the presence of laccase and positively charged gold nanoparticles caused by polyphenols. The developed chemical system is based on the ability of these substances to delay the oxidation of green indocyanine in the presence of laccase. When the fluorophore is mixed with the laccase, its fluorescence shows a rapid decrease, which can be attributed to the catalytic effect of the enzyme on the oxidation of the fluorophore. However, this effect is delayed in the presence of a phenolic compound in a manner proportional to the concentration of the polyphenols. This biosensor allowed for catechol a LOD of 0.01 μM and a linear range of 0.08–5 μM.

Laccases and phenol reaction products present significant optical characteristics in UV and visible range that are routinely adopted for the development of spectrophotometric methods for measuring laccase activity [56]. These optical features can be suitably exploited for biosensing applications since laccase interacting with different phenols shows different optical absorption spectra. These differences can be useful for increasing the specificity of laccase-based biosensors [57]. To take advantage of the above-cited spectral properties, it is necessary to have optically transparent matrices for enzyme immobilization. Sol–gel technology can be a good choice for fabricating matrices for laccase immobilization with suitable chemical stability, optical transparency, and porosity [58,59]. The changes occurring in the optical absorption spectra of laccase reaction products at 425, 375, and 400 nm have been used to determine hydroquinone, resorcinol, and catechol concentrations, respectively (see Figure 10).

Owing to the slow response time of the hydroquinone–laccase reaction, the proposed optical biosensor was used for resorcinol and catechol. Linear ranges up to 1.4 and 0.2 mM and an LOD of 4.5 and 0.6 μM were evidenced for resorcinol and catechol, respectively. This type of biosensor is characterized by larger linear ranges, significant sensitivities, and good LODs when compared to other biosensors employing laccase from *Trametes versicolor*. Tap water samples spiked with a known amount of catechol and resorcinol were also employed for testing this biosensing device with real samples.

Another interesting optical biosensor for continuous monitoring of phenolic species in water was proposed by Jędrichowska et al. by using laccase from *Cerrena unicolor* immobilized by physical adsorption in low-temperature co-fired ceramics (LTTC) [60].

LTTC technology is a well-known technique largely employed in the industry. This kind of material has excellent physical and chemical properties, and three-dimensional structures can be easily obtained in the LTTC supports [61]. Microfluidic systems for sampling micro- and nanoliter volumes and optoelectronic components can be integrated in a single LTCC multilayer substrate [62]. In their paper, Jędrichowska et al. describe all the different steps required for biosensor fabrication and the optimization of the different parameters [60]. Scanning electron microscopy was used for visualizing the laccase, which was positioned onto the chemically modified substrate and the morphology of the deposited layer. Sensing measurements were performed in a flow-through system by evaluating the optical absorbance changes occurring in response to various concentrations of standard laccase assay substrate 2,20-azino-bis (3-ethylbenzothiazoline-6-sulfonic acid)-ABTS. In addition, the role of different sensor parameters, such as flow rate, optical source characteristics, and reproducibility, was investigated. According to the authors, the use of LTTC technology allows the realization of optical sensors characterized by significant advantages in terms of the sensitivity, precision, linearity, and simplicity of construction [60].

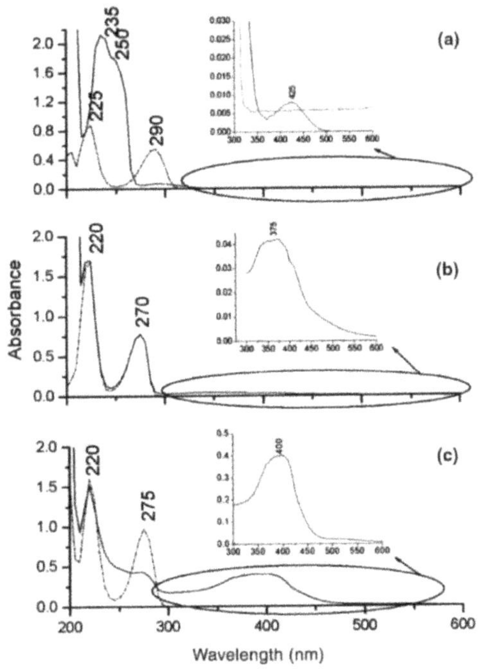

Figure 10. Experimental results obtained with an optical biosensor based on sol–gel immobilized laccase. (**a**) Absorption spectra of hydroquinone (dashed line) and laccase-reaction product (continuous line), (**b**) the same for resorcinol, (**c**) the same for catechol. In the different insets, the spectra in the 300–600 nm region. (Reprinted with permission from [59]).

Very recently, Cano-Raya et al. proposed a new laccase-based optical biosensor for catechol concentration determination [63]. Laccase from *Trametes Versicolor* is attached to anionic polyamide 6 (PA6) porous microparticles placed in a Pebax MH1657 polymer binder that includes MBTH that can produce a colored product when it interacts with the o-benzoquinone produced by the enzymatic reaction of catechol. The analyte concentration is estimated by measuring the absorbance at 500 nm. The proposed biosensor is characterized by an LOD of 11 μM and a linear range up to 118 μM of catechol and has been challenged with spiked natural water samples from rivers and springs, showing a recovery rate varying in the 97–108% interval.

3.1.2. Tyrosinase-Based Biosensors

Tyrosinase is a copper protein that catalyzes two successive reactions ([64] and references therein). In the first reaction, a hydroxyl group is added in the ortho position of a monophenolic compound, converting it into an o-diphenolic compound (monophenolase or cresolase activity). This diphenolic compound is subsequently oxidized into o-quinones by diphenolase or catecholase. Monophenols and diphenols are used by tyrosinase as a substrate. This enzyme can be found in different organisms, such as plants, animals, and fungi. The characteristics of this enzyme are different in size, sequence of amino acids, and glycosylation pattern [65]. Tyrosinase shows hydroxylase and catecholase activities due to histidine. In 2012 and 2017, two reviews prepared by Karim et al. [16] and Gui et al. [54] presented the optical biosensors reported in the literature.

In 1999, Russell and Burton [66] proposed a portable disposable biosensor using tyrosinase immobilized on a synthetic membrane for the detection and quantification of phenolic compounds in water. The enzyme produces changes of color in the solution, and these changes were found to be proportional to the phenolic substance concentration. The proposed biosensor is characterized by an LOD of 0.05 mg/L.

Using MBTH, Abdullah et al. developed also an optical biosensor based on the tyrosinase enzyme immobilized in a chitosan film similar to that previously described for the detection of phenol by laccase [67]. This biosensor exploits changes in the absorption spectra of the tyrosinase in the presence of MBTH and phenolic compounds. The authors investigated the response of this biosensor to different compounds and estimated the linear concentration range for 4-chlorophenol (2.5–50.0 μM), m-cresol (2.5–100.0 μM), and p-cresol (12.5–400.0 μM). The authors also reported interesting low LOD for the investigated phenolic substances.

Fiorentino et al. adopted a singular immobilization procedure in which ordered tyrosinase films deposited on an optical transparent support were immobilized by a "layer-by-layer" assembly, alternating the enzyme with the polycation polymer poly(dimethyldiallylammonium chloride) [68]. This procedure allowed a high loading of enzyme. The proposed biosensor was adopted for the detection of the o-diphenolic compound l-3,4-dihydroxyphenyl-alanine (l-DOPA) by means of absorption and fluorescence measurements. The developed device showed good repeatability and time stability. Using absorption measurements, an LOD equal to 23 μM and a linear response up to 350 μM was obtained; fluorescence measurements allow an LOD of 3 μM and a linear response in the range up to 10 μM.

Another example of biosensors exploiting a "layer-by-layer" immobilization was proposed by Alkasir et al. [69], who adopted this procedure for fabricating a colorimetric biosensor. The different layers are formed by chitosan and alginate polyelectrolytes deposited on a filter paper. The tyrosinase is embedded between these layers. This biosensor was used for the detection of phenol, BPA, catechol, and cresols [69]. The visual inspection of the color changes allows the detection of one of these substances. The color change is due to the specific binding of the quinone given by the enzymatic reaction to the multilayers of chitosan deposited on the paper. The digitalized approach was also used for more sensitive detection. The LOD was 0.86 ± 0.1 μg/L for each of the phenolic compounds studied. The proposed device showed very good time stability and was tested with real environmental samples.

Microarray-based biosensor systems were proposed by Jang et al. for the determination of phenol using CdSe/ZnS quantum dots [70].Microarrays were based on poly(ethylene glycol)(PEG) hydrogel. They were prepared by photopatterninga solution containing PEG diacrylate (PEG-DA), a photoinitiator, and tyrosinase. Tyrosinase and QDs were entrapped within the hydrogel microarrays because of a photo-induced crosslinking. The obtained hydrogel microarray was characterized by a fluorescent signal whose intensity linearly decreases phenol concentration. The detection limit of this biosensor is 1.0 μM [70]. As we said before, the surface plasmon resonance (SPR) approach allows the realization of a highly performing sensing scheme. Singh et al. [71] presented an SPR-based fiber-optic biosensor for the detection of phenolic compounds in an aqueous solution [71], based on the use of the wavelength interrogation approach. Differently from the mentioned angular interrogation approach, when wavelength interrogation is used, the wavelength of light is varied, and the angle of incidence is kept constant and greater than the critical angle (wavelength interrogation). In this method, light from a polychromatic source is coupled into the input end of the fiber, and the spectrum of the transmitted power at the output end of the fiber is recorded. A dip at a specific wavelength is observed in the transmitted spectrum (this is the resonance wavelength): its position depends on the refractive index of the sensing medium around the metal layer, and, hence, the shift of the dip is related to the number of molecules captured by the substrate. In the reported case, a silver film was deposited on the core of an optical fiber and tyrosinase from lyophilized mushroom powder was immobilized using the gel entrapment technique. The experimental setup of this wavelength interrogation SPR-based fiber-optic phenol biosensor is shown in Figure 11.

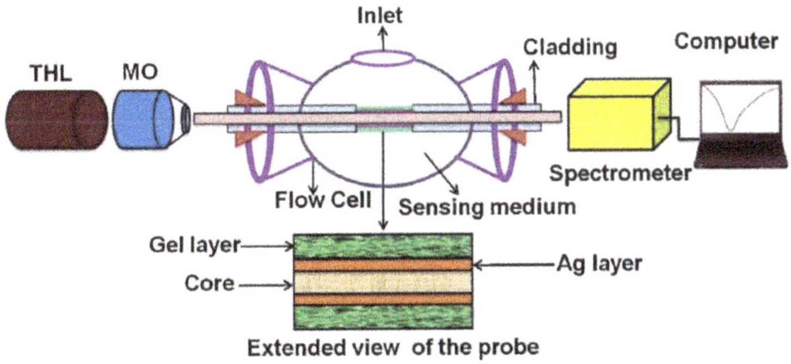

Figure 11. Schematic sketch of the experimental setup of an SPR-based fiber-optic phenol biosensor is shown. The fiber-optic probe is fixed in a small flow cell to enable the delivery and removal of aqueous samples of phenol around the sensing surface. Light from a tungsten–halogen lamp is coupled into the fiber. The spectrum of the transmitted power is recorded by using a spectrometer and a personal computer. (Reprinted with permission from [71]).

The fiber-optic probe was attached to a small flow cell in which the aqueous solutions can be delivered to the sensing element and removed. The developed biosensor was used for determining the concentration of different phenolic compounds (phenol, catechol, m-cresol, and 4-chlorophenol).

Aqueous samples of phenolic species with a variable concentration in the range of 0–1000 μM were examined. SPR spectra were collected for different concentrations, and the resonance wavelength showed a red shift when the concentration of the analyte increased. Representative calibration curves are reported in Figure 12.

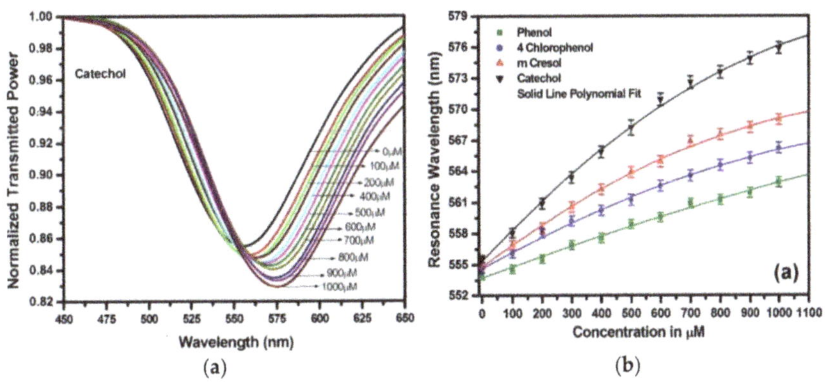

Figure 12. (**a**) Surface plasmon resonance spectra of a fiber-optic SPR probe for different concentrations of catechol, (**b**) Calibration curves obtained by measuring the variation in resonance wavelength for different concentrations of various phenol species. (Reprinted-adapted-with permission from [71]).

The LODs were evaluated for catechol, m-cresol, 4-chlorophenol, and phenol and resulted to be around 11, 17, 25, and 38 μM, respectively. The authors evidenced that the characteristics (high sensitivity, wider operating range, reusability, and reproducibility of results) of the developed sensor make it suitable for practical applications. In these cases, it would be also possible to take advantage of miniaturization properties, low costs, online monitoring and remote sensing potentiality, immunity to electromagnetic fields, and biocompatibility.

An SPR approach has also been adopted very recently by Hashim et al. for realizing a biosensor for phenol solution in which tyrosinase is immobilized on graphene oxide thin film. The device is characterized by a sensitivity of 0.00193 μM^{-1} and a LOD of 1 μM with a linear range up to 100 μM [72,73].

As is evident from the few examples of enzymatic biosensors described in this section, the interest in this category of devices is always alive. New ideas will probably be developed by exploiting the properties of nano enzymes, a new class of nanomaterials that has peculiar physicochemical properties. Nanozymes can imitate natural enzymes and show similar properties. Their reactions are effective, fast, and highly selective. These characteristics make nano enzymes exceptionally good candidates for the development of new sensing and monitoring applications [74].

3.2. *Immunosensors*

Immunosensors are sensing devices composed of an antigen or antibody coupled to a transducer that can evidence the binding of complementary species. Antibodies are proteins that are produced by mammals in response to foreign elements (bacteria, viruses, chemicals, etc.). The analyte detection is very specific and can allow concentration measurements.

SPR technology can be advantageously coupled with antibody immobilization, and various examples of SPR-based immunosensors for the detection and monitoring of low-molecular-weight analytes for environmental applications are described by Shankaran et al. [75]. For SPR immunosensor fabrication, biomolecules, as antigen or antibody, are adsorbed on the gold surface, and all the changes occurring for these molecules or the different interaction occurring processes can be studied. The binding between the antibody and the analyte causes a change in the refractive index that induces a shift in the resonance angle that can be registered as previously described. These shifts allow the determination of bound analyte concentration and the evaluation of the affinity between analyte and antibody and give information about their association or dissociation processes. In Ref. [75], the advantages of SPR immunoassay are exhaustively described, and the different SPR immunoassay format characteristics are discussed.

Dostalek et al. [76] presented an example of an SPR-based immunosensor for different endocrine disruptors and, in particular, for 4-nonylphenol, which is widely used as detergents in both domestic and industrial products. The sensing scheme and the working principles are clearly described in Ref. [76], and for 4-nonylphenol, an LOD of 0.26 ng/mL is estimated. The obtained calibration curve allows the determination of concentration up to 4.4 ng/mL. Analytes can be detected in 45-min cycles, including 30-min incubation of antibodies with samples. The sensor is regenerable. The LOD is relevant in comparison with the maximum admissible concentrations in drinking water currently permitted by the regulatory authorities in the USA and EU.

Long et al. [77] presented a highly sensitive and selective immunosensor for BPA detection that takes advantage of evanescent wave fiber-optic sensor and microfluidic technology for developing an all-fiber optofluidic-based bioassay platform [77]. In Figure 13, representative results are reported. The changes in fluorescence signal related to different BPA concentrations are used for realizing the calibration curve that shows a linear range between 0.5 μg/L and 1.0 μg/L and an LOD of 0.06 μg/L. This value is particularly appealing when compared with ELISA and amperometric biosensor performances. The biosensor was also tested with BPA-spiked samples, and the recovered data and the relative standard deviations were between 90–120% and 3.8–9.1%, respectively. The authors use the developed biosensing device also for investigating the BPA leaching of polycarbonate (PC) bottles of different brands. In fact, the residual and degraded BPA in this kind of bottle may migrate into food, especially at elevated temperatures for long periods [78]. The presented results show that the risk of BPA leaching from PC bottles is a real problem, and this approach can give a sensitive, rapid, on-site, real-time detection of BPA leaching.

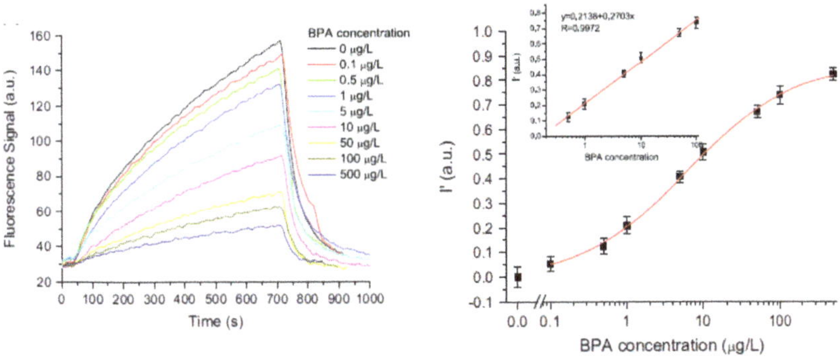

(**a**) (**b**)

Figure 13. (**a**) Typical fluorescence curves signals obtained at various concentrations of BPA. (**b**) Logarithmic calibration curves for determination of BPA obtained by using an optofluidics-based immunosensor. In the inset, the linear relationship between BPA concentration and fluorescence intensity is shown. (Reprinted-adapted-with permission from [77]).

3.3. *Receptor-Based Biosensors*

In 1998, Wright et al. [79] developed an SPR sensor with specific receptors for the detection of phenols in water. The authors synthesized some receptor molecules and immobilized them in gold or silver films. These films were deposited on glass slides mounted on a semicircular glass prism (refractive index 1.5151) and fastened to a small chamber connected to a peristaltic pump and waste reservoir. A p-polarized He: Ne laser was used for studying SPR response. The reported results evidenced the capability of this approach to detect and discriminate different phenolic species at low concentrations in aqueous media [79].

An example of this approach is represented by the device developed by Filik et al. for PAP employing CAL4 and reflectance measurements that we have previously described [25]. A typical result has been already reported in Figure 3. A linear calibration curve is obtained in the PAP concentration range of 0.5–5.5 ppm) with an LOD of 0.109 ppm. A response time of about 5 min is obtained for a stirred solution. The proposed sensor was also tested with several complex samples with spiked PAP, with recovered data ranging between 97 and 102%.

A very recent example of a receptor-based immunosensor has been developed by Conti et al. for BPA optical sensing by exploiting the luminescence emission of a new RuII complex that is able to bind BPA in an aqueous solution and to quench the luminescence emission of the core. The quenching effect is not remarkable, but the appropriately designed complexes can be used for determining BPA concentration in water. A linear calibration range up 50 µM BPA concentration has been obtained (see Figure 14) by using the luminescence quenching effect [80].

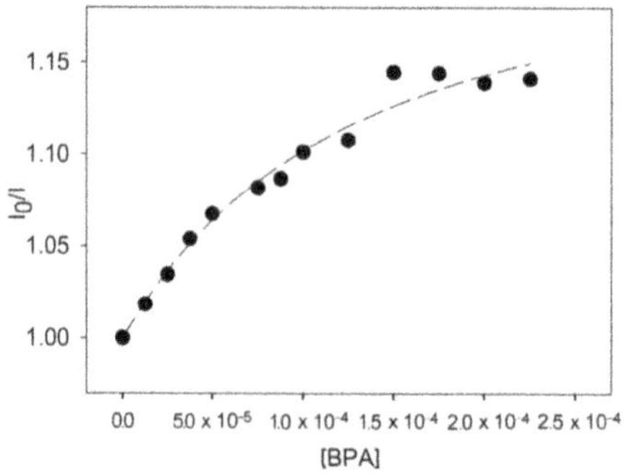

Figure 14. Calibration curve for different BPA concentration obtained by using the ratio of luminescence intensity of the [Ru (phen)2 L]2+ complex over luminescence intensity of the complex upon addition of increasing concentration. (Reprinted from [80] under Open Access conditions).

3.4. Nucleic Acids-Based Biosensors

Another very attractive class of biosensors is represented by the devices that exploit DNA as biosensing element. These biosensors can take advantage of the excellent stability of nucleic acids and the remarkable selectivity of the interaction of nucleic acids.

Yildirim et al. [81] proposed a portable, evanescent, wave fiber-optic DNA-based sensor for rapid, on-site detection of BPA with excellent sensitivity and selectivity. The authors covalently immobilized DNA on the optical fiber sensor surface. This biosensor uses an indirect competitive detection mode. For this sensing scheme, a pre-injection of BSA is used for avoiding the nonspecific binding to the sensor surface. After this step, the concentration of the remaining free aptamers becomes inversely proportional to that of BPA in the water sample. The sample solution is sent to the optical fiber sensor surface for allowing the generation of a useful fluorescence signal. The working parameters of the developed biosensors were investigated in detail. The authors reported a linear range for BPA from 2 nM to 100 nM with an LOD of 1.86 nM that becomes competitive with standard liquid chromatography detection results for BPA. Good reproducibility, stability, and selectivity for BPA detection were also demonstrated. The proposed sensor was successfully tested with wastewater samples [81].

Lim et al. [82] developed a palm-size NanoAptamer analyzer able to detect BPA at environmentally relevant concentrations (<1 ng/mL or ppb) with excellent sensing characteristics [82]. The presented biodevice uses a modified NanoGene assay [83] for BPA detection using magnetic beads for covalent bonding with a BPA-specific aptamer. After interaction with BPA, there is a decrease in the fluorescence signal that is proportional to the analyte concentration. The proposed biosensor showed a linear range for BPA from 0.0005 to 1 ng/mL. The BPA detection using this analyzer requires an incubation time of 30 min. This time can be positively compared with the time of other DNA aptamer methods for BPA detection that usually need incubation time lasting from 20 min to 8 h.

In 2019 Allsop et al. designed and fabricated another aptamer-based optical biosensor able to test BPA solutions in the concentration range from 10 nM to 1 fM. The presented device employs an array of gold nano-antennae that generate coupled localized surface plasmon (LSP) and are modified with an aptamer specifically for BPA detection [84]. The array of nano-antennae is assembled on a section of a standard telecommunication optical fiber. This configuration potentially enables multiplexing and remote sensing applications. Using a linear regression analysis, the authors can attain an extremely low LOD (330 ± 70 aM) that represents the lowest measured LOD.

3.5. Whole-Cells Biosensors

Optical microbial biosensors are devices that use microorganisms with an optical transducer to allow fast and accurate monitoring of the analytes of interest. Different researchers presented similar devices for applications in the field of environmental monitoring.

Mazhari and Agsar [23] proposed the use of *Streptomyces tuirus* DBZ39 synthesized on extracellular gold nanoparticles for the visual detection of phenol. The visual detection was improved by the addition of sodium sulphate, and the change of color occurred within 2 min. The proposed method was successfully tested with water samples from the effluents of fertilizer and distillery industries.

The same authors further exploited the properties of *Streptomyces tuirus* DBZ39 together with tyrosinase and gold nanoparticles for developing a paper biosensor for the detection of phenol from industrial water. The proposed biosensor can efficiently detect the changes in absorbance due to phenol presence thanks to the specific catalytic activity of the tyrosinase and the SPR contribution due to gold nanoparticles. This biosensor was tested with different types and quantities of phenolic constituents in various industrial effluents. The peculiar optical properties of gold nanoparticles increase the efficacy of tyrosinase for detecting phenol compounds [85].

3.6. Molecularly Imprinted Polymer-Based Sensors

Notwithstanding their synthetic origin, molecularly imprinted polymers (MIPs) are often considered biomimetic materials, and MIPs-based sensors are usually regarded as biosensors.

Griffete et al. exploit the characteristics of MIPs and photonic crystals for preparing a defect-embedded imprinted photonic polymer that is constituted by an ordered and interlinked three-dimensional microporous array [26]. In this structure, some nanocavities can interact with BPA using binding sites. Reflectance spectroscopy has been used to investigate the optical properties of the structure that are influenced by the interaction with BPA (see Figure 3 of Ref. [26]). The authors also demonstrate the selectivity and specificity of the developed MIP-based sensors for BPA solutions.

Taguchi et al. developed a slab-type optical waveguide (s-OWG) and fabricated a microfluidic system [86]. On this OWG, consecutive parallel gold and silver bands are deposited. These can generate two individual SPR signals because of the difference in resonant reflection spectra of these metals. MIP nanoparticles were used as a recognition element for the BPA compound. In Figure 15, the immobilization procedure for preparing MIP nanoparticles and BPA grafted to gold nanoparticles on the sensor chip for the binding of free BPA is described. Peak shifts of SPR spectra by the addition of free BPA into MIP with immobilized nanoparticles were observed for BPA concentration varying in the range 0.1–2000 mM.

The detection of BPA is also the aim of other MIP-based sensors [87,88]. In Ref. [88], the use of black phosphorus and hollow-core anti-resonant fiber allowed two orders of magnitude enhancement of sensitivity in a fluorescence-sensing scheme. The simulated LOD was 1.69 pM, according to the calibration curve based on the IUPAC definition. The sensor was tested with real samples, such as water (collected from a lake near the campus of Beijing University of Technology) and human blood (see Figure 5 of Ref. [88]).

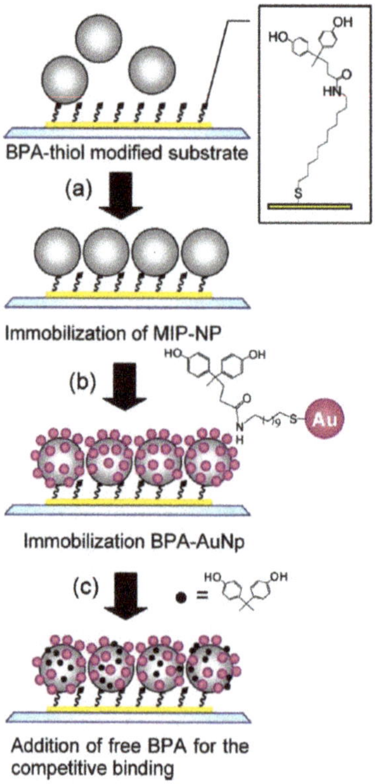

Figure 15. Schematic representation of the immobilization procedure of MIP-Np and BPA-Np on the sensor chip for the BPA detection. (Reprinted with permission from [86]. Copyright 2012 American Chemical Society).

4. Sensors

In addition to the sensing schemes using biological transducers for detecting phenols that we described until now, optical techniques allow the implementation of chemical and/or physical sensors in which a chemical or physical property of a specific analyte is converted into a measurable optical signal that is proportional to the concentration of the analyte of interest. These sensors are less common for phenolic species detection due to their low sensitivity and selectivity, but currently, the development of new nanomaterials, such as gold, silver, and other metal nanoparticles; nanotubes; and quantum dots enables a significant improvement of these characteristics [89].

4.1. Optical Chemical Sensors

A clear classification of the different types of chemical optical sensors is shown in Figure 2 of Ref. [90]. The simplest sensing schemes are based on direct and reagent-mediated spectroscopic techniques. This framework includes methods based on the variations induced in the absorption or fluorescence signals of suitably designed inorganic probes [91–93]. Another relevant class of sensors is related to the design and fabrication of optical fiber chemical sensors that we have already mentioned in Section 2 [32,33]. A representative example of these devices has been developed by Wang et al. using a plastic optical fiber, a polymer membrane, a gold mirror, and a TiO_2-based composite layer [38,94]. In particular, the relative variations of reflected light intensity are used as a working parameter. Phenol solutions at different concentrations were used for testing the sensor, and an LOD of $0.294 \cdot 10^{-3}$ mg/L was obtained with a high selectivity.

4.2. Nanostructure-Based Sensors

As previously mentioned, the development of nanotechnology allowed the design and development of new sensing devices for phenolic species detection [95]. In 2008 Nezhad et al. presented an indirect colorimetric method for the optical detection of phenolic compounds exploiting the SPR band shown by gold nanoparticles [96]. By using the changes in the absorbance signal (typically occurring in a few tens of seconds) the authors were able to detect low concentrations of hydroquinone, catechol, and pyrogallol. Linear ranges from $7.0 \cdot 10^{-7}$ to $1.0 \cdot 10^{-4}$ M, $6.0 \cdot 10^{-6}$ to $2.0 \cdot 10^{-4}$ M, and $6.0 \cdot 10^{-7}$ to $1.0 \cdot 10^{-4}$ M were, respectively, obtained for the three above-mentioned phenolic species. The proposed method was also successfully tested with tap and river water samples.

Gold nanoparticles were also used for BPA detection by various researchers; for example, Ma et al. synthesized a diazonium carrying ligand monolayer film on the nanoparticles with the diazonium ions exposed on their surface [97]. The BPA-diazonium interaction causes a BPA concentration-dependent color change that can be employed for concentration determination. Also, for BPA, the gold nanoparticles-SPR method allows a fast response time (4 min), a broad linear range (0.1–4 nM), and a low LOD (0.02 nM). Table 1 of the cited paper also reports an interesting comparison among different methods available for BPA sensing.

In the last several years, semiconductor quantum dots, nanocubes, and nanorods have also attracted great interest in sensing applications due to their appealing physical and chemical properties [98,99]. Very representative examples of this class of devices are reported in Refs. [100–104]. In particular, Jaiswal et al. proposed a fast synthesis route of doped carbon nitride quantum dots for the detection of hydroquinone by photoluminescence quenching. They obtained an LOD of 50 nM and a linear range from 12 to 57.5 μM [102].

4.3. Photonic Crystal Fiber Sensors

Recently, Frazao et al. reviewed optical sensing applications of photonic crystal fibers. They revised the different approaches based on fiber Bragg gratings, long-period gratings, and interferometric structures. In addition, the role of nonlinear effects and the main sensing schemes for gaseous and liquid compounds were also discussed [105]. Using a photocatalytic long-period fiber grating (PLPFG), a fiber Bragg grating (FBG), a polymer membrane, an ultraviolet light, and microchannels, Zhong et al. developed a lab-on-a-chip device for phenol concentration sensing [38]. This approach is characterized by easy and fast in situ use, low consumption of agents and reagents, low costs, and high sensitivity. The PLPFG component is a three-layer structure in which the refractive index of the cladding layer is less than that of the core, and both are less than the refractive index of the coating layer. This is a photocatalytic film for UV-visible-driven photocatalytic degradation of phenol. The LPFG enhances the evanescent wave absorption and shifts the central wavelength due to the interaction between the evanescent wave and the analyte. The developed sensor is characterized by a linear performance in a large range of phenol concentrations (7.5 μg/L to 100 mg/L). It can operate at pH values and temperatures ranging from 2.0 to 14.0 and from 10 °C to 48 °C, respectively.

A sensor for BPA and bisphenol S (BPS) based on photonic crystal technology was proposed by Niger et al. employing a dodecagonal photonic crystal fiber structure having a floral pattern in the first cladding layer [106]. Using this approach, the authors reported a relative sensitivity of 97.6% and 94.9%, respectively, for BPA and BPS.

4.4. Sensing Schemes Using Vibrational Spectroscopies

As described in Section 2, vibrational spectroscopies have been largely employed for designing and developing a huge number of devices for sensing different phenolic compounds. In 1984, the pioneering work of Marley et al. was devoted to evaluating the possibility to use RS for the quantitative analysis of six compounds (phenol, o-chlorophenol, 2,4-dichlorophenol, 2,4,6-trichlorophenol, 2-chloro-5-methylphenol and

2-chloro-4-nitrophenol) in water [107]. The authors used two different methods for the quantitative determination of concentrations, namely peak area measurement and cross-correlation. Both methods were applied to data after Savitzky-Golay smoothing and after correction for internal standard fluctuations. Areas were measured on selected bands in the spectra, and cross-correlation operation was accomplished on a complete set of spectra for a given compound, eliminating frequency components from both the high and low ends of the set. LODs ranged from 0.3 ppm to 100 ppm, depending on the compound.

The use of properly designed nanostructured substrates has allowed the implementation of a certain number of sensing devices for BPA [108–118] by means of SERS. Typically, these devices exploit silver or gold nanoparticles that can be functionalized with organic groups to enhance Raman signal intensity and, consequently, the sensitivity for BPA or other phenolic substances concentration determination. In particular, Roschi et al. employed silver nanoparticles functionalized with thiolated-cyclodextrin (CD-SH) for the detection of bisphenols (BPs) A, B, and S. Using multivariate analysis of the SERS data, the LOD for BPs was estimated at about 10^{-7} M, in the range of the tens of ppb (see Figure 16) [112].

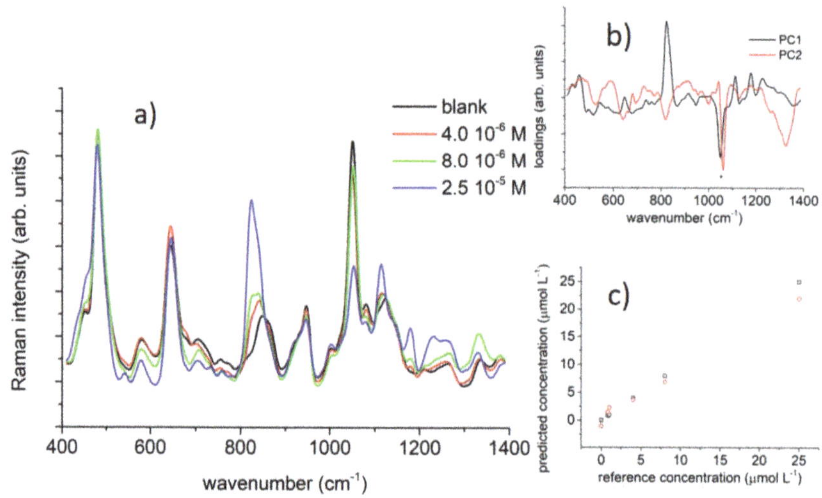

Figure 16. (**a**) BPA SERS spectra as a function of the BPA molar concentration. (**b**) Loadings for the first two components in the PCA analysis on the BPA SERS data. (**c**) Calibration curve as obtained by the partial least square regression analysis on SERS data. (Reprinted from [112] under Open Access conditions).

5. Conclusions

The danger of phenolic compounds and their tendency to remain present in the environment motivates the intense search for new methods for their detection and measurement of their concentrations. In this paper, we intended to highlight the relevant role that optical techniques can play in this framework. For this reason, we focused our attention on UV-vis fluorescence, reflectance, and absorption experimental approaches, on the different uses of optical fibers and Bragg gratings and on SPR and vibrational spectroscopies methods. In addition, we revised some representative applications of these techniques in developing new sensors for some phenolic compounds especially present in water.

In Table S2, we summarized the most relevant working parameter of these optical sensors. In particular, their sensitivity, linear range, LOD, and response time have been reported. From the inspection of this Table, it is evident that the largest number of optical sensing schemes among those discussed in this review has been proposed for bisphenol A and phenol. More conventional optical techniques, such as absorption and UV-vis fluorescence, are still adopted, often using innovative biocomponents. However, it is also evident that more novel optical techniques, such as SPR, SERS, and SERRS (Surface-Enhanced-Resonant Raman Spectroscopy), with the help of sophisticated nanostructures, are making their way. Despite the difficulty of a rigorous comparison due to the different experimental conditions, it can be seen that the use of advanced new biocomponents and new optical techniques is offering ever more performing working parameters.

Although limited to the most common techniques and not particularly complete, this review certainly confirms the important role played by optical techniques for the development of sensitive, fast, and easy-to-use biosensors and physical and chemical sensing schemes for phenolic species monitoring.

Supplementary Materials: The following are available online at https://www.mdpi.com/article/10.3390/s21227563/s1, Table S1: Permissible concentration limits for different phenolic compounds, Table S2: Relevant working parameters for the optical sensors for the determination of phenolic compounds in environmental applications discussed in the present paper.

Author Contributions: Conceptualization, I.D. and M.L.; methodology, I.D. and M.L.; validation, I.D., N.D. and M.L.; formal analysis, I.D., N.D. and M.L.; writing—original draft preparation, I.D., N.D. and M.L.; writing—review and editing, I.D. and M.L.; visualization, I.D.; supervision, M.L.; project administration, M.L. All authors have read and agreed to the published version of the manuscript.

Funding: This research received no external funding.

Institutional Review Board Statement: Not applicable.

Informed Consent Statement: Not applicable.

Data Availability Statement: Not applicable.

Conflicts of Interest: The authors declare no conflict of interest.

References

1. Mainali, K. Phenolic compounds contaminants in water: A glance. *Curr. Trends Civ. Struct. Eng.* **2020**, *4*, 1–3. [CrossRef]
2. Sun, R.; Wang, Y.; Ni, Y.; Kokot, S. Spectrophotometric analysis of phenols, which involves a hemin–graphene hybrid nanoparticles with peroxidase-like activity. *J. Hazard. Mater.* **2014**, *266*, 60–67. [CrossRef]
3. Appendix A to 40 CFR, Part 423–126 Priority Pollutants. U.S. Environmental Protection Agency. Available online: https://www3.epa.gov/region1/npdes/permits/generic/prioritypollutants.pdf (accessed on 2 November 2021).
4. Priority Substances and Certain Other Pollutants according to Annex II of Directive 2008/105/EC. European Commission. Available online: http://data.europa.eu/eli/dir/2008/105/oj (accessed on 2 November 2021).
5. Boudet, A.M. Evolution and current status of research in phenolic compounds: Review. *Phytochem* **2007**, *68*, 2722–2735. [CrossRef]
6. Gami, A.A.; Shukor, M.Y.; Khalil, K.A.; Dahalan, F.A.; Khalid, A.; Ahmad, S.A. Phenol and its toxicity. *J. Environ. Microbiol. Toxicol.* **2014**, *2*, 11–24.
7. Michalowicz, J.; Duda, W. Phenols-sources and toxicity: Review. *Pol. J. Environ. Stud.* **2007**, *16*, 347–362.
8. Wasi, S.; Tabrez, S.; Ahmad, M. Toxicological effect of major environmental pollutants: An overview. *Environ. Monit. Assess.* **2013**, *185*, 2585–2593. [CrossRef]
9. Frye, C.A.; Bo, E.; Calamandrei, G.; Calzà, L.; Dessì-Fulgheri, F.; Fernández, M.; Fusani, L.; Kah, O.; Kajta, M.; Le Page, Y.; et al. Endocrine disrupters: A review of some sources, effects, and mechanisms of actions on behaviour and neuroendocrine systems. *J. Neuroendocrinol.* **2012**, *24*, 144–159. [CrossRef]
10. Schug, T.T.; Janesick, A.; Blumberg, B.; Heindel, J.J. Endocrine disrupting chemicals and disease susceptibility. *J. Steroid Biochem. Mol. Biol.* **2011**, *127*, 204–215. [CrossRef]
11. Kang, J.H.; Kondo, F.; Katayama, Y. Human exposure to bisphenol A. *Reprod. Toxicol.* **2006**, *226*, 79–89. [CrossRef]
12. Vandenberg, L.N.; Hauser, R.; Marcus, M.; Olea, N.; Welshons, W.V. Human exposure to bisphenol A (BPA). *Reprod. Toxicol.* **2007**, *24*, 139–177. [CrossRef]
13. Calafat, A.M.; Kuklenyik, Z.; Reidy, J.A.; Caudill, S.P.; Ekong, J.; Needham, L.L. Urinary concentrations of bisphenol A and 4-Nonylphenol in a human reference population. *Environ. Health Perspect.* **2005**, *113*, 391–395. [CrossRef] [PubMed]
14. Calafat, A.M.; Ye, X.; Wong, L.Y.; Reidy, J.A.; Needham, L.L. Exposure of the U.S. population to bisphenol A and 4-tertiary-octylphenol: 2003–2004. *Environ. Health Perspect.* **2008**, *116*, 39–44. [CrossRef] [PubMed]
15. Dekant, W.; Volkel, W. Human exposure to bisphenol A by biomonitoring: Methods, results and assessment of environmental exposures. *Toxicol. Appl. Pharmacol.* **2008**, *228*, 114–134. [CrossRef]

16. Karim, F.; Fakhruddin, A.N.M. Recent advances in the development of biosensor for phenol: A review. *Rev. Environ. Sci. Biotechnol.* **2012**, *11*, 261–274. [CrossRef]
17. Rodionov, P.V.; Veselova, I.A.; Shekhovtsova, T.N. Optical sensors for determining phenolic compounds with different structures. *J. Anal. Chem.* **2013**, *68*, 931–941. [CrossRef]
18. Portaccio, M.; Menale, C.; Diano, N.; Serri, C.; Mita, D.G.; Lepore, M. Monitoring production process of Cisplatin-loaded PLGA nanoparticles by FT-IR microspectroscopy and univariate data analysis. *J. Appl. Polym. Sci.* **2015**, *132*, 41305. [CrossRef]
19. Camerlingo, C.; Portaccio, M.; Tatè, R.; Lepore, M.; Delfino, I. Fructose and pectin detection in fruit-based food products by surface-enhanced Raman spectroscopy. *Sensors* **2017**, *17*, 839. [CrossRef]
20. Ricciardi, V.; Portaccio, M.; Piccolella, S.; Manti, L.; Pacifico, S.; Lepore, M. Study of SH-SY5Y cancer cell response to treatment with polyphenol extracts using FT-IR spectroscopy. *Biosensors* **2017**, *7*, 57. [CrossRef]
21. Camerlingo, C.; Verde, A.; Manti, L.; Meschini, R.; Delfino, I.; Lepore, M. Graphene-based Raman spectroscopy for pH sensing of X-rays exposed and unexposed culture media and cells. *Sensors* **2018**, *18*, 2242. [CrossRef]
22. Di Meo, V.; Crescitelli, A.; Moccia, M.; Sandomenico, A.; Portaccio, M.; Lepore, M.; Ruvo, M.; Galdi, V.; Esposito, E. Pixeled metasurface for multiwavelength detection of vitamin D. *Nanophotonics* **2020**, *9*, 3921–3930. [CrossRef]
23. Mazhari, B.B.Z.; Agsar, D. Detection of phenols from industrial effluents using streptomyces mediated gold nanoparticles. *Indian J. Mat. Sci.* **2016**, *2016*, 6937489. [CrossRef]
24. Bayram, A.; Horzum, N.; Metin, A.U.; Kılıç, V.; Solmaz, M.E. Colorimetric bisphenol-A detection with a portable smartphone-based spectrometer. *IEEE Sens. J.* **2018**, *18*, 5948–5955. [CrossRef]
25. Filik, H.; Aksua, D.; Apaka, R.; Sener, I.; Kılıc, E. An optical fibre reflectance sensor for p-aminophenol determination based on tetrahydroxycalix[4]arene as sensing reagent. *Sens. Actuat. B* **2009**, *136*, 105–112. [CrossRef]
26. Griffete, N.; Frederich, H.; Maître, A.; Schwob, C.; Ravaine, S.; Carbonnier, B.; Chehimi, M.M.; Mangeney, C. Introduction of a planar defect in a molecularly imprinted photonic crystal sensor for the detection of bisphenol A. *J. Colloid Interface Sci.* **2011**, *364*, 18–23. [CrossRef]
27. Bishnu, P.; Popma, T. Optical Waveguide Sensors. In *Handbook of Optical Sensors*; Santos, J.L., Farahi, F., Eds.; CRC Press: Boca Raton, FL, USA, 2014; pp. 1–43.
28. Krohn, D.A.; MacDougall, T.; Mendez, A. *Fiber Optic Sensors: Fundamentals and Applications*; SPIE Press: Bellingham, WA, USA, 2014.
29. Milvich, J.; Kohler, D.; Freude, W.; Koos, C. Surface sensing with integrated optical waveguides: A design guideline. *Opt. Express* **2018**, *26*, 19885–19906. [CrossRef]
30. Del Villar, I.; Matias, I.R. *Optical Fibre Sensors: Fundamentals for Development of Optimized Devices*; IEEE Press Series on Sensors; Wiley-IEEE Press: Oxford, UK, 2020.
31. Floris, I.; Adam, J.M.; Calderón, P.A.; Sales, S. Fiber optic shape sensors: A comprehensive review. *Opt. Lasers Eng.* **2021**, *139*, 106508. [CrossRef]
32. Pospíšilová, M.; Kuncová, G.; Trögl, J. Fiber-optic chemical sensors and fiber-optic biosensors. *Sensors* **2015**, *15*, 25208–25259. [CrossRef] [PubMed]
33. Wang, X.; Wolfbeis, O.S. Fiber-optic chemical sensors and biosensors (2015–2019). *Anal. Chem.* **2020**, *92*, 397–430. [CrossRef]
34. Tow, K.H.; Chow, D.M.; Vollrath, F.; Dicaire, I.; Gheysens, T.; Thevenaz, L. Exploring the use of native spider silk as an optical fiber for chemical sensing. *J. Lightwave Technol.* **2018**, *36*, 1138–1144. [CrossRef]
35. Campanella, C.E.; Cuccovillo, A.; Campanella, C.; Yurt, A.; Passaro, V.M.N. Fibre bragg grating based strain sensors: Review of Technology and Applications. *Sensors* **2018**, *18*, 3115. [CrossRef] [PubMed]
36. Lo Presti, D. Fiber Bragg gratings for medical applications and future challenges: A Review. *IEEE Access* **2020**, *8*, 156863–156888. [CrossRef]
37. Zhong, N.; Chen, M.; Chang, H.; Zhang, T.; Wang, Z.; Xin, X. Optic fiber with Er^{3+}: $YAlO_3/SiO_2/TiO_2$ coating and polymer membrane for selective detection of phenol in water. *Sens. Actuators B Chem.* **2018**, *273*, 1744–1753. [CrossRef]
38. Zhong, N.; Chen, M.; Wang, Z.; Xin, X.; Li, B. Photochemical device for selective detection of phenol in aqueous solutions. *Lab Chip* **2018**, *18*, 1621–1632. [CrossRef]
39. Daniyal, W.M.E.M.M.; Fen, Y.W.; Fauzi, L.N.I.M.; Hashim, H.S.; Ramdzan, N.S.M.; Omar, N.A.S. Recent advances in surface plasmon resonance optical sensors for potential application in environmental monitoring. *Sens. Mater.* **2020**, *32*, 4191–4200. [CrossRef]
40. Yesudasu, V.; Pradhan, H.S.; Pandya, R.J. Recent progress in surface plasmon resonance-based sensors: A comprehensive review. *Heliyon* **2021**, *7*, e06321. [CrossRef] [PubMed]
41. Kraft, M. Vibrational spectroscopic sensors fundamentals, instrumentation and applications. In *Optical Chemical Sensor*; Baldini, F., Chester, A., Homola, J., Martellucci, S., Eds.; In *NATO Science Series II: Mathematics, Physics and Chemistry*; Springer: Dordrecht, The Netherlands, 2006; Volume 224, pp. 117–155.
42. Baker, M.J.; Trevisan, J.; Bassan, P.; Bhargava, R.; Butler, H.J.; Dorling, K.M.; Fielden, P.R.; Fogarty, S.W.; Fullwood, N.J.; Heys, K.A.; et al. Using fourier transform IR spectroscopy to analyze biological materials. *Nat. Protoc.* **2014**, *9*, 1771–1791. [CrossRef]
43. Butler, H.J.; Ashton, L.; Bird, B.; Cinque, G.; Curtis, K.; Dorney, J.; Esmonde-White, K.; Fullwood, N.J.; Gardner, B.; Martin-Hirsch, P.L.; et al. Using raman spectroscopy to characterize biological materials. *Nat. Protoc.* **2016**, *11*, 664–687. [CrossRef]

44. Neubrech, F.; Huck, C.; Weber, K.; Pucci, A.; Giessen, H. Surface-Enhanced infrared spectroscopy using resonant nanoantennas. *Chem. Rev.* **2017**, *117*, 5110–5145. [CrossRef]
45. Di Meo, V.; Caporale, A.; Crescitelli, A.; Janneh, M.; Palange, E.; De Marcellis, A.; Portaccio, M.; Lepore, M.; Rendina, I.; Ruvo, M.; et al. Metasurface based on cross-shaped plasmonic nanoantennas as chemical sensor for surface-enhanced infrared absorption spectroscopy. *Sens. Actuators B Chem.* **2019**, *286*, 600–607. [CrossRef]
46. Pilot, R.; Signorini, R.; Fabris, L. Surface-enhanced raman spectroscopy: Principles, substrates, and applications. In *Metal Nanoparticles and Clusters*; Deepak, F., Ed.; Springer: Berlin/Heidelberg, Germany, 2018.
47. Borisov, S.M.; Wolfbeis, O.S. Optical biosensors. *Chem. Rev.* **2008**, *108*, 423–461. [CrossRef]
48. Yashas, S.R.; Shivakumara, B.P.; Udayashankara, T.H.; Krishna, B.M. Laccase biosensor: Green technique for quantification of phenols in wastewater (A review). *Orient. J. Chem.* **2018**, *34*, 631–637.
49. Claus, H. Laccases: Structure, reactions, distribution. *Micron* **2004**, *35*, 93–96. [CrossRef] [PubMed]
50. Dennison, C. Investigating the structure and function of cupredoxins. *Coord. Chem. Rev.* **2005**, *249*, 3025–3054. [CrossRef]
51. Arregui, L.; Ayala, M.; Gómez-Gil, X.; Gutiérrez-Soto, G.; Hernández-Luna, C.E.; Herrera de Los Santos, M.; Levin, L.; Rojo-Domínguez, A.; Romero-Martínez, D.; Saparrat, M.C.N.; et al. Laccases: Structure, function, and potential application in water bioremediation. *Microb. Cell. Fact.* **2019**, *8*, 200–232. [CrossRef]
52. Rodríguez-Delgado, M.M.; Alemán-Nava, G.S.; Rodríguez-Delgado, J.M.; Dieck-Assad, G.; Martínez-Chapa, S.O.; Barceló, D.; Parra, R. Laccase-based biosensors for detection of phenolic compounds. *Trends Anal. Chem.* **2015**, *74*, 21–45. [CrossRef]
53. Abdullah, J.; Ahmad, M.; Heng, L.Y.; Karuppiah, N.; Sidek, H. An optical biosensor based on immobilization of laccase and MBTH in stacked films for the detection of catechol. *Sensors* **2007**, *7*, 2238–2250. [CrossRef]
54. Sanz, J.; de Marcos, S.; Galbán, J. Autoindicating optical properties of laccase as the base of an optical biosensor film for phenol determination. *Anal. Bioanal. Chem.* **2012**, *404*, 351–359. [CrossRef]
55. Andreu-Navarro, A.; Fernández-Romero, J.M.; Gómez-Hens, A. Determination of polyphenolic content in beverages using laccase, gold nanoparticles and long wavelength fluorimetry. *Anal. Chim. Acta* **2012**, *713*, 1–6. [CrossRef]
56. Cantarella, G.; d'Acunzo, F.; Galli, C. Determination of laccase activity in mixed solvents: Comparison between two chromogens in a spectrophotometric assay. *Biotechnol. Bioeng.* **2003**, *82*, 395–398. [CrossRef]
57. Xu, F. Oxidation of phenols, anilines, and benzenethiols by fungal laccases: Correlation between activity and redox potentials as well as halide inhibition. *Biochemistry* **1996**, *35*, 7608–7614. [CrossRef]
58. Delfino, I.; Portaccio, M.; Della Ventura, B.; Manzo, G.; Mita, D.G.; Lepore, M. Optical properties of sol-gel immobilized Laccase: A first step for its use in optical biosensing. In *Optical Sensing and Detection II*; International Society for Optics and Photonics: Washington, DC, USA, 2012; Volume 8439.
59. Lepore, M.; Portaccio, M. Optical detection of different phenolic compounds by means of a novel biosensor based on sol-gel immobilized laccase. *Biotechnol. Appl. Biochem.* **2017**, *64*, 782–792. [CrossRef]
60. Jędrychowska, A.; Malecha, K.; Cabaj, J.; Sołoducho, J. Laccase biosensor based on low temperature co-fired ceramics for the permanent monitoring of water solutions. *Electrochim. Acta* **2015**, *165*, 372–382. [CrossRef]
61. Thelemann, T.; Fischer, M.; Groß, A.; Müller, J. LTCC-based fluidic components for chemical applications. *J. Microelectron. Electron. Pack.* **2007**, *4*, 167. [CrossRef]
62. Baeza, M.; López, C.; Alonso, J.; López-Santin, J.; Álvaro, G. Ceramic microsystem incorporating a microreactor with immobilized biocatalyst for enzymatic spectrophotometric assays. *Anal. Chem.* **2010**, *82*, 1006–1011. [CrossRef]
63. Cano-Raya, C.; Dencheva, N.V.; Braz, J.F.; Malfois, M.; Denchev, Z.Z. Optical biosensor for catechol determination based on laccase-immobilized anionic polyamide 6 microparticles. *J. Appl. Polym. Sci.* **2020**, *137*, e49131. [CrossRef]
64. Gul, I.; Ahmad, M.S.; Naqvi, S.M.S.; Hussain, A.; Wali, R.; Farooqi, A.A.; Ahmed, I. Polyphenol oxidase (PPO) based biosensors for detection of phenolic compounds: A Review. *J. App. Biol. Biotechnol.* **2017**, *5*, 72–85.
65. Claus, H.; Decker, H. Bacterial tyrosinases. *Syst. Appl. Microbiol.* **2006**, *29*, 3–14. [CrossRef]
66. Russell, I.M.; Burton, S.G. The development of an immobilized enzyme bioprobe for the detection of phenolic pollutants in water. *Anal. Chim. Acta* **1999**, *389*, 161–170. [CrossRef]
67. Abdullah, J.; Ahmad, M.; Karuppiah, N.; Heng, L.Y.; Sidek, H. Immobilization of tyrosinase in chitosan film for an optical detection of phenol. *Sens. Actuators B Chem.* **2005**, *114*, 604–609. [CrossRef]
68. Fiorentino, D.; Gallone, A.; Fiocco, D.; Palazzo, G.; Mallardi, A. Mushroom tyrosinase in polyelectrolyte multilayers as an optical biosensor for o-diphenols. *Biosens. Bioelectron.* **2010**, *25*, 2033–2037. [CrossRef]
69. Alkasir, R.S.; Ornatska, M.; Andreescu, S. Colorimetric paper bioassay for the detection of phenolic compounds. *Anal. Chem.* **2012**, *84*, 9729–9737. [CrossRef] [PubMed]
70. Jang, E.; Son, K.J.; Kim, B.; Koh, W.G. Phenol biosensor based on hydrogel microarrays entrapping tyrosinase and quantum dots. *Analyst* **2010**, *135*, 2871–2878. [CrossRef]
71. Singh, S.; Mishra, S.K.; Gupta, B.D. SPR based fibre optic biosensor for phenolic compounds using immobilization of tyrosinase in polyacrylamide gel. *Sens. Actuators B Chem.* **2013**, *186*, 388–395. [CrossRef]
72. Hashim, H.S.; Fen, Y.W.; Omar, N.A.S.; Daniyal, W.M.E.M.M.; Saleviter, S.; Abdullah, J. Structural, optical and potential sensing properties of tyrosinase immobilized graphene oxide thin film on gold surface. *Optik* **2020**, *212*, 164786–164796. [CrossRef]

73. Hashim, H.S.; Fen, Y.W.; Sheh Omar, N.A.; Abdullah, J.; Daniyal, W.M.E.M.M.; Saleviter, S. Detection of phenol by incorporation of gold modified-enzyme based graphene oxide thin film with surface plasmon resonance technique. *Opt. Express* **2020**, *28*, 9738–9752. [CrossRef]
74. Cavalcante, F.T.T.; de, A. Falcão, I.R.; da S. Souza, J.E.; Rocha, T.G.; de Sousa, I.G.; Cavalcante, A.L.G.; de Oliveira, A.L.B.; de Sousa, M.C.M.; dos Santos, J.C.S. Designing of nanomaterials-based enzymatic biosensors: Synthesis, properties, and applications. *Electrochem* **2021**, *2*, 149–184. [CrossRef]
75. Shankaran, D.R.; Gobi, K.V.; Miura, N. Recent advancements in surface plasmon resonance immunosensors for detection of small molecules of biomedical, food and environmental interest. *Sens. Actuators B Chem.* **2007**, *121*, 158–177. [CrossRef]
76. Dostálek, J.; Přibyl, J.; Homola, J.; Skládal, P. Multichannel SPR biosensor for detection of endocrine-disrupting compounds. *Anal. Bioanal. Chem.* **2007**, *389*, 1841–1847. [CrossRef]
77. Long, F.; Zhu, A.; Zhou, X.; Wang, H.; Zhao, Z.; Liu, L.; Shi, H. Highly sensitive and selective optofluidics-based immunosensor for rapid assessment of Bisphenol A leaching risk. *Biosens. Bioelectron.* **2014**, *55*, 19–25. [CrossRef]
78. Li, X.; Ying, G.G.; Su, H.C.; Yang, X.B.; Wang, L. Simultaneous determination and assessment of 4-nonylphenol, bisphenol A and triclosan in tap water, bottled water and baby bottles. *Environ. Int.* **2010**, *36*, 557–562. [CrossRef]
79. Wright, J.D.; Oliver, J.V.; Nolte, R.J.M.; Holder, S.J.; Sommerdijk, N.A.J.M.; Nikitin, P.I. The detection of phenols in water using a surface Plasmon resonance system with specific receptors. *Sens. Actuators B Chem.* **1998**, *51*, 305–310. [CrossRef]
80. Conti, L.; Mummolo, L.; Romano, G.M.; Giorgi, C.; Giacomazzo, G.E.; Prodi, L.; Bencini, A. Exploring the ability of luminescent metal assemblies to bind and sense anionic or ionizable analytes a Ru(phen)2bipy-based dizinc complex for bisphenol A (BPA) recognition. *Molecules* **2021**, *26*, 527. [CrossRef] [PubMed]
81. Yildirim, N.; Long, F.; He, M.; Shid, H.C.; Gu, A.Z. A portable optic fiber aptasensor for sensitive, specific and rapid detection of bisphenol-A in water samples. *Environ. Sci. Process Impacts* **2014**, *16*, 1379–1386. [CrossRef]
82. Lim, H.J.; Chu, B.; Son, A. Detection of bisphenol A using palm-size NanoAptamer analyzer. *Biosens. Bioelectron.* **2017**, *94*, 10–18. [CrossRef]
83. Kim, G.Y.; Wang, X.; Ahn, H.; Son, A. Gene quantification by the NanoGene assay is resistant to inhibition by humic acids. *Environ. Sci. Technol.* **2011**, *45*, 8873–8880. [CrossRef] [PubMed]
84. Allsop, T.D.P.; Neal, R.; Wang, C.; Nagel, D.A.; Hine, A.V.; Culverhouse, P.; Castañón, J.D.A.; Webb, D.J.; Scarano, S.; Minunni, M. An ultra-sensitive aptasensor on optical fibre for the direct detection of bisphenol A. *Biosens. Bioelectron.* **2019**, *135*, 102–110. [CrossRef] [PubMed]
85. Mazhari, B.B.Z.; Agsar, D.; Prasad, M.V.N.A. Development of paper biosensor for the detection of phenol from industrial effluents using bioconjugate of Tyr-AuNps mediated by novel isolate *Streptomyces tuirus* DBZ39. *J. Nanomat.* **2017**, *2017*, 1352134. [CrossRef]
86. Taguchi, Y.; Takano, E.; Takeuchi, T. SPR sensing of bisphenol A using molecularly imprinted nanoparticles immobilized on slab optical waveguide with consecutive parallel Au and Ag deposition bands coexistent with bisphenol A-immobilized Au nanoparticles. *Langmuir* **2012**, *28*, 7083–7088. [CrossRef]
87. Xiong, Y.; Ye, Z.; Xu, J.; Liu, Y.; Zhang, H. A microvolume molecularly imprinted polymer modified fiber-optic evanescent wave sensor for bisphenol A determination. *Anal. Bioanal. Chem.* **2014**, *406*, 2411–2420. [CrossRef]
88. Qiao, P.; Wang, X.H.; Gao, S.; Yin, X.; Wang, Y.; Wang, P. Integration of black phosphorus and hollow-core anti-resonant fiber enables two-order magnitude enhancement of sensitivity for bisphenol A detection. *Biosens. Bioelectron.* **2020**, *149*, 111821–111829. [CrossRef]
89. Ge, L.; Li, S.P.; Lisak, G. Advanced sensing technologies of phenolic compounds for pharmaceutical and biomedical analysis. *J. Pharm. Biomed. Anal.* **2020**, *179*, 112913–112926. [CrossRef] [PubMed]
90. McDonagh, C.; Burke, C.S.; MacCraith, B.D. Optical chemical sensors. *Chem. Rev.* **2008**, *108*, 400–422. [CrossRef] [PubMed]
91. Tang, Y.; Lee, D.; Wang, J.; Li, G.; Yu, J.; Lin, W.; Yoon, J. Development of fluorescent probes based on protection-deprotection of the key functional groups for biological imaging. *Chem. Soc. Rev.* **2015**, *44*, 5003–5015. [CrossRef] [PubMed]
92. Sun, Q.; Yang, S.H.; Wu, L.; Yang, W.C.; Yang, G.F. A highly sensitive and selective fluorescent probe for thiophenol designed via a twist-blockage strategy. *Anal. Chem.* **2016**, *88*, 2266–2272. [CrossRef]
93. Li, Y.; Su, W.; Zhou, Z.; Huang, Z.; Wu, C.; Yin, P.; Li, H.; Zhang, Y. A dual-response near-infrared fluorescent probe for rapid detecting thiophenol and its application in water samples and bio-imaging. *Talanta* **2019**, *199*, 355–360. [CrossRef]
94. Wang, Z.; Zhong, N.; Chen, M.; Chang, H.; Zhong, D.; Wu, Y.; Liu, H.; Xin, X.; Zhao, M.; Tang, B.; et al. Photochemical reflective optical fiber sensor for selective detection of phenol in aqueous solutions. *Appl. Opt.* **2019**, *58*, 2091–2099. [CrossRef]
95. Ganesan, M.; Nagaraaj, P. Quantum dots as nanosensors for detection of toxics: A literature review. *Anal. Methods* **2020**, *12*, 4254–4275. [CrossRef] [PubMed]
96. Nezhad, M.R.H.; Alimohammadi, M.; Tashkhourian, J.; Razavian, S.M. Optical detection of phenolic compounds based on the surface plasmon resonance band of Au nanoparticles. *Spectrochim. Acta Part A Mol. Biomol. Spectrosc.* **2008**, *71*, 199–203. [CrossRef] [PubMed]
97. Ma, Y.; You, F.; Promthaveepong, K.; Li, N. Ultrasensitive bisphenol A sensing based on responsive plasmonic nanoparticles. *Sens. Actuators B Chem.* **2017**, *245*, 369–374. [CrossRef]
98. Frasco, M.F.; Chaniotakis, N. Semiconductor quantum dots in chemical sensors and biosensors. *Sensors* **2009**, *9*, 7266–7286. [CrossRef] [PubMed]

99. Lesiak, A.; Drzozga, K.; Cabaj, J.; Bański, M.; Malecha, K.; Podhorodecki, A. Optical sensors based on II-VI quantum dots. *Nanomaterials* **2019**, *9*, 192. [CrossRef]
100. Yan, X.; Hu, T.; Wang, L.; Zhang, L.; Su, X. Near-infrared fluorescence nanoprobe for enzyme-substrate system sensing and in vitro imaging. *Biosens. Bioelectron.* **2016**, *79*, 922–929. [CrossRef]
101. Nayak, R.; Ali, F.A.; Mishra, D.K.; Ray, D.; Aswald, V.K.; Sahoo, S.K.; Nanda, B. Fabrication of CuO nanoparticle: An efficient catalyst utilized for sensing and degradation of Phenol. *J. Mater. Res. Technol.* **2020**, *9*, 11045–11059. [CrossRef]
102. Jaiswal, A.; Kumar, A.; Prakash, R. Facile synthesis of doped C_xN_y QDs as photoluminescent matrix for direct detection of hydroquinone. *Spectrochim. Acta Part A Mol. Biomol. Spectrosc.* **2021**, *246*, 119019–119027. [CrossRef]
103. Kumar, R.; Umar, A.; Kumar, R.; Chauhan, M.S.; Al-Hadeethi, Y. ZnO–SnO_2 nanocubes for fluorescence sensing and dye degradation applications. *Ceram. Int.* **2021**, *47*, 6201–6210. [CrossRef]
104. Ponnusamy, R.; Venkatesan, R.; Samal, R.; Kandasamy, M.; Gandhiraj, V.; Chakraborty, B.; Rout, C.S. Bifunctional WO3 microrods decorated RGO composite as catechol sensor and optical limiter. *Appl. Surf. Sci.* **2021**, *536*, 147669. [CrossRef]
105. Frazão, O.; Santos, J.L.; Araũjo, F.M.; Ferreira, L.A. Optical sensing with photonic crystal fibers. *Laser Photon. Rev.* **2008**, *2*, 449–459. [CrossRef]
106. Niger, M.; Hasin, T.F. Detection of harmful chemical compounds in plastics with highly sensitive photonic crystal fiber with higher nonlinear coefficient. In Proceedings of the 2019 IEEE International Conference on Signal Processing, Information, Communication & Systems (SPICSCON), Dhaka, Bangladesh, 28–30 November 2019.
107. Marley, N.A.; Mann, C.K.; Vickers, T.J. Determination of phenols in water using Raman spectroscopy. *Appl. Spectrosc.* **1984**, *380*, 540–543. [CrossRef]
108. Marks, H.L.; Pishko, M.V.; Jackson, G.W.; Coté, G.L. Rational design of a bisphenol a aptamer selective surface-enhanced Raman scattering nanoprobe. *Anal. Chem.* **2014**, *86*, 11614–11619. [CrossRef] [PubMed]
109. Bakar, N.A.; Salleh, M.M.; Umar, A.A.; Shapter, J.G. Design and measurement technique of surface-enhanced Raman scattering for detection of bisphenol A. *Adv. Nat. Sci. Nanosci. Nanotechnol.* **2017**, *8*, 025008–025012. [CrossRef]
110. Qiu, L.; Liu, Q.; Zeng, X.; Liu, Q.; Hou, X.; Tian, Y.; Wu, L. Sensitive detection of bisphenol A by coupling solid phase microextraction based on monolayer graphene-coated Ag nanoparticles on Si fibers to surface enhanced Raman spectroscopy. *Talanta* **2018**, *187*, 13–18. [CrossRef] [PubMed]
111. Wang, C.Y.; Zeng, Y.; Shen, A.G.; Hu, J.M. A highly sensitive SERS probe for bisphenol A detection based on functionalized Au@Ag nanoparticles. *Anal. Methods* **2018**, *10*, 5622–5628. [CrossRef]
112. Roschi, E.; Gellini, C.; Ricci, M.; Sanchez-Cortes, S.; Focardi, C.; Neri, B.; Otero, J.C.; López-Tocón, I.; Smulevich, G.; Becucci, M. Surface-enhanced Raman spectroscopy for bisphenols detection: Toward a better understanding of the analyte–nanosystem interactions. *Nanomaterials* **2021**, *11*, 881. [CrossRef]
113. Li, D.; Li, D.W.; Fossey, J.S.; Long, Y.T. Portable surface-enhanced Raman scattering sensor for rapid detection of aniline and phenol derivatives by on-site electrostatic preconcentration. *Anal. Chem.* **2010**, *82*, 9299–9305. [CrossRef] [PubMed]
114. Han, X.X.; Pienpinijtham, P.; Zhao, B.; Ozaki, Y. Coupling reaction-based ultrasensitive detection of phenolic estrogens using surface-enhanced resonance Raman scattering. *Anal. Chem.* **2011**, *83*, 8582–8588. [CrossRef] [PubMed]
115. Zeng, Z.; Tang, D.; Liu, L.; Wang, Y.; Zhou, Q.; Su, S.; Hu, D.; Han, B.; Jin, M.; Ao, X.; et al. Highly reproducible surface-enhanced Raman scattering substrate for detection of phenolic pollutants. *Nanotechnology* **2016**, *27*, 455301–455319. [CrossRef]
116. Pompeu, D.R.; Larondelle, Y.; Rogez, H.; Abbas, O.; Pierna, J.A.F.; Baeten, V. Characterization and discrimination of phenolic compounds using Fourier transform Raman spectroscopy and chemometric tools. *Biotechnol. Agron. Soc. Environ.* **2018**, *22*, 13–28.
117. Carreira-Casais, A.; Montes-García, V.; Pastoriza-Santos, I.; Prieto, M.Á.; Simal-Gandara, J.; Pérez-Juste, J. Multiple SERS detection of phenol derivatives in tap water. *Proceedings* **2021**, *70*, 88. [CrossRef]
118. Liu, C.; Hu, J.; Biswas, S.; Zhu, F.; Zhan, J.; Wang, G.; Tung, C.H.; Wang, Y. Surface-enhanced Raman scattering of phenols and catechols by a molecular analogue of titanium dioxide. *Anal. Chem.* **2020**, *92*, 5929–5936. [CrossRef]

MDPI
St. Alban-Anlage 66
4052 Basel
Switzerland
www.mdpi.com

Sensors Editorial Office
E-mail: sensors@mdpi.com
www.mdpi.com/journal/sensors

www.ingramcontent.com/pod-product-compliance
Lightning Source LLC
Chambersburg PA
CBHW040741250926
51850CB00007BB/1376
* 9 7 8 3 0 3 6 5 9 9 1 7 5 *